Mobile Information Communication Technologies Adoption in Developing Countries:

Effects and Implications

Ahmed Gad Abdel-Wahab
Mansoura University, Egypt

Ahmed Ahmed A. El-Masry
Plymouth University, UK

INFORMATION SCIENCE REFERENCE
Hershey · New York

Director of Editorial Content:	Kristin Klinger
Director of Book Publications:	Julia Mosemann
Acquisitions Editor:	Lindsay Johnston
Development Editor:	Joel Gamon
Publishing Assistant:	Milan Vracarich Jr.
Typesetter:	Keith Glazewski
Production Editor:	Jamie Snavely
Cover Design:	Lisa Tosheff

Published in the United States of America by
Information Science Reference (an imprint of IGI Global)
701 E. Chocolate Avenue
Hershey PA 17033
Tel: 717-533-8845
Fax: 717-533-8661
E-mail: cust@igi-global.com
Web site: http://www.igi-global.com

Library of Congress Cataloging-in-Publication Data

Mobile information communication technologies adoption in developing countries : effects and implications / Ahmed Gad Abdel-Wahab and Ahmed Ahmed A. El-Masry, editor.
p. cm.
Includes bibliographical references and index.
Summary: "This book reviews different approaches and methodologies used in dealing with issues related to mobile ICTs, and presents successful examples mobile ICT adoption in developing countries, addressesing the impact of culture on mobile ICT adoption and deployment"--Provided by publisher.
ISBN 978-1-61692-818-6 (hbk.) -- ISBN 978-1-61692-820-9 (ebook) 1. Telecommunication--Developing countries. 2. Mobile communication systems--Developing countries. 3. Information technology--Developing countries. I. Abdel-Wahab, Ahmed Gad. II. El-Masry, Ahmed Ahmed A., 1965-
HE8635.M66 2011
384.5'3091724--dc22
2010027731

British Cataloguing in Publication Data
A Cataloguing in Publication record for this book is available from the British Library.

Editorial Advisory Board

Table of Contents

Preface xiv

Acknowledgment xvii

Section 1
M-Technology and Communications

Chapter 1
The Critical Mass of Wireless Communications: Differences between Developing and Developed Economies 1
Kaisu Puumalainen, Lappeenranta University of Technology, Finland
Lauri Frank, University of Jyväskylä, Finland
Sanna Sundqvist, Lappeenranta University of Technology, Finland
Anni Tuppura, Lappeenranta University of Technology, Finland

Chapter 2
The Competitive Growth Pattern of Mobile Telecommunications in Korea 18
Moon-Soo Kim, Hankuk University of Foreign Studies (HUFS), Korea
Sungjoo Lee, Ajou University, Korea

Chapter 3
Mobile Information Communication Technologies and Construction Project Management: Indian Scenario Case Study 36
Vanita Ahuja, Project Management Consultant, Teacher and Trainer, India

Chapter 4
Requirements Engineering in the ICT4D Domain 52
Kristina Pitula, Concordia University, Canada
Daniel Sinnig, Concordia University, Canada
Thiruvengadam Radhakrishnan, Concordia University, Canada

Section 2
M-Technology Applications

Chapter 5
In Defence of Mobile Technologies: Exploring the Socio-Technological Dimensions of M-Learning ... 67
Ayse Kok, University of Oxford, UK

Chapter 6
Mobile Learning in China ... 79
Yuqin Yang, Nanyang Technological University, Singapore
Qiyun Wang, Nanyang Technological University, Singapore

Chapter 7
Blended Mobile Learning in Developing Nations and Environments with Variable Access: Three Cases ... 91
Susan Smith Nash, AAPG (American Association of Petroleum Geologists) & University of Oklahoma, USA

Chapter 8
Mobile Technologies and Rich Media: Expanding Tertiary Education Opportunities in Developing Countries ... 103
Trish Andrews, The University of Queensland, Australia
Robyn Smyth, The University of New England, Australia
Belinda Tynan, The University of New England, Australia
Andrew Berriman, The University of New England, Australia
Deborah Vale, The University of New England, Australia
Richard Caladine, The University of Wollongong, Australia

Chapter 9
The Role of M-Government in Western China Development ... 117
Jesper Schlæger, University of Copenhagen, Denmark

Chapter 10
Exploring Civil Servant Resistance to M-Government: A Story of Transition and Opportunities in Turkey ... 134
Ronan de Kervenoael, Sabanci University, Turkey & Aston University, UK
Mark Palmer, Aston University, UK
N. Meltem Cakici, Gediz University, Turkey

Section 3
M- Technology: Economic and Social Effects

Chapter 11
Mobile Telephony and Economic Growth in Developing Economies 161
Heli Virta, Lappeenranta University of Technology, Finland
Kaisu Puumalainen, Lappeenranta University of Technology, Finland
Anni Tuppura, Lappeenranta University of Technology, Finland

Chapter 12
Understanding Mobile Phone Usage While Driving: Mini-Bus and Taxi Drivers' Experiences in Istanbul.. 173
Ronan de Kervenoael, Sabanci University, Turkey & Aston University, UK
Canan Devletkusu, Dogus University, Turkey

Chapter 13
Mobile Technology and the Gulf Society: Perception and Attitude 195
Khaled A. Sabry, Hamdan Bin Mohammed e-University, UAE
Ahmed Al-Nakeeb, Al-Ain University, UAE
Khalid W. Alrawi, Al-Ain University, UAE

Chapter 14
BlogWall: Promoting Social Communication through Mobile Technology in Sri Lanka.................. 205
Adrian David Cheok, National University of Singapore, Singapore
Owen Noel Newton Fernando, National University of Singapore, Singapore
Nimesha Ranasinghe, National University of Singapore, Singapore
Kening Zhu, National University of Singapore, Singapore
Chamari Edirisinghe, National University of Singapore, Singapore

Chapter 15
The Role of Information Communication Technologies within the Field of Communication for Social Change 218
Jan Servaes, University of Massachusetts, USA

Chapter 16
Health Hazards of Mobile Information Communication Technologies.. 237
Sohayla M. Attalla, Mansoura University, Egypt

Compilation of References 252

About the Contributors 289

Index.. 298

Detailed Table of Contents

Preface xiii

Acknowledgment xvi

Section 1
M-Technology and Communications

This section provides the reader with an introduction and background of m-technology and contains four chapters. It starts from Chapter 1 and ends with Chapter 4. Chapter 1 deals with the critical mass of wireless communications in developing and developed economies. Chapter 2 explores the competitive growth pattern of mobile telecommunications in Korea. Chapter 3 investigates mobile information communication technologies and construction project management with case study from India. The fourth chapter presents for the requirements engineering in the ICT4D domain.

Chapter 1
The Critical Mass of Wireless Communications: Differences between Developing and Developed Economies 1
Kaisu Puumalainen, Lappeenranta University of Technology, Finland
Lauri Frank, University of Jyväskylä, Finland
Sanna Sundqvist, Lappeenranta University of Technology, Finland
Anni Tuppura, Lappeenranta University of Technology, Finland

This chapter provides an introduction and background on mobile telecommunication diffusion and then moves in to the economic development and critical mass point in the innovation's diffusion process and then economic development and diffusion of wireless communications. The chapter also presents for data analysis and main results. The chapter ends with conclusions and suggestions for future research.

Chapter 2
The Competitive Growth Pattern of Mobile Telecommunications in Korea 18
Moon-Soo Kim, Hankuk University of Foreign Studies (HUFS), Korea
Sungjoo Lee, Ajou University, Korea

This chapter provides an introduction and an overview of Korean mobile telecommunications addressing some issues such as technology market in Korea. The chapter also presents for data, analysis model, diffusion models, and LVC model for empirical analysis. The chapter also presents for strategic and policy implications and the success factors of Korean mobile communications market. The chapter ends with concluding remarks.

Chapter 3
Mobile Information Communication Technologies and Construction Project Management:
Indian Scenario Case Study 36
Vanita Ahuja, *Project Management Consultant, Teacher and Trainer*, *India*

This chapter provides an introduction of construction projects management and an overview of mobile communications and construction industry addressing technology issues and management issues and then it provides an overview of Indian construction industry. The chapter also presents for a case study of renovation and expansion of Indira Gandhi international airport– New Delhi, India. It discusses the case study results. Then the chapter ends with conclusion and recommendations for future research.

Chapter 4
Requirements Engineering in the ICT4D Domain.. 52
Kristina Pitula, Concordia University, Canada
Daniel Sinnig, Concordia University, Canada
Thiruvengadam Radhakrishnan, Concordia University, Canada

This chapter starts with a detailed introduction of requirements engineering and ict4d context. The chapter also presents for storytelling and how it constitutes. Then the chapter compares storytelling with conventional elicitation techniques. In the chapter, the authors also present for incorporating 'SDS' in requirements gathering process. Then they move to requirements elicitation, requirements abstraction, and requirements specification and validation. The chapter also demonstrates for validation of the SDS approach with a preliminary experiment and then presents for the next generation e-tool. The chapter ends with a summary.

Section 2
M-Technology Applications

Section two mainly focuses on m-technology applications and consists of six chapters. It starts from Chapter 5 and ends with Chapter 10. Chapter 5 demonstrates some defences of mobile technologies with exploring the socio-technological dimensions of m-learning. While Chapter 6 deals with mobile learning in China, Chapter 7 explores m-learning in developing nations and environments with variable access. Chapter 8 presents for mobile technologies and rich media-expanding tertiary education opportunities in developing countries. The last two chapters in this sections deal with m-government. Chapter 9 presents for the role of m-government in Western China development, and Chapter 10 explores civil servant resistance to m-government in Turkey.

Chapter 5
In Defence of Mobile Technologies: Exploring the Socio-Technological Dimensions of M-Learning 67
Ayse Kok, University of Oxford, UK

This chapter starts with a detailed introduction of mobile technologies as a tool for revolutionizing education, transforming the traditional ways of learning and teaching into 'anytime' and particularly, 'anyplace' education. Then the chapter provides a brief background and definition of m-learning. In the chapter, the author presents for data analysis and some methodological issues related to m-learning. Then the author discusses research some gaps and future research directions. The chapter ends with conclusion and points for future research.

Chapter 6
Mobile Learning in China 79
Yuqin Yang, Nanyang Technological University, Singapore
Qiyun Wang, Nanyang Technological University, Singapore

This chapter starts with an introduction of mobile learning and provides some statistics on mobile users in china. Then the authors define m-learning based on mobile devices, the context of learning experiences and learning environments, and the context of e-learning. The chapter presents for educational policies and initiatives and telecommunications policies and reforms in China. It also discusses the challenges in adopting m-learning such as illiteracy, poor infrastructure, digital divide, and political/social climate. The chapter details m-learning projects in Chinese universities and companies. The chapter ends with conclusion and future research directions and challenges.

Chapter 7
Blended Mobile Learning in Developing Nations and Environments with Variable Access: Three Cases 91
Susan Smith Nash, AAPG (American Association of Petroleum Geologists) & University of Oklahoma, USA

This chapter starts with an introduction and background of m-learning and claims that m-learning using laptops or desktop computers connected to the Internet offers robust, web-based solutions that can be multimedia rich, and effective synchronous and asynchronous collaboration. The author presents for three cases of blended mobile learning in developing countries. Case one: almost no internet access away from home base or centre, case Two: limited access and case three: good urban or "base camp" connectivity for downloads; good cell phone coverage for sending basic text messages. The chapter ends with conclusion and future research recommendations.

Chapter 8
Mobile Technologies and Rich Media: Expanding Tertiary Education Opportunities in Developing Countries 103
Trish Andrews, The University of Queensland, Australia
Robyn Smyth, The University of New England, Australia
Belinda Tynan, The University of New England, Australia
Andrew Berriman, The University of New England, Australia
Deborah Vale, The University of New England, Australia
Richard Caladine, The University of Wollongong, Australia

This chapter details for mobile technologies and rich media and the authors claim that m-technology and rich media can enable greater opportunities for situated and personal learning in both real and virtual higher education contexts in developing countries. The chapter provides some statistics on mobile phone usage in developing countries. The chapter discusses factors influencing adoption of mobile technologies such as infrastructure, affordability, and government policy. The chapter also discusses the key Issues relating to use of mobile technologies and rich media and considerations for tertiary education in developing countries. Then the chapter moves into discussing the implications for m-learning practice in developing countries and forecasting issues for higher education institutions for adopting mobile technology and Rich Media in m-learning.

Chapter 9
The Role of M-Government in Western China Development 117
Jesper Schlæger, University of Copenhagen, Denmark

This chapter discusses the role of mobile government in western china development. The chapter presents for digital governance model and then analysing mobile government in western china. The chapter explores some ideas of service-oriented mobile government and how corruption affects m-government. The author concludes the chapter with some directions for future research.

Chapter 10
Exploring Civil Servant Resistance to M-Government: A Story of Transition and Opportunities in Turkey 134
Ronan de Kervenoael, Sabanci University, Turkey & Aston University, UK
Mark Palmer, Aston University, UK
N. Meltem Cakici, Gediz University, Turkey

This chapter introduces for civil servant resistance to m-government in Turkey. The authors provide a detailed background and definition of m-government in general and in Turkey in particular. The chapter presents for empirical study and data analysis. The authors presents for important results and different concepts of resistance to m-government in Turkey such as functional resistance, ideological resistance, market driven resistance and geographic resistance.

Section 3
M- Technology: Economic and Social Effects

Section three provides the readers with some economic and social effects of m-technology. It consists of six chapters. It starts with Chapter 11 and ends with Chapter 16. Chapter 11 presents for the connection between mobile telephony and economic growth in developing economies. Chapter 12 explores mobile phone usage while driving with empirical study on Mini-bus and taxi drivers' experiences in Istanbul, Turkey. Chapter 13 demonstrates to perception and attitude on mobile technology in the Gulf countries with case study from UAE. Chapter 14 presents for BlogWall system as it promotes social communication through mobile technology in Sri Lanka. While Chapter 15 presents the role of information communication technologies within the field of communication for social change, Chapter 16 confirms the health hazards of mobile information communication technologies.

Chapter 11
Mobile Telephony and Economic Growth in Developing Economies 161
Heli Virta, Lappeenranta University of Technology, Finland
Kaisu Puumalainen, Lappeenranta University of Technology, Finland
Anni Tuppura, Lappeenranta University of Technology, Finland

In this chapter, the authors assess how mobile phone penetration impacts on economic growth in developing economies. They present for a good literature review, then their mode specification, dataset, and method of estimation. They demonstrate their results using the GMM technique. The chapter ends with conclusion and ideas for future research.

Chapter 12
Understanding Mobile Phone Usage While Driving: Mini-Bus and Taxi Drivers' Experiences
in Istanbul... 173
Ronan de Kervenoael, Sabanci University, Turkey & Aston University, UK
Canan Devletkusu, Dogus University, Turkey

The chapter starts with an introduction of how dominant dangerous behaviour in the absence of an enforced legal framework is often described as becoming routine practice for many mobile phone users. The authors point out how advanced motorisation affects road traffic safety mechanisms in Turkey. The authors present for a detailed literature review on the topic under study and then the methodology used in their study. The chapter discusses the main results and then presents for conclusion and implications of the study.

Chapter 13
Mobile Technology and the Gulf Society: Perception and Attitude 195
Khaled A. Sabry, Hamdan Bin Mohammed e-University, UAE
Ahmed Al-Nakeeb, Al-Ain University, UAE
Khalid W. Alrawi, Al-Ain University, UAE

This chapter introduces of mobile phone technology and its effect on attitudes and perceptions of mobile users in Gulf countries. Then the authors explore the advancements of mobile technology and the gulf region. The authors present for research objectives and methodology and then the survey results in UAE. Then the chapter ends with discussion and future work.

Chapter 14
BlogWall: Promoting Social Communication through Mobile Technology in Sri Lanka 205
Adrian David Cheok, National University of Singapore, Singapore
Owen Noel Newton Fernando, National University of Singapore, Singapore
Nimesha Ranasinghe, National University of Singapore, Singapore
Kening Zhu, National University of Singapore, Singapore
Chamari Edirisinghe, National University of Singapore, Singapore

This chapter starts with an introduction of Mobile Technology and its social effects. The authors then discuss Mobile culture in developing countries. The authors also discuss mobile culture and Poetry culture in Sri Lanka. They introduce the Blogwall system. The authors then presents for the user evaluation and acceptance of BlogWall. The chapter also discusses the results of the user evaluation and ends with a discussion and conclusion.

Chapter 15
The Role of Information Communication Technologies within the Field of Communication for Social Change 218
Jan Servaes, University of Massachusetts, USA

The chapter starts with summarizing the field of communication for development and social change addressing development paradigms, communication paradigms and research priorities. The author moves to detail information and communication technologies for development and social change focusing on technology and culture, technology transfer and consequences for policymaking. Then the author discusses ICTs and the millennium development goals. The chapter ends with conclusion.

Chapter 16
Health Hazards of Mobile Information Communication Technologies 237
Sohayla M. Attalla, Mansoura University, Egypt

The chapter discusses how mobile phones affect health through radio frequency. The chapter starts with what RF energy is and how it can affect the body. The chapter points out health hazards of cell phone base stations and neurobehavioral deficits. The author discusses the health hazards of cell phone base stations and human head. The author points out some hazards such as thermal effects, cancer, and audiovestibular functions. The author also discusses the effects of cell phone stations on parotid gland, metallic implants in the body, chemical and hormonal, Blood Brain Barrier, Genotoxic effects,

Sleep and EEG effects and pregnancy. The author extends his analysis to the effect of mobile phones on children.

Compilation of References 252

About the Contributors 289

Index 298

Preface

Now more than ever, the mobile technology field is growing with emerging research and new discoveries that expand to all walks of life. Mobile technology may have its greatest impact in the developing world, because it brings telephony to districts that had never been reached before. With the introduction of third generation mobile network services and the convergence of mobile and traditional internet services, Mobile will be one of the key enablers for achieving competitive advantages in developing countries.

Titled *Mobile Information Communication Technologies Adoption in Developing Countries*, the target audience of this book will be composed of researchers and professionals working in the field of mobile information and communication technology, investors, health care organizations, telecommuters, international traders and businessmen, students and their families, governments, bank managers, and economy as a whole.

The book contains three sections:

- Section One: M-technology and communications,
- Section Two: M-technology applications and,
- Section Three: M-technology: Economic and social effects

Section one **M-Technology and Communications** contains four chapters, first chapter entitled *The Critical Mass of Wireless Communications: Differences between Developing and Developed Economies*, by Kaisu Puumalainen, Lauri Frank, Sanna Sundqvist, and Anni Tuppura. It identifies and analyzes the timing and level of critical mass in the development of market penetration for wireless communications. The findings suggest considerable differences between developing and developed countries. The second chapter entitled *The Competitive Growth Pattern of Mobile Telecommunications in Korea*, by Moon-Soo Kim and Sungjoo Lee. Through the empirical study on Korean mobile market, the authors investigate not only the effects on demand diffusion patterns by the competition between technologies and operators, but also strategic implications for the service providers and the policy-makers.

The third chapter entitled *Mobile Information Communication Technologies and Construction Project Management – Indian Scenario: Case Study* by Vanita Ahuja. This chapter presents a case study of mobile communications adoption at a major construction project in India and further discusses the factors leading to sustainable mobile communication adoption by construction project teams. The fourth chapter entitled "*Requirements Engineering in the ICT4D Domain*" by Kristina Pitula, Daniel Sinnig and Thiruvengadam Radhakrishnan. The authors propose a requirements management process especially suited for ICT4D projects. The process supports both bottoms-up and top-down development.

Section two **M-Technology Applications** contains six chapters. The first chapter entitled *In Defence of Mobile Technologies: Exploring the Socio-Technological Dimensions of M-Learning* by Ayse Kok, The aim of this chapter is to improve understanding of the expanding use of mobile phones for the delivery of the learning experience in developing countries. The second chapter entitled *Mobile Learning in China*, by Yuqin Yang and Qiyun Wang. In this chapter; the authors present an overview of the development of m-learning in China, including the construction of m-learning infrastructure, and research projects conducted by universities and companies.

The third chapter entitled *Blended Mobile Learning in Developing Nations and Environments with Variable Access: Three Cases* by Susan Smith Nash. The author presents an overview of the experience of implementing mobile technology solutions in developing nations in conditions of limited Internet access, challenging logistics, and informal learning settings. The fourth chapter entitled *Mobile Technologies and Rich Media-Expanding Tertiary Education Opportunities in Developing Countries* by Trish Andrews, Robyn Smyth, Belinda Tynan, Andrew Berriman, Deborah Vale and Richard Caladine. This chapter focuses on the emerging possibilities and issues arising from the rapid adoption of mobile technologies for learning in tertiary and higher education contexts in developing countries. In particular, it explores the implications for developing nations of the rapid proliferation of mobile device.

The fifth chapter entitled *The Role of M-Government in Western China Development*, by Jesper Schlæger. This chapter describes the content of ideas, institutions, and technologies of m-government to understand how fitting these levels, has led to such an improvement in governance. The sixth chapters entitled *Exploring Civil Servant Resistance to M-Government: A Story of Transition and Opportunities in Turkey*, by Ronan de Kervenoael, Mark Palmer, and N. Meltem Cakici. Drawing on the resistance and mobility literature, this chapter investigates how civil servants' behaviours, in an emerging country technological environment, through their everyday practice, react and resist the influence of m-government transition.

Section three **M-Technology: Economic and Social Effects** contains six chapters. The first chapter entitled *Mobile Telephony and Economic Growth in Developing Economies*, by Heli Virta, Kaisu Puumalainen, and Anni Tuppura. The chapter investigates the influence of mobile phone penetration on economic growth in developing economies. The results suggest that extensive mobile cellular network coverage facilitates economic development in developing countries. The second chapter entitled *Understanding Mobile Phone Usage While Driving: Mini-Bus and Taxi Drivers' Experiences in Istanbul* by Ronan de Kervenoael and Canan Devletkusu. This chapter claims that dominant dangerous behavior in the absence of enforced legal framework is being deployed and has become routine for many m-users. This chapter adopts a qualitative case study approach (20 cases) to examine the public transport drivers' motives, logic and legitimacy processes.

The third chapter entitled *Mobile Technology and the Gulf Society: Perception and Attitude*, by Khaled Sabry, Ahmed Al-Nakeeb, and Khalid Alrawi. This chapter reviews and explores mobile technology growth in the Gulf region with particular focus on the use of mobile phones in the UAE. It further explores, through a snapshot survey, people's perception, attitude, and possible implications of the technology on their behavior. The fourth chapter entitled *BlogWall: Promoting Social Communication through Mobile Technology in Sri Lanka*, by Adrian David Cheok, Owen Noel Newton Fernando, Nimesha Ranasinghe, Kening Zhu, and Chamari Edirisinghe. Taking into account the speedy progress of the mobile technology, especially the SMS, the evolution of the Sri Lankan way of living the authors explore the possibility to use mobile technology, especially SMS (Short Message Service), to promote social interactivity in Sri Lankan community using Blogwall system.

The fifth chapter entitled *The Role of Information Communication Technologies Within the Field of Communication for Social Change*, by Jan Servaes. The author indicates that there is a lot of talk about the 'newness' of mobile and wireless information and communication technologies (ICTs) these days. What is so 'new' about them? And in what way will they solve the still unresolved problems of poverty, inequality and information divides in the world? This chapter takes a bird's eye perspective and presents a number of observations regarding the role of ICTs within the field of Communication for Development and Social Change (CDSC). The last chapter entitled *Health Hazards of Mobile Information Communication Technologies* by Sohayla Attalla. The author states that in this age, it is very difficult not to have technology. But with technology, come certain hazard Inhabitants living nearby mobile phone base stations are at risk for developing neuropsychiatric problems and some changes in the performance of neurobehavioral functions either by facilitation or inhibition. Therefore, revision of standard guidelines for public exposure to RER from mobile phone base station antennas and using of NBTB for regular assessment and early detection of biological effects among inhabitants around the stations are recommended.

The diverse coverage of mobile information communication technologies adoption in developing countries in this book will contribute to a better understanding of all topics, research, and discoveries in this developing, significant field of study. Furthermore, the contributions included in this book will be fuelling the research initiatives in emerging fields. We hope that you will find the discussion about present day reality and future challenges of mobile information communication technologies adoption in developing countries as useful as we hope it to be.

Ahmed Gad Abdel-Wahab
Mansoura University, Egypt

Ahmed Ahmed A. El-Masry
Plymoth University, UK

Acknowledgment

In editing this book, we have been fortunate to have had valuable comments on various draft chapters from our editorial Advisory members. We should like to thank Dr. R. Rena; Professor J. Hoff; Dr. R. Bons; Professor C. k. Ayo; Professor R. Munro; Professor M Falch; Dr. S. S. Nash; and Dr. R.K. Kamat. In addition, many thanks are due to our reviewers. Finally we should like to thank commissioning people at IGI Global, especially Beth Ardner, Assistant Director of Marketing for her enthusiastic and excellent support especially in the first and second stages of the book and Mr. Joel Gamon for his excellent assistance at the last stage of the book.

Ahmed Gad Abdel-Wahab
Mansoura University, Egypt

Ahmed Ahmed A. El-Masry
Plymoth University, UK

Section 1
M–Technology and Communications

Chapter 1
The Critical Mass of Wireless Communications:
Differences between Developing and Developed Economies

Kaisu Puumalainen
Lappeenranta University of Technology, Finland

Lauri Frank
University of Jyväskylä, Finland

Sanna Sundqvist
Lappeenranta University of Technology, Finland

Anni Tuppura
Lappeenranta University of Technology, Finland

ABSTRACT

The aim of this chapter is to identify and analyze the timing and level of critical mass in the development of market penetration for wireless communications. The authors assume that critical mass is fulfilled at the point when the acceleration of the diffusion process is at its maximum. In practice, this point is determined by estimating the diffusion function and calculating the second derivative with respect to time. The diffusion of mobile subscription is modelled using the Bass diffusion model. The penetration levels and points of time of critical mass of 209 countries or areas of the world are identified and subsequently subjected to regression analysis against population and economic characteristics. The findings suggest considerable differences between developing and developed countries.

INTRODUCTION

During the last two decades mobile telephone technology has diffused all over the world. In addition to the drastic impact it has exercised on the habits and styles of everyday communication, it has had a noticeable impact on business and entrepreneurship. Diffusion of telecommunications is desirable since by decreasing the costs of transactions and speeding up the propagation and dissemination of knowledge it exerts a posi-

DOI: 10.4018/978-1-61692-818-6.ch001

tive economic influence. Mobile technology may have its greatest impact in the developing world, because it brings telephony to districts fixed-line telephones formerly never reached. However, as data has not been available until recently, empirical evidence on mobile telephony impact or diffusion in developing countries is scarce.

Like many other modern technologies mobile telephony has positive network effects. Economides (1991) concludes that network effects occur when "The buyer of the last unit of a good has a higher benefit than the buyer of the first because the sale of the earlier units has created some benefits in a related dimension". The existence of network effects makes forecasting the success of telecommunication services less easy and reliable, especially when traditional forecasting methods are applied (Schoder, 2000). Network effects can significantly influence the adoption and hence the diffusion of goods and services (Church and Gandal, 1993; Katz and Shapiro, 1985; Witt, 1997). One of the consequences of network effects is the existence of a critical mass point in the innovation's diffusion process. The rate of adoption does not take off until a critical mass of adopters has been reached, but becomes self-sustaining thereafter (Mahler and Rogers, 1999). Despite the importance of the critical mass and the fact that telecommunications has been one of the main application areas of innovation diffusion studies (Meade and Islam, 2006), existing research on this topic is scarce and there are no established guidelines on how to identify the critical mass point.

BACKGROUND

Mobile telecommunication diffusion has been studied generally in a number of recent papers (Grajek and Kretschmer, 2009; Jang et al., 2005; Botelho and Pinto, 2004; Lim et al., 2003; Gruber 2001; Gruber and Verboven, 2001) but only a few of them focus on developing countries (e.g., Singh, 2008) or, especially, on comparing the diffusion in developed and developing countries (Rouvinen, 2006). However, there are some studies that report different patterns and types of use of mobile phones in low-income economies compared to high-income economies (e.g., Hahn and Kibora, 2008; James and Versteeg, 2007). Moreover, the magnitude of this innovation may be greater in geographic areas,where the fixed telephone provision has a very limited presence or is technically unreliable (Hahn and Kibora, 2008). Thus, the diffusion of mobile telecommunications may exhibit very different features in the least developed or developing countries compared to developed countries.

The diffusion of products with associated positive network effects – such as mobile phones – depends on whether the technology gains enough users or adopters to become self-sustaining. If the technology does not reach the critical mass of adopters, it languishes or can be marginalized by competing technologies. An ability to predict the point of time when critical mass will be achieved may have positive consequences for gadget producers and service providers and may also offer an effective planning tool for governments. In case of telecommunications, for instance, strategic investment to speed up otherwise slow rates of adoption may pay off and telecommunications can positively influence economic growth. However, the existing research on the critical mass or takeoff time is scarce and no conclusive guidelines as how to identify the critical mass are available. Even the definitions of the phenomena are somewhat confusing in the literature, as some authors equate the concepts of critical mass and takeoff (Kim and Kim, 2007), while others maintain takeoff occurs earlier than critical mass (Wiorkowski and Gylys, 2006). Most of the earlier studies in the field of marketing define the phenomenon as takeoff, apply it for consumer durables in a limited set of developed countries, and identify the takeoff point visually from the sales curve over time (Golder and Tellis 1997, Agarwal and Bayus

2002, Tellis et al 2003, Haapaniemi and Mäkinen 2007). Applications for telecommunications are rare with only few exceptions (Kim and Kim 2004, Kim and Kim 2007, Lim et al 2003). Even where these studies identify the time of takeoff, they do not explicitly discuss the penetration level of the critical mass and the analysis is limited to a small number of countries.

In this study, we aim to fill in the gaps in previous research by a) identifying the time and level of critical mass from the diffusion parameter estimates for wireless subscriptions in more than 200 developed and developing countries, and b) explaining the cross-country differences in critical mass points through regression analysis of the countries' population and economic characteristics.

ECONOMIC DEVELOPMENT AND CRITICAL MASS

Network Effects and Critical Mass in the Diffusion of an Innovation

Network effects create a so-called critical mass point in the innovation's diffusion process. The rate of adoption of interactive innovations does not take off in the familiar S-shape curve until a critical mass of adopters has been reached (Mahler and Rogers, 1999). In other words, if critical mass is not fulfilled, the diffusion will fail. In the early phase of diffusion the network seems unattractive to potential adopters, because there are only few users. Thus, the network does not create enough utility for potential adopters to join the network. Only after the critical mass is reached is the interactive innovation perceived as valuable by potential adopters (Mahler and Rogers, 1999). At that point, the network has enough users to be attractive to other potential users, and the diffusion process becomes self-sustaining. A telecommunications service provider introducing

Figure 1. Diffusion without (solid) and with (dotted) network effects

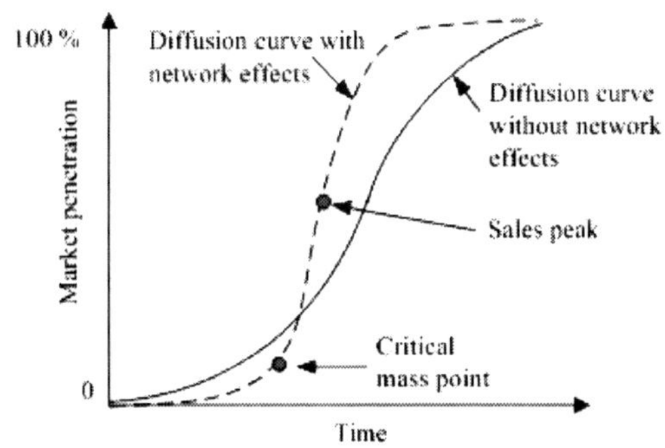

new interactive services often faces formidable problems in getting the innovation to reach critical mass, but is thereafter able to benefit from the self-sustaining diffusion process (Mahler and Rogers, 1999). Thus, it is very valuable to identify the critical mass in order to have an innovation which enjoys network effects diffused successfully. A comparison of innovations with and without network effects shows how the latter is seen to diffuse more slowly until a critical mass of adopters is reached (see Figure 1).

Critical mass also may be defined as the minimal number of adopters of an interactive innovation for the further rate of adoption to be self-sustaining (Mahler and Rogers, 1999). It can also be interpreted as the turning point between positive and negative returns to adoption (Markus, 1990a). The critical mass point in the diffusion process is generally expected to occur approximately between 10 and 20% adoption (Rogers, 1983; Valente, 1995). Weiber (1995) explained that the left-skewed diffusion curve and the critical mass at the early stages of diffusion were a result of market-related factors. He argued that these factors first hamper the rate of diffusion but later, after reaching critical mass, accelerate the diffusion process. As mentioned above, it has been proposed that critical mass does not only speed up the rate of adoption but also may cause the collapse of an innovation (e.g. Markus, 1990b).

Figure 2. Deriving the critical mass point from the s-curves second derivative

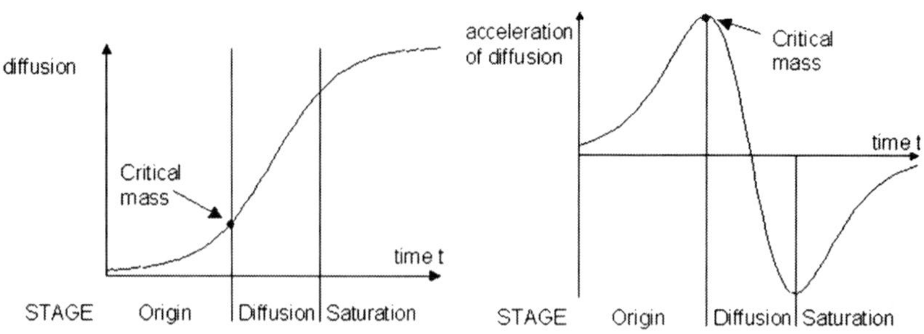

Formally, the point of time of critical mass achievement may be extracted from the diffusion curve as follows. Griliches (1957) suggests three phases from the sigmoid aggregate diffusion curve: (1) origin, (2) diffusion and (3) saturation. These phases can be identified, for example, using the sigmoid shaped curve's second time derivatives' extreme points as cut points for the phases. Figure 2 shows the derivation of stages for the lifecycle of an innovation: In the beginning, the innovation diffuses slowly. At this point, the innovation is new and in its first stage. As Figure 2 shows, the first stage is cut by the maximum of the second derivative with respect to time (the acceleration), which is the moment when the increase of the innovation's adopters is at its maximum. Thus, it may also be considered as the point of critical mass fulfillment. This kind of approach has been used by Lim et al (2003) and Kim and Kim (2007) while methods based on visual examination of the diffusion curve have been proposed by e.g. Tellis et al (2003) and Agarwal and Bayus (2002).

Economic Development and Diffusion of Wireless Communications

Geographic, socio-economic, demographic, and cultural differences of countries, which are uncontrollable by the firm, are likely to influence the way in which a product or service diffuses over time. The present study assesses the effects of economic (GDP and income inequality) and demographic (population and population density) variables on the level and timing of critical mass. The wealth of a country, which is often operationalized as GDP per capita, has a positive effect on the diffusion process in reducing the time before a country adopts as well as in speeding up the diffusion within the country (e.g., Jha and Majumdar 1999; Madden et al. 2004; Koski and Kretschmer, 2005) . It is suggested that income inequality hinders diffusion as the internal communication structures within such countries are more fragmented. The effect of population density on innovation diffusion should generally be positive, as it should increase word-of-mouth communication. However, the effect of population density in the case of wireless communications could be the opposite, as sparsely populated areas are more likely to benefit from wireless technology as a means of communication instead of face-to-face or fixed telephone (e.g., Koski and Kretschmer, 2005). Figure 3 summarizes the discussion above.

Data Collection and Measures

The population, population density, GDP per capita, and wireless communications data were obtained from the Euromonitor's GMID. The database included 201 countries or areas from seven regions: Asia 43, Australasia 2, Eastern Europe 19, South America 43, Middle East and Africa 67, North America 2, and Western Europe

Figure 3. Relationships of interest

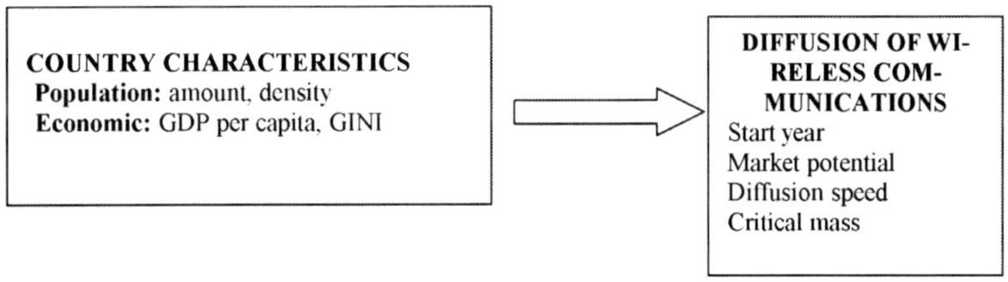

25. The wireless communications data was collected as the number of subscribers per population from the year 1980 to the year 2008. Population and GDP values are taken from the year 1995 (in order to represent the average time when the diffusion of wireless communications started) and a recent situation from the year 2005. The income inequality measures (Gini coefficients) were obtained from the United Nations UNDP (2007) Human Development Report 2007/08, but these were only available for 126 countries. Larger coefficients mean less equal income distribution within the country. The wireless technologies used, the numbers of wireless operators, and average revenues per user were obtained from World Cellular Information Service (http://www.wcisdata.com/).

The countries were classified according to the level of economic development using the World Bank definitions. Low income countries were those 47 countries with GNI per capita in year 2007 less than 935 \$. Lower middle category consisted of 51 countries in the \$ 936-3705 range. The upper middle category consisted of 38 countries within the \$ 3706-11455 range, and finally there were 65 high income countries with GNI per capita in excess of \$11456.

At the first stage of analysis the Bass (1969) diffusion model (Equation 1) was fitted to each country.

$$N(t) = \frac{m\left(1 - e^{-(p+q)t}\right)}{1 + \frac{q}{p} e^{-(p+q)t}} \tag{1}$$

where at any time t, N(t) is the cumulative number of adopters, and m is the total market potential for the new product. Bass calls the constants p and q the coefficients of innovation and imitation, respectively. Coefficient p captures the proportional adoptions due to mass media, and coefficient q represents the adoptions due to interpersonal communications. These have also been called the coefficients of external and internal influence, respectively (Mahajan, Muller and Wind, 2000). The timing of the sales peak is the maximum value of the first derivative of (1), i.e. the non-cumulative number of adopters, and can be estimated from the innovation and imitation coefficients using Equation 2.

$$t^* = -\frac{1}{p+q} \ln \frac{p}{q} \tag{2}$$

The timing of the critical mass point was identified based on the maximum point of the second derivative of (1), (see Equation 3).

$$t** = -\frac{1}{p+q} \ln \left[(2+\sqrt{3}) \frac{p}{q} \right] \tag{3}$$

Critical time t** was further inserted into the Bass equation (1) in order to calculate the level of penetration at the critical mass point. The calendar year of the critical mass was computed simply by adding t** to the start year of diffusion. Furthermore, the critical mass penetration level was divided by the estimated market potential.

Table 1. Mean values of country characteristics in income categories

Variable	Low Income n=35-47	Lower Middle n=45-51	Upper Middle n=33-38	High n=61-65	Total n=174-201	ANOVA F
Population 1995 mean *(Std.Dev.)*	20570.38 *(30993.94)*	58638.81 *(210925.06)*	19543.53 *(37903.71)*	15010.15 *(38793.64)*	28237.32 *(111406.37)*	1.74
Pop. Density 1995 mean *(Std.Dev.)*	90.75 *(150.63)*	95.34 *(131.48)*	105.11 *(134.67)*	645.01 *(2087.48)*	270.14 *(1201.24)*	3.09**
GDP/capita 1995 mean *(Std.Dev.)*	201.14 *(184.24)*	815.72 *(561.72)*	2376.69 *(1564.30)*	20470.29 *(14351.30)*	7258.79 *(12194.42)*	81.28***
Pop.growth 1990-2005 mean *(Std.Dev.)*	.45 *(.14)*	.25 *(.19)*	.17 *(.19)*	.21 *(.22)*	.27 *(.21)*	20.88***
GDP/cap. growth 1990-2005 mean *(Std.Dev.)*	58.14 *(268.79)*	270.28 *(1222.76)*	1152.14 *(5199.01)*	7.15 *(34.210)*	300.09 *(2358.56)*	1.97
Gini mean *(Std.Dev.)*	42.64 *(8.37)*	44.13 *(10.83)*	43.90 *(10.16)*	32.69 *(5.42)*	40.84 *(9.98)*	11.65***

*** p< 0.01, ** p< 0.05, *p<.10

Results

The Bass model was estimated using SPSS software and its non-linear least squares regression analysis. The estimation succeeded in 174 countries, but failed to converge and yielded implausible or non-significant estimates in 27 countries. The estimation success rate was significantly related to development category, as it varied from 75% in low income category to 94% in high income category (Chi Square=9.003 with three degrees of freedom, p=.029). The descriptive information about the country characteristics can be seen in Table 1. A simple one-way analysis of variance was initially used to explore differences in the variables across the levels of development, and the ANOVA F statistics and the associated significance level for the equality of category means is shown in the last column of Table 1.

The descriptive information about the diffusion parameter estimates can be seen in Table 2.

The average start year of wireless communications in the world is 1993, but the development categories follow each other with an average two-year lag starting from 1990 in the high income countries to 1996 in the low income countries. The factual penetration levels at the end of year 2008 are also vastly different across the four income groups: in low income group the average penetration level is currently 19% of the population, in lower middle income group about 55%, in upper middle group about 85%, and in high income group already as high as 107% of the population.

According to the Bass model estimates, the high income countries have already almost achieved the final penetration level, 115% of the population. Upper middle income category countries are expected to saturate at 108% penetration

Table 2. Mean values of diffusion variables in income categories

Variable	Low Income n=35-47	Lower Middle n=45-51	Upper Middle n=33-38	High n=61-65	Total n=174-201	ANOVA F
Start year mean *(Std.dev.)*	1996.15 *(3.26)*	1994.08 *(3.43)*	1992.24 *(2.39)*	1990.51 *(2.43)*	1993.06 *(3.62)*	37.40***
m mean *(Std.dev.)*	.34 *(.33)*	.78 *(.41)*	1.09 (.44)	1.15 *(.29)*	.88 *(.47)*	42.81***
p mean *(Std.dev.)*	.004 *(.010)*	.002 *(.007)*	.003 *(.007)*	.005 *(.008)*	.004 *(.008)*	.94
q mean *(Std.dev.)*	.65 *(.19)*	.60 *(.20)*	.72 *(.70)*	.54 *(.25)*	.61 *(.37)*	1.99
Penetr. 08 mean *(Std.dev.)*	.19 *(.16)*	.55 *(.30)*	.86 *(.29)*	1.07 *(.28)*	.69 *(.43)*	108.64***
CM time mean *(Std.dev.)*	9.90 *(3.54)*	10.75 *(3.26)*	10.73 *(2.77)*	9.10 *(2.88)*	10.00 *(3.16)*	3.21**
CM penetr. mean *(Std.dev.)*	.07 *(.07)*	.16 *(.09)*	.23 *(.09)*	.23 *(.06)*	.18 *(.10)*	38.51***
CM year mean *(Std.dev.)*	2005.72 *(1.35)*	2004.79 *(1.33)*	2002.91 *(2.04)*	1999.58 *(2.15)*	2002.79 *(3.11)*	114.57***
Peak time mean *(Std.dev.)*	12.09 *(3.52)*	13.12 *(3.44)*	13.33 *(3.23)*	11.97 *(2.89)*	12.55 *(3.26)*	1.99
CM pen. / m mean *(Std.dev.)*	.21 *(.02)*	.21 *(.01)*	.21 *(.01)*	.20 *(.01)*	.21 *(.01)*	2.12*

*** $p< 0.01$, ** $p< 0.05$, *$p<.10$

while the estimated market potential in the lower middle category is 78% of the population. The low income countries stand out most clearly, with the expected saturation level at only one-third of the population. The estimated innovation coefficients are close to zero, with larger variance within the development categories than between them. The innovation coefficient being close to zero is typical for products with direct network effects. The imitation coefficients range from about 0.5 in the high income countries to 0.7 in the upper middle income group.

According to the Bass model estimates, the critical mass has already been achieved in all income groups. The high income countries achieved it slightly before the turn of the millennium, followed by upper middle category countries around 2003, lower middle income countries around 2005, and finally low income countries around 2006. The average time to achieve the critical mass was about ten years, while the annual adoptions peaked between 12 and 13 years after the start of the diffusion. The penetration level at the critical mass point varied from the low of 7% in low income countries to the high of 23% in the high income category. It appears that the critical mass penetration level for the diffusion of wireless communications is about one-fifth of the market saturation penetration level regardless of the level of development.

Figure 4 summarizes the estimated diffusion curves in the various development categories. The category-level diffusion curves are calculated

Figure 4. Estimated median diffusion curves for country categories

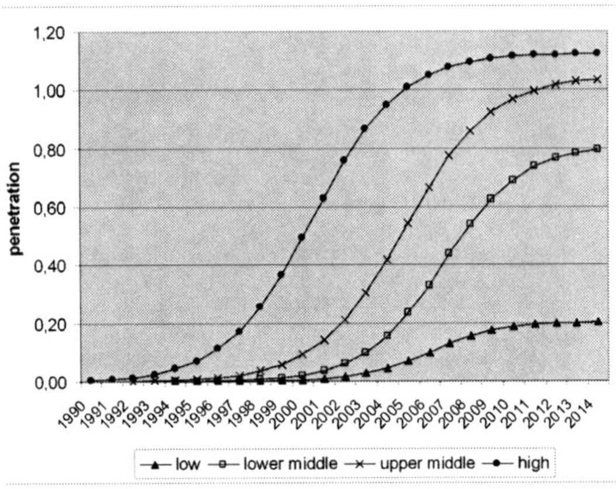

from the median values estimated for individual countries.

As a further illustration of the similarities and differences across the four development categories, we examine more closely the diffusion cases in Tanzania (low income), China (lower middle income), Brazil (upper middle income) and Finland (high income). Table 3 shows the values for our key variables in these countries.

Finland

Finland is a very sparsely populated high-income country, where distribution of income among the population is among the most equal in the world. It was the world leader in the adoption of mobile communications, starting with the NMT technology in 1982, followed by the GSM standard. The competition among operators was allowed from 1993, and the critical mass of adopters was achieved in 1996, when one fifth of the population had adopted wireless communications. The innovation coefficient (p) is above and the imitation coefficient (q) below the average in the reference group, this may be due to the slightly higher GDP per capita. The average revenues per user have remained at a high level in Finland due to extensive use of additional services like SMS messages, multimedia messaging, and Internet connection over the mobile handset.

Brazil

Brazil is a large country in terms of both population and area, the population density is low. GDP per capita is high but unequally distributed. The mobile phone diffusion in Brazil has started in 1992, and it has taken about 11 years to achieve the critical mass. This is near the average of the upper middle income countries which is 10.7 years. The diffusion coefficient for the market potential is about the same as in the reference group on average. The innovation coefficient is above and imitation coefficient below the average. This could be due the above average Gini coefficient of Brazil, which indicate less equal income distribution. The wireless communications in Brazil is characterized by the application of very many different technologies, and this may have slowed down the diffusion in the beginning as evidenced by the much lower innovation coefficient compared to Finland. The competition

Table 3. Values of key variables in the case countries

Country	start year	m	p	q	penetr 2008	cm_t	cm_ penetr	popden 1995	GDP_PC 1995	Gini 1995
Finland	1990	1.19	0.0136	0.31	1.17	5.58	0.21	16.7	27685.4	22.8
Brazil	1992	1.02	0.0017	0.36	0.71	11.09	0.21	19.1	2381.0	61
China	1992	0.55	0.0014	0.43	0.47	10.24	0.11	128.5	730.0	35.9
Tanzania	1995	0.72	0.0002	0.54	0.31	12.07	0.15	33.8	94.8	-

Figure 5. Diffusion in Finland

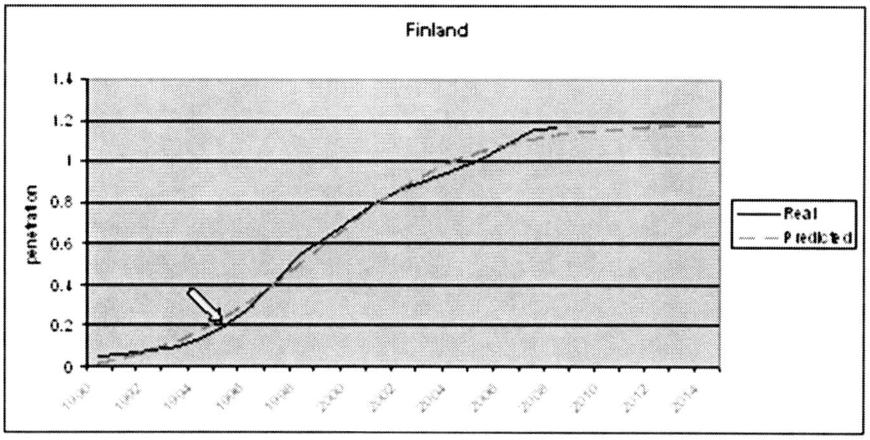

emerged in 1995, i.e. three years from the start, but achieving the critical mass took much longer. After the emergence of competition among wireless service operators, the average revenues per user (ARPU) have declined drastically, from about 70 USD per month in 1998 to 10 USD per month in 2003. After the critical mass was reached in 2003, the ARPU has remained stable.

China

With its enormous population and fast economic growth, China has long been seen as the country offering the most potential for wireless services. In China the mobile phone diffusion has started in 1992, which is two years earlier than in the lower middle income countries on average. The critical mass of users was achieved in about ten years (2002) at about 11% penetration rate. The estimated eventual market potential is only about 55% of the population, which is clearly smaller than the corresponding estimates in higher-income countries. However, as the current penetration rate is already 46%, it is hard to believe that the growth would stabilize at such a low ultimate penetration. All three diffusion coefficients for

Figure 6. Diffusion in Brazil

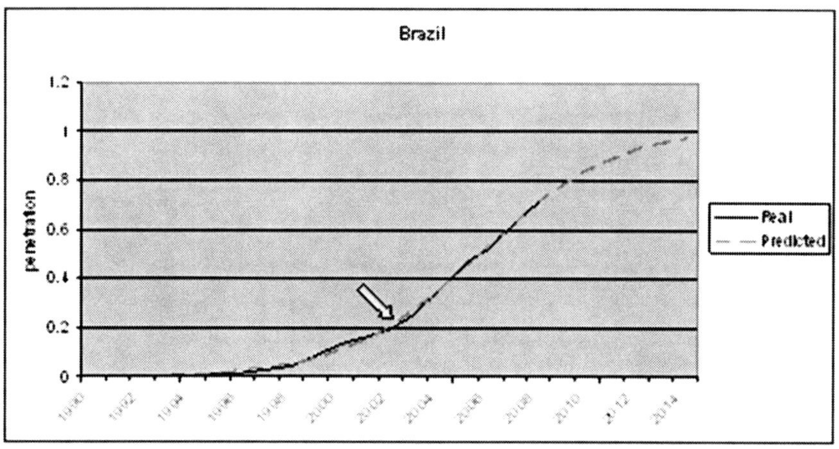

Figure 7. Diffusion in China

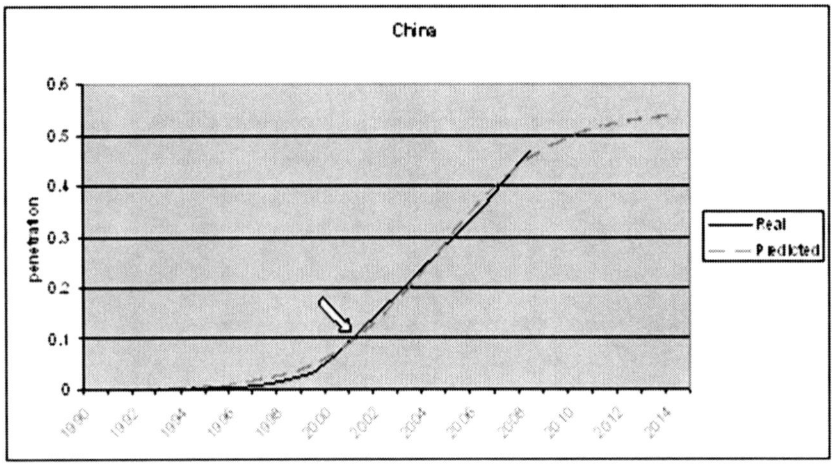

China are below the reference group average. The average revenues per user in China have developed in a similar manner as in Brazil, and stabilized at around ten USD per month.

Tanzania

In Tanzania the diffusion of mobile phones has started in 1995, which is a bit earlier than in low income countries on average. The achievement of the critical mass of users has taken about 12 years, which is longer than in the low income countries on average. One reason for this may be that the GDP per capita in 1995 Tanzania has been below the low income country group mean. From the diffusion model coefficients the market potential (m) has been greater and both innovation and imitation coefficients lower compared to the reference group. Since 1995, Tanzania has utilized two competing technologies until 2005, and competition among operators emerged at the end of the millennium. In 2003, there was a rapid increase in adoptions and at the same time the average revenues per user declined from 30

Figure 8. Diffusion in Tanzania

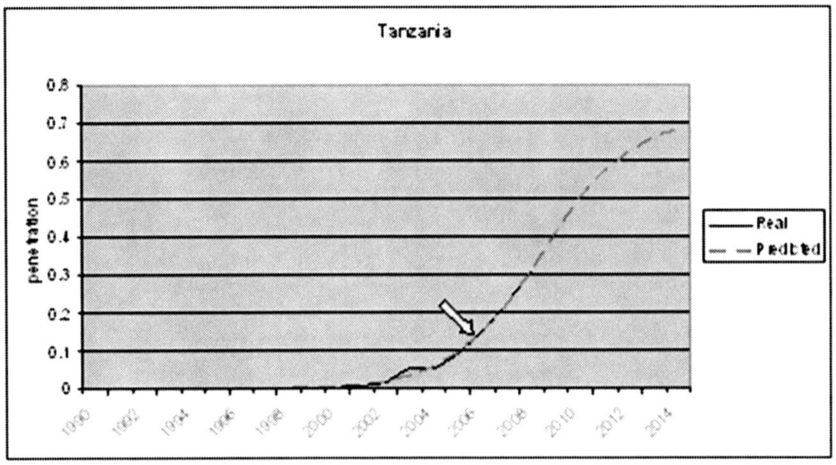

Table 4. Linear regression results: Start year, current penetration, and estimated sales peak as dependents

	Start year		Penetr. 08		Peak time	
R square (adj.)	.204 (.177)		.486 (.465)		.616 (.598)	
F	7.617***		22.351***		34.037***	
Independents	B	t	B	t	B	t
Start year	n.a.	n.a.	-.513***	-6.943	-.849***	-12.584
Population 1995	.015	.189	-.024	-.363	.094	1.561
Pop. Density 95	-.059	-.712	.110	1.646	-.025	-.410
GDP per capita 1995	-.434***	-4.746	.169**	2.099	-.414***	-5.645
GINI	.009	.100	-.164**	-2.259	.174**	2.619
*** p< 0.01, ** p< 0.05, *p<.10, B: standardized regression coefficient						

to 20 USD per month. The recent levels of ARPU are about the same as in middle income countries China and Brazil.

The effects of demographic and economic characteristics on the diffusion of wireless communications were tested by linear regression. Table 4 shows that 20.4% of the variation in the start year can be accounted for by population and economic factors. Of these factors, GDP per capita is the only significant predictor with a standardized regression coefficient of -.434. This confirms the assumption that wealthier countries have adopted wireless communications earlier than less wealthy countries. The model for actual penetration level in 2008 is also significant (R Square .486), showing the obvious result that wealthier countries have higher penetration levels, but also that when the effect of adoption timing and wealth are taken into account, the current penetration is higher in countries that have more equal distribution of income within the country. The time taken for the annual adoptions to peak is significantly shorter for later adopter countries, confirming the earlier results by e.g. Rouvinen (2006) and Gruber and Verboven (2001). Wealthier countries, and those with more equal income distribution, also achieve the sales peak faster.

Table 5 shows the linear regression analysis results for the estimated Bass diffusion model parameters. The models for explaining the variation in market potential (m) and innovation coefficient (p) are significant with R squares around

Table 5. Linear regression results: Bass diffusion model parameters as dependents

	m		p		q	
R square (adj.)	.308 (.276)		.359 (.328)		.135 (.094)	
F	9.453***		11.859***		3.302***	
Independents	B	t	B	t	B	t
Start year	-.477***	-5.272	.625***	7.167	.121	1.195
Population 1995	.058	.717	-.002	-.024	-.110	-1.215
Pop. Density 95	.101	1.227	.010	.125	-.008	-.086
GDP per capita 1995	.088	.890	.489***	5.158	-.279***	-2.534
GINI	-.034	-.379	.185**	2.151	-.291**	-2.923
*** p< 0.01, ** p< 0.05, *p<.10, B: standardized regression coefficient						

Table 6. Linear regression results: Critical mass variables as dependents

	CM time		CM penetr.		CM year		CM pen. / m	
R square (adj.)	.610 (.592)		.314 (.282)		.633 (.616)		.399 (.370)	
F	33.190***		9.715***		36.617***		14.060***	
Independents	**B**	**t**	**B**	**t**	**B**	**t**	**B**	**t**
Start year	-.828***	-12.177	-.526***	-5.834	.269***	4.080	-.540***	-6.403
Population 1995	.016	.262	.045	.557	.002	.030	-.061	-.815
Pop. Density 95	-.027	-.438	.102	1.249	-.027	-.457	.019	.243
GDP per capita 1995	-.584***	-7.900	.003	.035	-.627***	-8.750	-.680***	-7.408
GINI	.069	1.038	-.047	-.532	.014	.210	-.250***	-3.003
*** p< 0.01, ** p< 0.05, *p<.10, B: standardized regression coefficient								

.30 for both models. The model for imitation coefficient (q) explains only about 14 percent of the variation across countries. Earlier starters have higher market potentials and larger innovation coefficients. Population factors are not at all significant, but the economic factors have some significant effects. A country's wealth is positively related to the innovation coefficient, but negatively to the imitation coefficient. These coefficients imply that adoptions due to external information are relatively more significant in rich countries, whereas adoptions due to the imitation effect are relatively more important in poorer countries. When the adoption timing and income level are taken into account, the equality of within-country income distribution reduces the importance of external information and enhances the imitation effect.

The regression analysis results for the critical mass are shown in Table 6. The time taken from the start of the diffusion to critical mass was only weakly related to GDP per capita. Wealthier countries achieve critical mass faster. The calendar year of critical mass occurrence is naturally very strongly associated to wealth, as both start year and time taken to critical mass occur sooner in wealthier countries. Wealth also has a significant positive effect on the penetration level where critical mass occurs. Critical mass occurs at a higher absolute penetration in more developed countries, but as the last column of Table 6 shows, critical mass penetration is a smaller share of the ultimate penetration. Furthermore, the significant negative impact of the Gini coefficient implies that critical mass occurs at a smaller relative percentage of market potential when income is unequally distributed within the country.

FUTURE RESEARCH DIRECTIONS

Although our study covered a wider range of countries than any of the previous studies on the diffusion of wireless communications, and provided new knowledge about the achievement of the critical mass in developing as well as developed countries, it is not without limitations. First of all, we only applied the Bass diffusion model to estimate the market potential and innovation and imitation coefficients. While the Bass model has gained wide acceptance, especially in the field of marketing studies, other epidemic diffusion models like the logistic and Gompertz functions are also often applied in diffusion studies, especially in the field of economics. As the results indicated that the innovation coefficients were generally very close to zero, the simpler logistic function could also be used in future research.

Secondly, after estimating the parameters of the diffusion model, we applied a cross-sectional

OLS regression to explain the variation in the diffusion parameters. Particularly when the timing variables (peak time and takeoff time) are of interest, hazard models with time-varying covariates would provide a more rigorous way to test the relationships.

Thirdly, we used a very limited number of explanatory variables, as only two basic population indicators and two basic income indicators were examined. Further research might benefit from the inclusion of also socio-cultural demand-side covariates and particularly supply-side indicators, e.g. competition, payment policies and standards. It might also be beneficial to analyze the level of usage instead of just the number of users, as suppliers' revenues are normally very much usage-based rather than subscription-based (Grajek and Kretschmer, 2009).

The results for eventual market potential in low-income countries also raised concern that the diffusion model may be grossly underestimating the real potential. The mean market potential for the low income countries was about 33% of the population, but the median value was as low as 20%. For example, our estimate for market potential in India was 33%, whereas the study by Singh (2008) predicts a penetration of 71% in India by the year 2015. Gruber (2001) studied the diffusion of mobile communications in Central and Eastern Europe, and using data from 1990-1997 predicted the eventual market potential to be about 20% of the population. This has proven to be a gross underestimation, since the 2008 penetration levels in these countries range from the low of 69% for Albania to the high of 155% for Estonia. It may well be the case that when estimating with a limited number of observations, the market potential estimates tend to be biased downward. The existence and magnitude of this potential bias would definitely deserve further analysis.

Finally, even though our study was among the first to explicitly study the level of penetration at the critical mass point, the results of the critical mass point were determined by the estimated diffusion parameters. The estimated penetration at critical mass is thus heavily dependent on correct estimation of the market potential. More explicit evaluation of the critical mass penetration/market potential ratio deserves closer attention in further studies.

CONCLUSION

The results showed tremendous differences in the diffusion process and pace of adoption between developed and developing countries. This is partly due to the well-documented lead-lag effect in international diffusion, i.e. that in many developed countries the diffusion has taken off years earlier than in developing countries, which has resulted in quicker diffusion processes for the later group (Ganesh et al. 1997). Moreover, while the new technology may be relatively more expensive for an average user in a developing country, the growth of utility may be greater (in cases where the mobile phone is the first workable solution for long-distance communication) compared to that experienced by an average user in a developed country. This may have hastened the adoption of the innovation in developing countries.

With data covering more than 200 countries we were able to show that size and density of population within a country play a minor role in the diffusion of wireless communication. Income appears to be the main driver of demand. Higher GDP per capita promotes an early start to adoption, higher ultimate market penetration, faster diffusion, and faster achievement of the critical mass in terms of both time and share of the ultimate market potential. The results are in line with the 184-country study by Dekimpe et al (1998), but in contrast with the results from OECD countries by Jang et al (2005) and from EU countries by Gruber and Verboven (2001). The contrasting results about the role of wealth in our global-scale study, compared with previous studies, concentrated on a limited set of countries and clearly demonstrates the need for further global studies in order to understand the diffusion patterns in less developed countries. The least developed countries clearly stand out from the middle and high development categories, as the Bass model predictions indicate that the level of mobile penetration will saturate at only about one-fifth to one-third of the population. It seems that if many of these countries remain marginalized in terms of economic growth, the majority of population will remain outside the reach of wireless communications.

REFERENCES

Agarwal, R., & Bayus, B. L. (2002). The market evolution and sales takeoff of product innovations. *Management Science*, *48*(4), 1024–1041. doi:10.1287/mnsc.48.8.1024.167

Bass, F. (1969). A new product growth for model consumer durables. *Management Science*, *15*(5), 215–225. doi:10.1287/mnsc.15.5.215

Botelho, A., & Pinto, L. C. (2004). The diffusion of cellular phones in Portugal. *Telecommunications Policy*, *28*, 427–437. doi:10.1016/j.telpol.2003.11.006

Church, J., & Gandal, N. (1993). Complementary network effects and technological adoption. *International Journal of Industrial Organization*, *11*(2), 239–260. doi:10.1016/0167-7187(93)90006-X

Dekimpe, M. G., Parker, P. M., & Sarvary, M. (1998). Staged estimation of international diffusion models: An application to global cellular telephone adoption. *Technological Forecasting and Social Change*, *57*(1-2), 105–132. doi:10.1016/S0040-1625(97)00085-1

Economides, N. (1991). Compatibility and the creation of shared networks . In Guerin-Calvert, M. E., & Wildman, S. S. (Eds.), *Electronic services networks: A business and public policy challenge*. New York.

Ganesh, J., Kumar, V., & Subramaniam, V. (1997). Learning effect in multinational diffusion of consumer durables: An exploratory investigation. *Journal of the Academy of Marketing Science*, *25*(3), 214–228. doi:10.1177/0092070397253003

GMID. (n.d.). Retrieved from http://portal.euromonitor.com/PORTAL/tandc.aspx

Golder, P. N., & Tellis, G. J. (1997). Will it ever fly? Modeling the takeoff of really new consumer durables. *Marketing Science*, *16*(3), 253–270. doi:10.1287/mksc.16.3.256

Grajek, M., & Kretschmer, T. (2009). Usage and diffusion of cellular telephony, 1998-2004. *International Journal of Industrial Organization*, *27*, 238–249. doi:10.1016/j.ijindorg.2008.08.003

Griliches, Z. (1957). Hybrid corn: An exploration in the economics of technological change. *Econometrica*, *25*, 501–522. doi:10.2307/1905380

Gruber, H. (2001). Competition and innovation: The diffusion of mobile telecommunications in central and Eastern Europe. *Information Economics and Policy*, *13*, 19–34. doi:10.1016/S0167-6245(00)00028-7

Gruber, H., & Verboven, F. (2001). The diffusion of mobile telecommunications services in the European Union. *European Economic Review*, *45*, 577–588. doi:10.1016/S0014-2921(00)00068-4

Haapaniemi, T. P., & Mäkinen, S. J. (2007). Determining the takeoff point in adoptions of innovations: A comparison of content and discrimination analysis. *International Journal of Technology Marketing*, *2*(1), 65–80. doi:10.1504/IJTMKT.2007.011586

Hahn, H. P., & Kibora, L. (2008). The domestication of the mobile phone. Oral society and new ICT in Burkina Faso. *The Journal of Modern African Studies*, *46*, 87–109. doi:10.1017/S0022278X07003084

James, J., & Versteeg, M. (2007). Mobile phones in Africa: How much do we really know? *Social Indicators Research*, *84*, 117–126. doi:10.1007/s11205-006-9079-x

Jang, S.-L., Dai, S.-C., & Sung, S. (2005). The pattern and externality effect of diffusion of mobile telecommunications: The case of the OECD and Taiwan. *Information Economics and Policy*, *17*, 133–148. doi:10.1016/j.infoecopol.2004.05.001

Jha, R., & Majumdar, S. K. (1999). A matter of connections: OECD telecommunications sector productivity and the role of cellular technology diffusion. *Information Economics and Policy*, *11*, 243–269. doi:10.1016/S0167-6245(99)00017-7

Katz, M. L., & Shapiro, C. (1985). Network effects, competition, and compatibility. *The American Economic Review*, *75*(3), 424–440.

Kim, M.-S., & Kim, H. (2004). Innovation diffusion of telecommunications: General patterns, diffusion clusters and differences by technological attribute. *International Journal of Innovation Management*, *8*(2), 223–241. doi:10.1142/S136391960400099X

Kim, M.-S., & Kim, H. (2007). Is there early take-off phenomenon in diffusion of IP-based telecommunications services? *The International Journal of Management Science*, *35*, 727–739.

Koski, H., & Kretschmer, T. (2005). Entry, standards and competition: Firm strategies and the diffusion of mobile telephony. *Review of Industrial Organization*, *26*, 89–113. doi:10.1007/s11151-004-4085-0

Lim, B.-L., Choi, M., & Park, M.-C. (2003). The late take-off phenomenon in the diffusion of telecommunication services: Network effect and the critical mass. *Information Economics and Policy*, *15*, 537–557. doi:10.1016/S0167-6245(03)00037-4

Madden, G., Coble-Neal, G., & Dalzell, B. (2004). A dynamic model of mobile telephony subscription incorporating a network effect. *Telecommunications Policy*, *28*, 133–144. doi:10.1016/j.telpol.2003.12.002

Mahajan, V., Muller, E., & Wind, Y. (Eds.). (2000). *New-Product Diffusion Models*. Boston: Kluwer Academic Press.

Mahler, A., & Rogers, E. M. (1999). The diffusion of interactive communication innovations and the critical mass: The adoption of telecommunications services by German Banks. *Telecommunications Policy*, *23*(10-11), 719–740. doi:10.1016/S0308-5961(99)00052-X

Markus, M. L. (1990a). Critical mass contingencies for telecommunications consumers . In Carnevale, M., Lucertini, M., & Nicosia, S. (Eds.), *Modeling the innovation: Communications, automation and information* (pp. 103–112). Amsterdam: Elsevier.

Markus, M. L. (1990b). Toward a critical mass theory of interactive media . In Fulk, J., & Steinfield, C. W. (Eds.), *Perspectives on new information technology in organizations*. Newbury Park, CA: Sage.

Meade, N., & Islam, T. (2006). Modelling and forecasting the diffusion of innovation – A 25-year review. *International Journal of Forecasting*, *22*, 519–545. doi:10.1016/j.ijforecast.2006.01.005

Rogers, E. M. (1983). *Diffusion of innovations* (3rd ed.). New York: Free Press.

Rouvinen, P. (2006). Diffusion of digital mobile telephony: Are developing countries different? *Telecommunications Policy*, *30*, 46–63. doi:10.1016/j.telpol.2005.06.014

Schoder, D. (2000). Forecasting the success of telecommunication services in the presence of network effects. *Information Economics and Policy*, *12*, 181–200. doi:10.1016/S0167-6245(00)00006-8

Singh, S. K. (2008). The diffusion of mobile phones in India. *Telecommunications Policy*, *32*, 642–651. doi:10.1016/j.telpol.2008.07.005

Tellis, G. J., Stremersch, S., & Yin, E. (2003). The international takeoff of new products: The role of economics, culture, and country innovativeness. *Marketing Science*, *22*(2), 188–208. doi:10.1287/mksc.22.2.188.16041

United Nations Human Development Report 2007/08 (2008). Retrieved from http://hdr.undp.org/en/media/HDR_20072008_EN_Complete.pdf

Valente, T. W. (1995). *Network models of the diffusion of innovations*. Creskill, NJ: Hampton Press.

Weiber, R. (1995). Systemguter und klassische diffusiontheorie: Elemente einer diffusionstheorie kristische masse-systeme . In Stoetzer, M.-W., & Mahler, A. (Eds.), *Die diffusion von innovationen in der telekommunikation*. Berlin: Springer.

Wiorkowski, J. J., & Gylys, V. A. (2006). An empirical real-time test for takeoff with applications to cellular telephony. *Review of Marketing Science,* 4, Article 1.

Witt, U. (1997). “Lock-in” vs. “Critical masses” - Industrial change under network effects. *International Journal of Industrial Organization*, *15*, 753–773.

World Bank. (n.d.). Retrieved from http://go.worldbank.org/K2CKM78CC0

ADDITIONAL READING

Allen, D. (1988). New telecommunications services: Network externalities and critical mass. *Telecommunications Policy*, *13*, 255–264.

Antonelli, C. (1993). Investment and adoption in advanced telecommunications. *Journal of Economic Behavior & Organization*, *20*(2), 227–246. doi:10.1016/0167-2681(93)90092-4

Bewley, P. P., & Fiebig, D. (1988). Flexible logistic growth model with applications in telecommunications. *International Journal of Forecasting*, *4*(2), 177–192. doi:10.1016/0169-2070(88)90076-3

Burki, A. A., & Aslam, S. (2000). The role of digital technology and regulations in the diffusion of mobile phones in Asia. *Pakistan Development Review*, *39*(4), 741–748.

Dekimpe, M. G., Parker, P. M., & Sarvary, M. (2000). ’Globalization’: Modeling technology adoption timing across countries. *Technological Forecasting and Social Change*, *63*(1), 25–42. doi:10.1016/S0040-1625(99)00086-4

Frank, L., & Sundqvist, S. (2004). The effect of human development and technology achievement on the diffusion of wireless communications. In: Y. Hosni (Ed.): *Management of Technology: Internet Economy: Opportunities and Challenges for Developed and Developing Regions of the World.* Elsevier Science, 327-336.

Ganesh, J., & Kumar, V. (1996). Capturing the cross-national learning effect: An analysis of an industrial technology diffusion. *Journal of the Academy of Marketing Science*, *24*(4), 328–337. doi:10.1177/0092070396244004

Gatignon, H., Eliashberg, J., & Robertson, T. S. (1989). Modeling multinational diffusion patterns: and efficient methodology. *Marketing Science*, *8*, 231–247. doi:10.1287/mksc.8.3.231

Geroski, P. A. (2000). Models of technology diffusion. *Research Policy*, *29*(4-5), 603–625. doi:10.1016/S0048-7333(99)00092-X

Gort, M., & Klepper, S. (1982). Time paths in the diffusion of product innovations. *The Economic Journal*, *92*, 630–653. doi:10.2307/2232554

Helsen, K., Jedidi, K., & DeSarbo, W. S. (1993). A New approach to country segmentation utilizing multinational diffusion patterns. *Journal of Marketing*, *57*(October), 60–71. doi:10.2307/1252219

Ilonen, J., Kämäräinen, J.-K., Puumalainen, K., Sundqvist, S., & Kälviäinen, H. (2006). Toward automatic forecasts for diffusion of innovations. *Technological Forecasting and Social Change*, *73*(2), 182–198. doi:10.1016/j.techfore.2004.11.005

Kiiski, S., & Pohjola, M. (2002). Cross-country diffusion of the internet. *Information Economics and Policy*, *14*(2), 297–310. doi:10.1016/S0167-6245(01)00071-3

Kumar, V., Ganesh, J., & Echambadi, R. (1998). Cross-national diffusion research: What do we know and how certain are we? *Journal of Product Innovation Management*, *15*, 255–268. doi:10.1016/S0737-6782(97)00082-9

Kumar, V., & Krishnan, T. (2002). Multinational diffusion models: An alternative framework. *Marketing Science*, *21*(3), 318–330. doi:10.1287/mksc.21.3.318.139

Mahajan, V., & Muller, E. (1994). Innovation diffusion in a borderless global market: Will the 1992 unification of the European community accelerate diffusion of new ideas, products, and technologies? *Technological Forecasting and Social Change*, *45*, 221–235. doi:10.1016/0040-1625(94)90047-7

Mahajan, V., Muller, E., & Bass, F. (1990). New product diffusion models in marketing: A review and directions for research. *Journal of Marketing*, *54*, 1–26. doi:10.2307/1252170

Maiorano, F., & Stern, J. (2007). Institutions and telecommunications infrastructure in low and middle-income countries: The case of mobile telephony. *Utilities Policy*, *15*, 165–181. doi:10.1016/j.jup.2007.04.001

Puumalainen, K. (2004). Predicting the global diffusion of internet usage – effects of market environment. In: Y. Hosni and T.Khalil (Eds.): *Internet Economy: Opportunities and Challenges for Developed and Developing Regions of the World*. Elsevier Science, 15-26.

Puumalainen, K., & Sundqvist, S. (2005). Global diffusion of telecommunications innovations – a cross-cultural review. *International Journal of Technology Marketing*, *1*(1), 22–36. doi:10.1504/IJTMKT.2005.008123

Puumalainen, K., Sundqvist, S., & Huiskonen, J. (2007). Managing uncertainty in forecasting the diffusion of innovations in telecommunications. *Journal of Scientific and Industrial Research*, *66*(4), 299–305.

Sultan, F., Fraley, J. U., & Lehmann, D. R. (1990). A meta-analysis of applications of diffusion models. *JMR, Journal of Marketing Research*, *37*, 70–77. doi:10.2307/3172552

Sundqvist, S., Frank, L., & Puumalainen, K. (2005). The Effects of Country Characteristics, Cultural Similarity and Adoption Timing on the Diffusion of Wireless Communications. *Journal of Business Research*, *58*(1), 107–110. doi:10.1016/S0148-2963(02)00480-0

Takada, H., & Jain, D. (1991). Cross-national analysis of diffusion of consumer durable goods in pacific rim countries. *Journal of Marketing*, *55*, 48–54. doi:10.2307/1252237

Chapter 2
The Competitive Growth Pattern of Mobile Telecommunications in Korea

Moon-Soo Kim
Hankuk University of Foreign Studies (HUFS), Korea

Sungjoo Lee
Ajou University, Korea

ABSTRACT

This study focuses on two perspectives from the competition between technologies and the change of market structure in terms of the number of operators in mobile service market. On the promise that the competition between the heterogeneous technologies as well as between the mobile service operators is a core driver of rapid diffusion, a diffusion model that can describe such two types of competition is proposed and further applied to the Korean mobile service case to examine the demand diffusion patterns of service and its characteristics. Through the empirical study on Korean mobile market, we investigate not only the effects on demand diffusion patterns by the competition between technologies and operators, but also strategic implications for the service providers and the policy-makers.

INTRODUCTION

Economic growth in such countries as Japan (Jorgenson & Nomura, 2005), Korea (Shin & Park, 2007), and Finland (Jalava & Pohjola, 2007), which are technically advanced in Information and Communications Technology (ICT), is dominated by investments and productivity growth in ICT, both for industries and the economy as a whole. Especially, Korea is well-known for its fast diffusion of mobile subscribers and technological advances in the industry. And thus it is worth understanding the factors that have driven such phenomena in Korea.

Mobile communication services in Korea started in the 1960s, which were based on a manual switching system for a division of government agencies in Seoul and a few other areas in the National Capital region. The services became importance in the 1980s as the needs of mobile communications grew rapidly. To respond to such needs, the Korean government established Korea Mobile Communication Services (KMCS) in March 1984. It provided a cellular service, taking

DOI: 10.4018/978-1-61692-818-6.ch002

the whole responsibility of mobile communications in Korea. The industry structure, however, was changed from monopoly to duopoly when KMCS was privatized (renaming as SK Telecom in 1998), and the Sinsegi Telecom was selected as a new service operator by the government in 1994. Later, in 1996, the government allowed another technology, PCS (Personal Communications Services) to enter the industry to facilitate its competition. As a result, three new operators including KTF, LG Telecom, and Hansol PCS began to provide PCS services. Since then, the two types of services have been fast-growing through active and aggressive marketing and now one cellular and two PCS operators exist after two M&As. Likewise, two factors, 'the competition between two technologies – inter-technology competition' and 'the competition between operators – intra-technology competition' are regarded as a core driver in the growth of Korean mobile communications industry. Therefore, this study focuses on two perspectives, one concerning the competition between technologies and the other concerning the changes of market structure in terms of the number of service providers in the market, in other words, the competition between operators.

And so, this chapter purposes to examine the Korean mobile service market by adopting a diffusion model incorporating the two factors and getting implications from empirical analysis. It is feasible and meaningful to study the Korean case since the growth of the industry is remarkable and the market structure is simple to analyze. Actually, theoretically and empirically diverse studies have been carried out addressing an explosive growth and a great success in the Korean mobile telecommunication (Choi et al., 2001; Kim et al., 2006; Kim & Kim, 2004), but most of them didn't consider market and technology together, although the combination of two factors could be a main driver for rapid diffusion (Lee et al., 2009; Kim et al., 1999). For the purpose of analysis, a LVC (Lotka-Volterra Competition) model, a famous competitive diffusion model, was adopted and applied to five successive periods, distinguished by an event which caused important changes in market structure such as new entrants, exits, and M&A from Dec. 1993 to Feb. 2003. The research findings are expected to help us investigate the effects of competition on demand diffusion patterns and further to provide strategic implications with mobile service operators and policy-makers in the mobile industry.

OVERVIEW OF KOREAN MOBILE TELECOMMUNICATIONS

Technology

Korean mobile telecommunications services are divided into two types according to the technology applied; one is a cellular service using 800 MHz band and the other is Personal Communications Services (PCS) based on 1,800 MHz band. Looking into the industry, the overall competition patterns are summarized as the competition between cellular operators and PCS operators, where the former is an early entrant dominating the industry and the latter is a late follower with relatively small size. Hence, we regard cellular service as the first generation technology and PCS as the second generation technology. The existing studies that highlight the competition in the industry take the similar approach by treating the competition between cellular operators and PCS operators as competition between two different technologies (Choi et al., 2001).

In case of mobile telecommunications services in Korea, cellular and PCS are not substituted as the second generation technology is not necessarily better than the first generation technology. On the contrary, some argue that cellular is superior to PCS in quality and price since the 1800 MHz band for cellular is more advantageous than the

Figure 1.

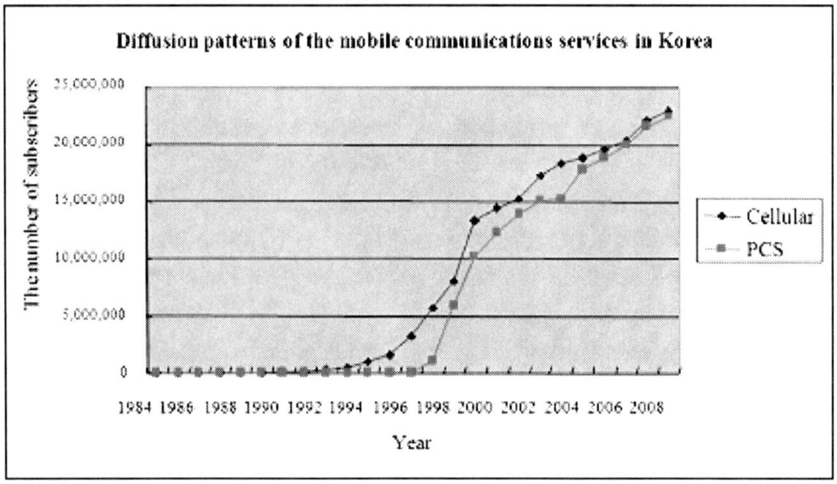

800 MHz band for PCS in terms of frequency characteristics and investment of base stations. Nonetheless, when a new ICT service is introduced in a market, it usually has remarkable advantages compared to the existing services. Likewise, at the introduction of PCS, the article from the electronic newspaper (http://www.etnews.co.kr/) on the 1st of October, 1997 insists that PCS has better performance than cellular. Theoretically, PCS technology has potential to develop various value-added services since it is based on the newest technology. Also, the sound quality might be better when using PCS than cellular because the baseband of PCS is 2.3 times as wide as that of cellular, which enables to split sound into 13,000 pieces to transfer digital voice as contrasted with cellular splitting sound into 8,000 pieces.

Considering such issues, this study takes the competition approach to analyzing the relations between technologies, especially focusing on cellular and PCS, and collected data on the number of subscribers to each of them. On the other hand, as the Korean mobile telecommunications service industry has only a few operators, their competition and M&A process will significantly affect the market, and so gathered data required to reflect the number of operators in a model.

Market

The Korean mobile telecommunications service was launched by the mobile phone for use in a car, so-called a car phone, in May 1984. The price of a car phone including subscription charge was about 4.1 million won at that time, which was almost worth of a compact car made in Korea. Thus the initial number of mobile phone subscriber was no more than 2,658. The number of subscribers was explosively increased in the past 25 years, however, 1 million in 1996, 10 million in 1998 and almost 46 million by early 2009 that jumped up 17,300 times since 1984 and reached its saturation point in 2007, considering 49 million people, the population of South Korea (see Figure 1). Figure 1 shows the number of subscribers to the mobile communications services where both cellular and PCS services seem to grow rapidly. The data was acquired from the ETRI (Electronic and Telecommunications Research Institute) website (http://www.etri.re.kr).

In this study, the period from 1993 to 2003 is the main focus, which has dynamism in market structure, and the period is divided into five periods according to the major events that caused the dynamism. *The first period (T1)* raging from

Table 1.The distinctions and descriptions of five periods

Period	Duration	Competition type	Description
T1	December 1993 ~ December 1996	No competition	Monopoly of cellular technology - SKT existed (Cellular)
T2	January 1997 ~ September 1997	Inter-technology competition	Entrant of new cellular service provider - Sinsegi Telecom entered (Cellular)
T3	October 1997 ~ April 2001	Inter- and Intra-technology competition	Introduction of PCS services - LGT, KT and Hansol entered (PCS)
T4	May 2001 ~ November 2001	Inter- and Intra-technology competition	Competition weakened in PCS services - KTF annexed Hansol (PCS)
T5	December 2001 ~ September 2008	Inter- and Intra-technology competition	Competition weakened in cellular services - SKT annexed Sinsegi Telecom (Cellular)

December 1993 to December 1996 was when SK Telecom (Korea Mobile Communication Co. Ltd at that time) had monopolistically provided mobile service based on cellular technology with 800 MHz band, during which only cellular technology was diffused.

The second period (T2) covers from January to September 1997 when the second mobile service operator, Shinsegi Telecom turned to provide the same cellular service and the first operator had become privatized. Therefore, a duopolistic structure in market had set up and a competition within cellular technology begun.

The third period (T3) from October 1997 to April 2001 is characterized as the most competitive period where five operators were competing and thus is frequently used for empirical analysis. At the beginning of T3, a new technology, PCS with 1.8 GHz band was introduced and three operators including KTF, LGT and Hansol were chosen to provide PCS services. The Korean competition policy on mobile market, in general, succeeded in the rapid demand diffusion of mobile service through such competition means among 5 operators as mobile phone subsidies, abolition of contract-using period and diverse promotion policies, etc. Such those severe competition between mobile operators, however, has caused to be getting worse their financial situation that resulted in M&A between them. Based on the M&A, the fourth and fifth periods are divided.

The fourth period (T4) corresponding to the period from May to November 2001 is when the competition within PCS technology was weakened as KTF merged Hansol PCS. The mobile service market was reaching the peak of its growth rate, where the growth was driven more by PCS subscribers than cellular ones.

Finally, *the fifth period (T5)* from December 2001 to September 2008 is the most recent one in which the market consists of three operators including SKT, KTF and LGT, after SKT merged Shinsegi Telecom. This three-operator structure has continued until now. Particularly in T5, the number of operators for cellular services was reduced to one and the Korean government started to impose legal controls on SKT, the only cellular service operator, in order to prevent its anti-competition practices with a significant market power. Actually, SKT had advantages over the other two operators in technology and market, as it entered the industry as first entrant and was dominating the market with more than 50% of market. In addition, general conception for service subscribers was that cellular technology was fundamentally superior to PCS. Table 1 summarizes the characteristics of each period.

This study applies the LVC model and estimate parameters for each period. Then, by analyzing the value of parameter together with the trend of value changes, the effect of technology competition and market structure on demand diffusion can be deduced. Here, the assumption that changes of parameter estimates result from changes of market structure is needed to justify the analysis. And the assumption is acceptable based on the claim that all marketing activities and institutional management influencing diffusion process depend on the number of competitors, a representative measure for market structure (Kim et al., 1999).

DATA AND MODEL

Although the most common model for diffusion analysis is Bass model (1969), LVC model is adopted in this research. In general, Bass model focuses on the diffusion and substitution of successive generations of technologies (Mahajan et al. 1996), while LVC model emphasized the diffusion and competition of several technologies (Lee et al., 2009). Since it is not the case that PCS, the late entrants is superior to Cellular, the early entrant nor is cellular service substituted by PCS, the LVC model is more appropriate. Then the LVC model was applied to the number of subscribers to the mobile communication services in Korea.

Data

Data in this study covers the period from 1993 to 2008. The PCS operators got down to providing their service in 1993 and both the competition between technologies and the market structure were stabilized in 2003. So, the period is enough to extract some meaningful implications for mobile telecommunications service market. And thus the monthly data for the number of subscribers to each operators and the number of operators in market were collected. In total, 172 monthly data for cellular from December 1993 to September 2008 was collected while 127 monthly data for PCS from October 1997 to September 2008 was collected from the website of the Ministry of Information and Communications (http://mic.go.kr).

Diffusion Models

In general, there are two approaches to model technology diffusion: (1) curve-fitting; and (2) simulation.

Curve-fitting model: The curve-fitting model is based on data, usually fitting time-series data to a mathematical S-shaped curve model. Again, it can be classified into three types according to the parameters used in the model (Teng et al., 2002). The first one, *internal influence model* assumes no interaction between individuals in a social system during diffusion. Here, the diffusion is mainly caused by the internal effects such as mass media and consulting firms (Fourt & Woodlock, 1960). The second one, *external influence model* is the other way round, which assumes that the diffusion happens only by the contacts between individuals such as the well-known effect, word of mouth (Manfield, 1961). The most representative models include logistic model and Gompertz-function. The final one is a *mixed influence model*, suggested by Bass (1969) and thus called Bass model. This model considers both internal and external influences and has been popular in marketing areas, being frequently used to forecast demand of durable goods. Basically, the diffusion can be regarded as a communication process and then the means to diffuse technology is called a communications channel. This model divides potential adopter of new technology into two groups, the one which is affected by mass-media channel (external influence) and the other by interpersonal channel (internal influence), where the formal is called innovator and the latter imitator. Bass model has been developed in a way to release its assumptions, for example, a model incorporating marketing mix (Kalish, 1985), a model for multi-generation products, a model in competition environments (Mahajan et al., 1993), a model for complimentary products and necessities, a model considering geographical effects (Mahajan & Peterson, 1979), a model assuming dynamic potential market (Kalish, 1985; Mahajan

& Peterson, 1978) or constraint on supply (Jain & Rao, 1990).

Recently, various other models have been suggested to model the interactive and complex diffusion process among technologies, such as technological generation change, technology substitution and technology competition. The variants of Bass model or Lotka-Volterra Competition (LVC) equations are commonly used for this purpose.

Simulation model: The simulation model is a method to carry out a computer simulation by modeling a diffusion mechanism. Like a curve-fitting model, this approach begins with finding internal and external factors that affect diffusion, designs a simple model to describe the relationship between the factors, simulates the model, and finishes with verifying the results. This model has an advantage over others in that it enables to identify the cause-effect relations and the dynamics of factors.

Basic LVC Model

Competition diffusion models primarily rely on evolutionary game theory where it is assumed that evolution really originates in the results of co-evolution, whether biological or technological evolution. Here, co-evolution is inter-related process where the behaviours of system members taken to adapt the environments or market can alter the structure of other members in the system. This evolutionary game theory has been widely applied to economics since the mid of 1970s. The model based on this theory is different from general diffusion model in that the adoption of technology occurs only when it is the best decision against the strategies of all other participants in the game. In addition, this model is marked by the fact that it is difficult to assume rational decision-making. Therefore, when the model is applied to technology evolutionary process, the competition between technologies will be severe as more technologies exist and cause an irrational cutthroat competition as the interaction with competing technologies affects technology diffusion process.

Among the competition diffusion models, one of the most famous ones is the LVC (Lotka-Volterra Competition) model (Dieckmann, 1995; Lotka, 1925). This model constitutes the simplest description of two interacting populations (Lotka, 1956), and has several advantages over other models. Firstly, although quite simple, the model can fully describe well-known economic principles such as competitive effects and the law of increasing returns. Secondly, it can account for the effects of group behaviors, as well as those of individual behavior as emphasized in previous diffusion theories, by modeling both the internal and external effects on the populations of two different species. Finally, it can be applied to illustrate not only a predator-prey system, but also various other types of competition systems (Modis, 1999), and some essential interaction features can be studied by this model (Arato, 2003). It is thus judged a very practical model and studies are increasingly adopting it to illustrate reality (Kim et al., 2006).

The Lotka-Volterra equations were initially suggested to describe the interaction of biological species competing for the same resources (Lotka, 1956), and then applied to display parasitic and symbiotic relations or emerging and declining competitors, allowing for an intuitive understanding of the factors that drive co-evolution (Bazykin, 1998). Later interesting analogies between biological ecology and technological development were drawn, and several researchers started to use the equations to model competing technologies, since when the model has attracted considerable attention in the technology diffusion areas (Bharagava, 1989; Porter et al., 1991). For example, Bharagava (1989) applied the LVC model to technology substitution process and tried to define substitution types based on parameter values. Porter et al. (1991) looked at the LVC equations to understand the meaning of parameters and proposed that the model can take similar forms with other functions such as linear functions, exponential functions,

Table 2.The interaction patterns between two technologies

a_3	b_3	Interaction types	Description
+	+	pure competition	The growth of each technology is interrupted by the other technology.
-(+)	+(-)	predator-prey	The growth of one technology is interrupted by the other technology, while the growth of the other technology is facilitated by the technology.
-	-	mutualism	Two technologies co-exist, promoting the growth of the other technology.
-(0)	0(-)	commensalism,	The growth of one technology is facilitated by the other technology, which the other technology gets no influence from the technology.
+(0)	0(+)	amensalism	The growth of one technology is interrupted by the other technology, which the other technology gets no influence from the technology.
0	0	neutralism	There is no interaction between two technologies.

logistic functions, and Gomperz functions by restricting parameter values.

In a similar vein, when applied to the diffusion areas, the LVC model enables one to examine the technological diffusion process though competition, and also to investigate relations between two technologies (Modis, 1997). Thus different technologies can interact with each other in a variety of modes - competition, predator-prey, mutualism, commensalism, amensalism and neutralism as well as just in simple 'competition' relationships (Modis, 1997). Pistorius & Utterback (1997) classified technological interaction types into three categories - pure competition, symbiosis and predator-prey – based on the parameter values. Of course, the use of LVC model is not restricted to the long-term evolution of technology. Farrell (1993) applied the model to describe the technology competition for a short period of time. The following equations (1) are the basic LVC equations to describe the relations between two competing technologies.

$$\begin{aligned} \frac{dX_1(t)}{dt} &= (a_1 - a_2X_1(t) - a_3X_2(t))X_1(t) \\ \frac{dX_2(t)}{dt} &= (b_1 - b_2X_2(t) - b_3X_1(t))X_2(t) \end{aligned} \quad (1)$$

Here, $X_i(t)$: the cumulative demands for technology i at time t (i=1,2)

a_1, b_1 : the influence by its own diffusion

a_2, b_2 : the influence by its own growth

a_3, b_3 : the influence by competing technology diffusion

The equations show the relationships between technologies 1 and 2 during their diffusion process on the assumption that individuals can move from one technology to the other at will. Here, the parameters a_1 and b_1 indicate the relative growth rate of each technology, while a_2 and b_2 represent an internal interaction of each technology, for instant, the internal tendency to reduce residents of a particular region if its population is beyond all limits. Finally, a_3 and b_3 mean the influence by one technology on the other one. Consequently, Kim et al. (2006) defined the interaction types according to the signs of a_3 and b_3 as shown in Table 2.

LVC Model for Empirical Analysis

Because the basic Lotka-Volterra equations are continuous forms, the transformation into a discrete form is necessary to perform the empirical study. The basic system of equations (1) is closely related to the following logistic equations (2) for two competing technologies, working in discrete time intervals (Leslie, 1958).

$$X_1(t+1) = \frac{\alpha_1 X_1(t)}{1+\beta_1 X_1(t)+\gamma_1 X_2(t)}$$
$$X_2(t+1) = \frac{\alpha_2 X_2(t)}{1+\beta_2 X_2(t)+\gamma_2 X_1(t)} \quad (2)$$

Here, α_i and β_i are the logistic parameters for the species X_i, which are the same ones used to model one technology without competition, while the coefficient of γ_i expresses the magnitude of the effect which each technology has on the rate of increase of the other. These equations (2) are discrete forms, but non-linear, and the non-linear results can be varied by changing initial values. To solve the problem and get reliable estimation results, the equations (2) are transformed into the following linear equations (3) by taking the reciprocal (Kim et al., 2006).

$$\frac{1}{X_1(t+1)} = c_1 + c_2\frac{1}{X_1(t)} + c_3\frac{X_2(t)}{X_1(t)}$$
$$\frac{1}{X_2(t+1)} = d_1 + d_2\frac{1}{X_2(t)} + d_3\frac{X_1(t)}{X_2(t)} \quad (3)$$

Then, the equations (4) express the relationships between the equations (1) and (3) (Lee et al., 2004).

$$a_1 = \ln c_2,\quad a_2 = \frac{c_1 \ln c_2}{c_2 - 1},\quad a_3 = \frac{c_3 \ln c_2}{c_2 - 1}$$
$$b_1 = \ln d_2,\quad b_2 = \frac{d_1 \ln d_2}{d_2 - 1},\quad b_3 = \frac{d_3 \ln d_2}{d_2 - 1} \quad (4)$$

Theoretically we can acquire the relationships between two competing technologies from the signs of a_3 and b_3, the coefficients of competition in the equations (1). If a_3 has positive value, it means that technology 2 has arrested the growth of technology 1. However, the equations (3) are used for the estimation (rather than equations (1)), the signs of a_3 and b_3 should be inferred from the relations between the parameters in equations (4). In the equations (4), the signs of a_3 (or b_3) and c_3 (or d_3) are the same because $\ln c_2/(c_2-1)$(or $\ln d_2/(d_2-1)$) always has a positive value in the range of $c_2 \geq 0$, $c_2 \neq 0$ ($d_2 \geq 0$, $d_2 \neq 0$). Thus, the signs of c_3 and d_3 can also be used to identify competition types, as shown in Table 2.

ANALYSIS RESULTS

For estimation of linear model, SAS 12.0 was used. In particular, MODEL procedure in SAS/ETS package was applied for SUR estimation and Gauss optimization. It should be noted that the data points for estimation are not sufficient for T4. And thus the estimates for the period from T3 to T4 (renamed as T4') instead of T4 were used for analysis. As the purpose of this study is to examine how the diffusion process is affected when a new market structure is formed, it is valid to analyze a set of data in which a new set of data is added to the exiting one.

Result of Parameter Estimation According to the Time Period

Table 3 shows the estimation results for parameters c_1, c_2, c_3, d_1, d_2, and d_3, which is mentioned in the equations (3), for five periods, from T1 to T4' and T5.

As a result, most parameters except c_3 are statistically significant at the significance level of 5%, and the R^2 values are also very high. Then, the parameters a_1, a_2, a_3, b_1, b_2, and b_3 in equations (1), are estimated by using the parameters listed in Table 3 and considering the relations between parameters according to equations (4). Table 4 shows the transformation results. The underlying assumption is that the parameters generated by the combination of significant parameters in Table 3 are also significant. According to the results, most parameters are significant except two a_3s, one b_2 and one b_3.

Table 3.Estimation results for parameters in the linear model

	T1	T2	T3	T4'(T3~T4)	T5
c_1	9.054E-9 (0.0469)	-3.2E-9 (0.63)	8.658E-9 (0.0276)	6.948E-9 (0.0414)	3.698E-9 (0.0046)
c_2	0.936556 (<.0001)	0.961336 (<.0001)	0.927253 (<.0001)	0.933013 (<.0001)	0.960955 (<.0001)
c_3	0	0	-5.39E-9 (0.0944)	-3.66E-9 (0.1699)	-2.06E-9 (0.0324)
d_1	0	0	2.724E-7 (<.0001)	2.654E-7 (<.0001)	1.658E-9 (0.3961)
d_2	0	0	1.83654 (<.0001)	1.844473 (<.0001)	1.000987 (<.0001)
d_3	0	0	-2.78E-7 (<.0001)	-2.79E-7 (<.0001)	-1.91E-9 (0.4694)
R^2	0.9993	0.9952	0.9972 0.9669	0.9973 0.9630	0.9981 0.9962

Note 1. The values in parenthesis refer to significance level.

a_2: In the case of cellular, the parameter indicates the impact of the increase in demand for itself

a_3: In the case of cellular, the parameter indicates the impact of the diffusion of demand for PCS which is competitive service of cellular

b_1: In the case of PCS, the parameter indicates the impact of the diffusion of demand for itself

b_2: In the case of PCS, the parameter indicates the impact of the increase in demand for itself

b_3: In the case of PCS, the parameter indicates the impact of the diffusion of demand for cellular which is competitive service of PCS

T1 Period

The period T1 was the time to start providing cellular services nationwide and exclusively by a company. During this period, the diffusion of cellular technology decreases the increasing rate of cellular service by encroaching on existing markets (a_1<0). This can be interpreted for two reasons; at the early stage, the cellular market potential is limited due to high-price and the active diffusion of technology cannot be possible in situations where the technology is dominated by SKT. However, interestingly, the growth of cellular service facilitates highly the demand dif-

Table 4.Transformation results for parameters in the original LVC model

	T1	T2	T3	T4'(T3~T4)	T5
a_1	-0.06555*	-0.03943*	-0.07553*	-0.06934*	-0.03983*
a_2	-9.35397E-09*	3.2635E-09	8.98908E-09*	7.19166E-09*	3.77213E-09*
a_3	0	0	-5.60E-09	-3.79E-09	-2.10E-09*
b_1	0	0	0.607883*	0.612194*	0.000987*
b_2	0	0	1.97943E-07*	1.92399E-07*	1.65718E-09
b_3	0	0	-2.02E-07*	-2.02E-07*	-1.91E-09

Note 1. * The parameter values generated by the combination of significant parameters at the significance level of 5%.

Note 2. a_1: In the case of cellular, the parameter indicates the impact of the diffusion of demand for itself

fusion based on internal interaction of the existing subscribers (a_2>0).

T2 Period

At the period T2, the competition within the homogeneous technology started by the entrance of ShinSegi Telecom in the market. The parameter values are increased from -0.066 to -0.039 because of the competition between firms, and it means that the introduction of competition in the homogeneous technology facilitates demand diffusion. But still the value of a_1 is (-), so it refers that cellular propagation failed to promote the demand of the technology. In other words, under the oligopoly of homogeneous technologies, technology diffusion did not happen actively. This acted as a motive the Korean government decided to introduce a new technology like PCS to such a slow market that was a typical facility-based competition policy.

T3 Period

During the period T3, PCS services were provided by three different service providers with the competitive diffusion of cellular technology, implying the competition of homogeneous technologies together with the heterogeneous technologies. The most severe competition was observed between five different service providers. Both cellular and PCS services are based on the IS-95 CDMA technology of Qualcomm and thus can be considered homogeneous system from the fundamental viewpoint. However, since the frequency range is different between cellular and PCS, and above all most users have thought that the two services are different, they should be considered as heterogeneous technologies.

Investigating the estimation results, the demand diffusion of cellular service seemed to depend more on the diffusion of PCS (a_3<0), although the estimation was not statistically significant, rather than cellular demand itself (a_1<0 and a_2>0) implying the demand is increased by an active marketing raised by the competition of heterogeneous technologies. In fact, cellular market showed great raise in 1999 due to the economic recovery, abolition of obligated subscription, and payment of mobile phone subsidy by the competition with PCS. As the results of such marketing activities, TTL service that targeted the young-generation recorded a great success in the late 1999. Consequently, the service contributed to expand the demand of mobile communication service from the old to the young or teenagers.

In case of PCS, on the other hand, its demand increases according to the diffusion of service possibly because of the word-of-mouth effect (b_1>0) although the demand decreases according to the increase of diffusion rate (b_2>0) representing an internal interaction between PCS subscribers. However, contrary to the cellular service with business purpose, PCS service with relatively inexpensive fee has been adopted by ordinary users, showing the high imitation effect. Therefore, when compared the increasing rate of service, a_1 shows comparably higher positive value. In fact, PCS service showed rapid growth by aggressive marketing and the collaboration with terminal providers that designed various phones with small size especially for PCS services. Particularly, the growth of PCS service was driven by the government policy to promote it. Also, the diffusion of PCS service in this period was facilitated by that of cellular service (b_3<0), while the diffusion of cellular service was not significantly related to the PCS services.

This period, the most severe competition between homogeneous and heterogeneous technologies, shows the coexistence patterns of PCS and cellular technologies (Kim et al., 2006), contributing to the expansion of mobile communication service. However, the patterns of excessive competition are also observed by the severe marketing among the service providers, which affects negatively on their management performance. As a result, payment of mobile phone subsidy is abolished in order to prevent the excessive competition. In addition, M&A between service providers has started as an alternative for excessive competition.

T4' Period

The competition of PCS technology was weakened under the circumstance of competition between homogeneous and heterogeneous technologies. Hansol was undertaken by KTF.

In case of cellular technology, the value of a_1 is increased compared to the T3 period (a_1(T3) < a_1(T4')), implying the demand of cellular technology is increased due to the weakened competition in PCS technology. In addition, some demand of PCS might be transferred to cellular service during the M&A of competitors. Actually, in June 2001 when the temporary regulation of market share was released[1], the waiting demand which was bound to the regulation seemed to show the positive effect on the demand increase of cellular technology. Compared to the T3 period, the value a_2 shows a decrease (a_2(T3) > a_2(T4')), indicating that the faster the diffusion rate of cellular technology is, the more the diffusion of cellular technology is impeded due to the government regulation to prevent the monopoly of cellular technology. Meanwhile, though the diffusion of PCS technology does not show any significant effect on cellular technology, the value of a_3 decreases compared to the previous period, meaning that the positive effect of diffusion of PCS service on cellular service has decreased.

In case of PCS, the value of b_1 is increased a little bit compared to the previous period (b_1(T3) < b_1(T4')), indicating that the imitation effect due to the communication channel or word of mouth effect is magnified under the circumstances of M&A of PCA operators. Unlike the cellular service, the PCS service was allowed to establish the strategy related to the direct-network effect such as a discount of on-net tariff. Meanwhile, the value of b_2 had also decreased compared to the T3 period (b_2(T3) > b_2(T4')), weakening the negative effects on the diffusion. In addition, the diffusion of PCS service was also promoted by that of cellular service (a_3<0).

The result of weakening the degree of competition in PCS technology had similar effects on the diffusion of both cellular technology and PCS technology, showing the possible co-existence of cellular and PCS service. However, the positive effect of PCS service diffusion on cellular service diffusion is decreased although the relevant parameters are not significant. This phenomenon seems to be based on the biased government regulation rather than the natural competition.

T5 Period

The competition degree of cellular technology is weakened due to the M&A of Shinsegi Telecom and SKT, resulting in the monopolistic patterns of cellular service, similar to those in the period T1.

In case of cellular technology, the value of a_1 had increased and value of a_2 had decreases, showing the similar patterns from T3 to T4'. The decrease of competitors of cellular technology shows the positive effect on the diffusion of itself. Meanwhile, the diffusion effect of PCS on cellular turned out to be significant and positive unlike those in T3 and T4' (a_3 <0) where the values of a_3 were insignificant. There is a possibility that some cellular demands were transferred to the PCS service under the circumstance of M&A in the cellular technology, which enabled customers to get more benefits. In addition, the government control on SKT was gradually deregulated with the growth of PCS and SKT started to employ marketing activities to attract customers.

On the other hand, in case of PCS service, the value of b_2 does not show the significant fluctuation whereas the value of b_1 decreased greatly compared to T4'. The rapid growth of PCS has slowed down and its technology might have become stagnant as the mobile communications market was matured in Korea. An interesting finding is that the decrease of competitions of cellular technology does not show significantly positive effects on the diffusion of PCS technology (see b_3). Actually, SKT became more powerful after M&A. Nevertheless, a_3 and b_3 for all periods take negative values, whether they are significant or not, which means that PCS and cellular technologies have promoted the growth of the other technology, showing the co-existence patterns.

STRATEGIC AND POLICY IMPLICATIONS

Use of LVC Model

We adopted LVC model as a main methodology for this research and found the model was more suitable for analyzing the telecommunication services than Bass substitution model because the fact that the PCS technology was introduced later than cellular technology does not guaranteed that PCS is superior to cellular, which is an essential condition to apply Bass substitution model. Especially, the debate on the superior service quality of cellular service to the PCS has continuously been discussed because of the discriminated sense of users. Along with this debate, the assertion of frequency re-distribution is even suggested by some parties based on the superiority of cellular technology to the PCS technology. However, there is another view that there is no significant difference in the actual service quality between cellular technology and PCS technology. This party regards the opinion of service quality difference as the customer perceived quality raised by the superior marketing capability of SKT. In other words, the assertion of PCS inferiority is raised by the biased market assessment originated from the relative success of cellular technology. As of now, the mobile communication market is expected to grow rapidly, compared to the saturated voice communication market. Therefore, preoccupying the mobile communication market is more imperative than achieving more reliable service quality. Under this situation, LVC model enabled us to analyze not only the diffusion patterns of each technology but also the relationship type of the technologies. Actually, the relationship does not show a uniform pattern but has been dynamic over time.

Competition Among Service Providers

As the result of this research, to stimulate the mobile communication, encouraging the entry and up-bringing of other service providers is thought to be rather helpful than encouraging the M&A of existing providers, since the competition between heterogeneous or homogeneous technologies may result in the market expansion. In addition, according to the estimation result using Bass model, expansion of potential market does not enabled by PCS, but enabled by the competition between cellular technologies. With this consideration together, it seems that the M&A of ShinSegi Telecom and SKT in December 2001, did not contribute the expansion of potential market from the perspective of technology diffusion. Therefore, the opinion of firm-division currently proposed in Korea is worth considering. In practice, most of M&A of cases between homogeneous companies resulted in a price increase (Farrel & Shaprio, 1990), which denotes the M&A of SKT and ShinSegi Telecom does not represent the effect of market expansion due to the price decrease. This result also can be interpreted that it is due to the impediment to the cellular market by government regulation to prevent the monopoly after the M&A of SKT and ShinSegi Telecom. Even these government regulations might be unavoidable attempts in order to activate the growth of PCS service, however, the regulation should be deliberatively considered if there is a chance of market impediment of whole mobile communication market. Kwon (2000) showed that there should be a more reliable evidence for efficiency raise to justify the M&A of SKT and ShinSegi Telecom than that of between newcomers. From this perspective, it is possible that the decision of Fair Trade Commission to allow the M&A might not be the best solution, even assuming the various constraints. In addition, according to the time-series analysis to identify the competition effects of homogeneous and heterogeneous technologies, PCS technology and cellular technology diffuse with the pattern of mutual coexistence. At first, the diffusion of cellular technology was not affected by that of PCS, whereas the diffusion of PCS was affected by that of cellular technology. However, the situation became the opposite recently.

Therefore, since the diffusion of each service has promoted the diffusion of other service, the regulations for the diffusion of cellular technology can be justified. If cellular technology is actually superior to the other technology, increasing the use of cellular technology to the maximum level is obviously considered as an effective and natural decision from the national viewpoint, and increasing the market share of cellular service compared to that of PCS service is rather desirable. However, the problematic thing from the government viewpoint is that there is only one service provider – SKT - to provide the cellular technology, and increasing market share of cellular technology, in the long run, can result the monopolistic pattern of the mobile communication market. However, even this situation happens, it should be thought that characteristics of mobile communications industry are naturally monopolistic if this result happens due to the competition by fair means.

Competition between Technologies

When investigating the time-series competition result at the significance level of 5%, we found cellular service was not affected by the fluctuation of PCS service whereas the increase of PCS was positively affected by the increase of cellular service for T_3 and T_4. In case of domestic mobile communications market, whole subscription demand explosively increased by the introduction of PCS service. Under this circumstance, cellular service which has been preoccupied the market showed the incremental increase of demand, whereas PCS service showed the radical increase of demand based on the reputation of mobile communication service built by cellular technology. Despite these results, the insignificant values of a_3, which denotes the effect of PCS service on cellular service, show that PCS service failed to take away the subscribers of cellular service with all the rapid increase of PCS service. Major causes of this phenomenon include the long pre-emption of cellular service for the mobile communication market, high brand loyalty due to the long pre-emption, high perceived quality of cellular service, low switching intention of customers using cellular service, and the difference of strategic and marketing capabilities between cellular and PCS service providers. In addition, the fact that PCS service has failed to impede the growth of cellular service in spite of the biased government regulation means that some of the prior government regulations might not be as effective as it should be. Nevertheless, the general patterns where the growth of cellular accelerates the growth of PCS could be the main drivers of development in the early market of mobile communications industry.

Meanwhile, contrary to the positive effect of decrease of competitors in PCS technology on the diffusion of cellular technology (as the growth rate has moved to the positive direction), decrease of competitors in cellular technologies at T_5 when the market was approaching its saturation does not seem to have positive effect on the growth of PCS. The growth rate of PCS has decreased and the positive effect from cellular has disappeared. During the time, PCS service was not influenced by cellular service while the growth of cellular service was promoted by the growth of PCS service. This denotes that the diffusion of PCS was not affected by the weakening of cellular technology, firstly because cellular technology providers might not be its direct competitor and secondly because the government regulation was reduced as the number of PCS service subscribers increased. Therefore, the diffusion patterns of Korean mobile telecommunications service is likely to be the result of the characteristics of service providers or government regulations, rather than the characteristics of cellular and PCS technology themselves. Especially, the mobile communication market has failed to naturally grow due to the government regulations concerning about the monopolistic operation of SKT, especially when SKT occupied more than 50% of market share as well as played as the only service provider in cellular technology.

Success Factors of Korean Mobile Communications Market

The success of Korean mobile Communications Market lies in several factors: customers, infrastructure, service providers, and policy. Firstly, the imitation effect was remarkable in buying mobile phone and subscribing a mobile service particularly among young generations in Korea, which facilitated the fast diffusion of the technology. Secondly, Korea is densely populated, which is a good condition for establishing the network infrastructure and operating it for mobile communications service. In addition, Korea has strength in IT and the relevant markets including the mobile communications market could have been influenced positively.

Above all, the political supports and the strategic efforts of service providers constructed the environments where two technologies and several firms in the market co-existed. The service providers have taken an effort on designing various mobile services liking to mobile Internet and introducing a brand-new phone simultaneously. Also, as one of their marketing activities, they have a business model that allows customers not to pay for new phones (they pay for the phones on behalf of customers) but earn money for the use of their service. As a result, the purchasing power of customer has increased and customers started to use the service not only for a call, which is a basic function of mobile communications service, but also for other benefits such as taking a picture, using the Internet, texting, and watching a TV. Finally, the potential market in Korea could be enlarged.

Especially, the government policy took a leading role in the fast diffusion of service. Since the mobile communications market is a government-dependent market due to the frequency allotment, more precise and reliable decision making is mandatory to investigate the appropriateness of these regulations. The Korean government made two different technologies having similar functions to compete with each other and several service providers to compete in the market. Accordingly, even when a market structure became monopoly in a particular technology market, the competition continues with the other technology, leading to the growth of the whole mobile communications market. The providers tried to offer more benefits for customers through price competition and active promotion.

CONCLUDING REMARKS

Korea has shown the outstanding diffusion rate of mobile communication services and now the market is almost saturated. To identify the drivers of success in Korea, this study tried to examine the competitive growth patterns of mobile telecommunication services in Korea on the two assumptions that firstly, the diffusion was driven by competition between technologies, and secondly, it was also affected by market structure. Taking the two assumptions into consideration, LVC equations were used to model the competitive diffusion process for PCS and Cellular, and the whole period of analysis was divided into five periods according to the changes of market structure. Research findings indicate that competition between PCS and cellular technology has positive effects on the diffusion of mobile communications services, with each facilitating the growth of the other in mutualism relationship. In addition, the difference in parameter values was observed over the periods, which means that the changes of market structure could affect the dynamic diffusion patterns in the market. These effects might be remarkable because the Korean mobile communications market has kept monopoly or oligopoly, only small changes in market structure may result in great effects on customers.

Interestingly, the technological competition in mobile communication services in Korea was mainly affected by the government regulations, for example, an asymmetry regulation on the dominating company of SKT rather than service quality. The government policy has affected the technology competition and market structure and then the policy results have affected the demand diffusion. Also, the characteristics of nation in need of mobile communications services might have taken a great role in fast-growing demands in Korea together with the active marketing efforts of operators continuously introducing new value-added services particularly for their own subscribers. Therefore, the effects of technological competition and market structure on demand diffusion might vary greatly by technology types and nations. Nevertheless, the implications from the research findings can be a useful source for policy- makings and strategy-developments in the ICT sector where various technologies with a short life-cycle are inter-related to each other, especially for emerging technologies such as VDSL, FTTH, or WiBro.

With all its meaningful implications, this research has some limitations and thus need further research. First, this research cannot consider the effect of various marketing activities and government regulations, which are regarded as an important factor for diffusion. In addition, this research divided the overall period into five sub-periods to apply the basic LVC model but a new LVC model using the period as a moderating variable can be developed and used. And thus future research is needed to develop a new diffusion model incorporating more parameters. Secondly, only LVC model was used for diffusion analysis in this research. LVC model was selected because it is the most suitable model for this research but there are several other models that can take complementary role for the model. Applying several models together to get more knowledge could be possible. Finally, it was interesting that the early entrant SKT has taken the dominant position during the whole diffusion process in spite of the government regulations. It seems that in the new market where market entrant is limited, the preoccupation of market is critical and the future research could address this issue in mobile telecommunications market.

REFERENCES

Arato, M. (2003). A famous nonlinear stochastic equation: Lotka-Volterra model with diffusion. *Mathematical and Computer Modelling*, *38*, 709–726. doi:10.1016/S0895-7177(03)90056-2

Bass, F. M. (1969). A new product growth model for consumer durables. *Management Science*, *15*, 215–227. doi:10.1287/mnsc.15.5.215

Bazykin, A. (1998). Nonlinear Dynamics of Interacting Populations . In Khibnik, A. I., & Krauskopf, B. (Eds.), *World Scientific series on nonlinear science (Series A, 11)*. River Edge, NJ: World Scientific.

Bhargava, S. C. (1989). Generalized Lotka-Volterra equations and the mechanism of technological substitution. *Technological Forecasting and Social Change*, *35*, 319–326. doi:10.1016/0040-1625(89)90068-1

Choi, S., Lee, M., & Chung, G. (2001). Competition in Korean mobile telecommunications market: Business strategy and regulatory environment. *Telecommunications Policy*, *25*, 125–138. doi:10.1016/S0308-5961(00)00075-6

Dieckmann, U., Marrow, P., & Law, R. (1995). Evolutionary cycling in predator-prey interactions: Population dynamics and the red queen. *Journal of Theoretical Biology*, *176*, 91–102. doi:10.1006/jtbi.1995.0179

Farrell, C. J. (1993). A theory of technological progress. *Technological Forecasting and Social Change*, *44*, 161–178. doi:10.1016/0040-1625(93)90025-3

Fourt, L. A., & Woodlock, J. W. (1960). Early prediction of market success for new grocery products. *Journal of Marketing*, *25*(October), 31–38. doi:10.2307/1248608

Jain, D., & Rao, R. C. (1990). Effect of price on the demand for durables. *Journal of Business & Economic Statistics*, *8*, 163–170. doi:10.2307/1391978

Jalava, J., & Pohjola, M. (2007). ICT as a source of output and productivity growth in Finland . *Telecommunications Policy*, *31*, 463–472. doi:10.1016/j.telpol.2007.05.011

Jorgenson, D., & Nomura, K. (2005). The industry origins of Japanese economic growth. *Journal of International Economics*, *19*, 482–542.

Kalish, S. (1985). A new product adoption model with price, advertising, and uncertainty. *Management Science*, *31*, 1569–1585. doi:10.1287/mnsc.31.12.1569

Kim, J., Lee, D., & Ahn, J. (2006). A dynamic competition analysis on the Korean mobile phone market using competitive diffusion model. *Computers & Industrial Engineering*, *51*(1), 174–182. doi:10.1016/j.cie.2006.07.009

Kim, M. S., & Kim, H. (2004). Innovation diffusion of telecommunications: general patterns, diffusion clusters and differences by technological attribute. *International Journal of Innovation Management*, *8*, 223–241. doi:10.1142/S136391960400099X

Kim, N., Bridges, E., & Srivastava, R. K. (1999). A simultaneous model for innovative product category sales diffusion and competitive dynamics. *International Journal of Research in Marketing*, *16*, 95–111. doi:10.1016/S0167-8116(98)00026-3

Lee, S., Kim, M., & Park, Y. (2009). ICT co-evolution and Korean ICT strategy: An analysis based on patent data. *Telecommunications Policy*, *33*(5-6), 253–271. doi:10.1016/j.telpol.2009.02.004

Lotka, A. J. (1925). *Elements of Physical Biology*. Baltimore: Williams and Wilkins.

Lotka, A. J. (1956). *Elements of Physical Biology. 1924, republished as*. Elements of Mathematical.

Mahajan, V., Muller, E., & Bass, F. M. (1996). Timing diffusion and substitution of successive generations of technological innovations: The IBM Mainframe Case. *Technological Forecasting and Social Change*, *51*, 109–132. doi:10.1016/0040-1625(95)00225-1

Mahajan, V., & Peterson, R. (1978). Innovation diffusion in a dynamic potential adopter population . *Management Science*, *24*, 1589–1597. doi:10.1287/mnsc.24.15.1589

Mahajan, V., & Peterson, R. A. (1979). First-purchase diffusion models of new-product acceptance. *Technological Forecasting and Social Change*, *15*, 127–146. doi:10.1016/0040-1625(79)90073-8

Mahajan, V., Sharma, S., & Buzzel, R. D. (1993). Assessing the impact of competitive entry on market expansion and incumbent sales. *Journal of Marketing*, *57*, 39–52. doi:10.2307/1251853

Manfield, E. (1961). Technical change and the rate of imitation. *Econometrica*, *29*, 741–766. doi:10.2307/1911817

Modis, T. (1997). Generic re-engineering of corporations. *Technological Forecasting and Social Change*, *56*, 107–118. doi:10.1016/S0040-1625(97)00076-0

Pistorius, C. W. I., & Utterback, J. M. (1997). Multi-mode interaction among technologies. *Research Policy*, *26*, 67–84. doi:10.1016/S0048-7333(96)00916-X

Porter, A. L., Roper, A. T., Mason, T. W., Rossini, F. A., & Banks, J. (1991). *Forecasting and Management of Technology*. New York: Wiley.

Shin, J., & Park, Y. (2007). Building the national ICT frontier: The case of Korea. *Information Economics and Policy*, *19*, 249–277. doi:10.1016/j.infoecopol.2007.01.003

Teng, J. T. C., Grover, V., & Guttler, W. (2002). Information technology innovations: general diffusion patterns and its relationships to innovation characteristics. *IEEE Transactions on Engineering Management*, *49*(1), 13–27. doi:10.1109/17.985744

ADDITIONAL READING

Bharagava, S. C. (1989). Generalized Lotka-Volterra equations and the mechanism of technological substitution. *Technological Forecasting and Social Change*, *35*, 319–326. doi:10.1016/0040-1625(89)90068-1

Choi, C., Kim, S., & Park, Y. (2007). A patent-based cross impact analysis for quantitative estimation of technological impact: The case of information and communication technology. *Technological Forecasting and Social Change*, *74*, 1296–1314. doi:10.1016/j.techfore.2006.10.008

Day, R. H. (1994). *Cycles Economic Dynamics: An Introduction to Dynamical Systems and Market Mechanisms*. Cambridge: MIT Press.

Dolan, R. J., Jeuland, A. P., & Muller, E. (1986). Models of New Product Diffusion-extension to Competition against Existing and Potential firms over time . In Mahajan, V., & Wind, Y. (Eds.), *Innovation diffusion models of new product acceptance* (pp. 117–149). Cambridge, MA: Ballinger Publishing.

Frieden, R. (2005). Lessons from broadband development in Canada, Japan, Korea and the United States. *Telecommunications Policy*, *29*, 595–61. doi:10.1016/j.telpol.2005.06.002

Gruber, H. (2001). Competition and innovation: The diffusion of mobile telecommunication in central and eastern Europe. *Information Economics and Policy*, *13*, 19–34. doi:10.1016/S0167-6245(00)00028-7

Kim, M. S. (2005). Household-use vs. business-use demand diffusion of telephone and Internet access service in Korea. *International Journal of Innovation and Technology Management*, *2*(1), 57–75. doi:10.1142/S0219877005000356

Ku, H., & Kim, J. (1997). An interim assessment of competition in the Korean international telephone service market. *Telecommunications Policy*, *21*(3), 265–274. doi:10.1016/S0308-5961(97)00006-2

Kumar, U., & Kumar, V. (1992). Technological innovation diffusion: The proliferation of substitution models and easing the user's dilemma. *IEEE Transactions on Engineering Management*, *39*(2), 158–168. doi:10.1109/17.141273

Lawrence, K. D., & Lawton, W. H. (1981). Applications of diffusion models: Some empirical results . In Wind, Y., Mahajan, V., & Cardozo, R. N. (Eds.), *New-product forecasting* (pp. 529–541). Lexington, MA: Lexington Books.

Lee, S., Gholami, R., & Tong, T. (2005). Time series analysis in the assessment of ICT impact at the aggregate level – Lessons and implications for the new economy. *Information & Management*, *42*, 1009–1022. doi:10.1016/j.im.2004.11.005

Leslie, P. (1958). A Stochastic model for studying the properties of certain biological systems by numerical methods. *Biometrika*, *45*, 16–31.

Morris, S., & Pratt, D. (2003). Analysis of the Lotka-Volterra competition equations as a technological substitution model. *Technological Forecasting and Social Change*, *70*, 103–133. doi:10.1016/S0040-1625(01)00185-8

Parker, P., & Gatignon, H. (1994). Specifying competitive effects in diffusion models: An empirical analysis. *International Journal of Research in Marketing*, *11*(1), 17–39. doi:10.1016/0167-8116(94)90032-9

Pianca, E. (1983). *Evolutionary Ecology*. NY: Harper and Row.

Pistorius, C., & Utterback, J. (1997). Multi-mode interaction among technologies. *Research Policy*, *26*, 67–84. doi:10.1016/S0048-7333(96)00916-X

Rogers, E. M. (1995). *Diffusion of Innovations*. NY: The Free Press.

Schoder, D. (2000). Forecasting the success of telecommunication services in the process of network effects. *Information Economics and Policy*, *12*, 181–200. doi:10.1016/S0167-6245(00)00006-8

Shin, J., Lee, W., & Park, Y. (2006). On the benchmarking method of patent-based knowledge flow structure: Comparison of Korea and Taiwan with USA. *Scientometrics*, *69*(3), 551–574. doi:10.1007/s11192-006-0169-y

Sohal, A., Moss, S., & Ng, L. (2001). Comparing IT success in manufacturing and service industries. *International Journal of Operations & Production Management*, *21*(1-2), 30–45. doi:10.1108/01443570110358440

Vicente, M., & López, A. (2006). Patterns of ICT diffusion across the European Union. *Economics Letters*, *93*, 45–51. doi:10.1016/j.econlet.2006.03.039

Young, P. (1993). Technological growth curves, a competition of forecasting models. *Technological Forecasting and Social Change*, *44*, 375–389. doi:10.1016/0040-1625(93)90042-6

KEY TERMS AND DEFINITIONS

Cellular Service: A service of analog or digital telecommunication in which a subscriber has a wireless connection from a mobile telephone to a relatively nearby transmitter using 800 MHz band. The transmitter's span of coverage is called a cell.

Diffusion: The process in which an innovation is communicated through certain channels over time among the members of a social system.

Interaction Pattern: A certain characteristic of action or influence of people or things on each other.

LVC Model: Lotka-Volterra Competition model.

Market Structure: A dispersion of the number of sellers in the market and/or relating to sellers' degree of product differentiation, the cost structure, the degree of vertical integration, etc.

PCS (Personal Communications Service): A wireless phone service similar to cellular service but emphasizing personal service and extended mobility using 1,800 MHz band.

ENDNOTE

1 The Korean government had controlled the market share of cellular technology with regulations.

Chapter 3
Mobile Information Communication Technologies and Construction Project Management:
Indian Scenario Case Study

Vanita Ahuja
Project Management Consultant, Teacher and Trainer, India

ABSTRACT

Construction project management requires effective and real time information and data communication between all the project team agencies. In present scenario when the associating agencies may be geographically separated, and for projects with large site areas or being executed in hazardous or remote sites, the required information and data communication can be achieved through use of mobile communications. But, research shows that the adoption of mobile communications by the construction industry is limited. This chapter presents the issues affecting adoption of mobile communications by the construction industry, the research background, a case study of mobile communications adoption at a major construction project in India and further discusses the factors leading to sustainable mobile communication adoption by construction project teams. Construction industry primarily comprises small and medium enterprises (SMEs). Thus, the issues are studied with focus on adoption of mobile communications by SMEs.

INTRODUCTION

Construction projects involve participation of multiple organizations including clients, project managers, architects, consultants, contractors, material suppliers etc. and successful construction project management requires effective intra and inter organization communication between all the project team organizations. Intra organization communication is in the form of communication within head office, between head office and site office, between site office and mobile site staff and sometimes between different sites when these sites are being managed by a common resource pool. Inter organization communication is in the

DOI: 10.4018/978-1-61692-818-6.ch003

form of communication between site and other organizations like consultants and material suppliers and also between head offices of different organizations. In present scenario, project team organizations may be geographically separated and required communication can be achieved by the use of Information communication technologies (ICT).

Construction project management comprises change management which requires effective communication and coordination between all the project agencies, resulting in efficient flow of information and documents to assist in decision-making, and the efficient flow of resources to maintain progress of development. Coordination tasks are often complicated by schedule pressure and productivity demands, worker fatigue, data loss during information exchange, misunderstandings because of poorly defined information, and iterative negotiation when unanticipated events (Magdic et al., 2004) or changes occur. This leads to poor access of the right information, at the right time which is required for efficient decision-making. In this environment, ICT and, especially wireless (Kondratova, 2004; Filos, 2002) and mobile communications would lead to overall efficient use of IT (Rebolj and Menzel, 2004) and required effective communication. Mobile computing consists of three major components: Hardware i.e computers which can be used indoors and outdoors while the user is in motion; Networks with sufficient bandwidth which can be accessed while in motion; and software i.e mobile applications supporting context-sensitivity and personalization (Rebolj and Menzel, 2004). Advantages of using mobile computing are in the ability to share data in real time, reduced rework and paperwork, ability to solve problems on site, construction of accurate databases by timely and continuous collection of data, increased quality of information and decrease in operational costs (Koseoglu and Bouchlaghem, 2006).

We are rapidly moving away from the Desktop and Laptop Web paradigms towards the Mobile Web paradigm, where mobile, smart devices such as the Smart phone, Pocket PC, PDA (Personal Digital Assistant), hybrid devices (such as phone-enabled PDAs or Pocket PCs), and wearable computers will become powerful enough to replace laptop computers in the field and will be widely used for real time communication of construction project information to project repositories or between project participants (Kondratova, 2004). Also, wireless LAN (WLAN) and virtual private networks (VPN) enable effective mobile data communication.

Table 1 indicates communication channels with respect to construction projects and as it is indicated, information as well as data communication between mobile site staff and other project nodes requires use of mobile communication technologies and if the sites are at remote locations, communication between site offices and other project nodes also requires mobile communication technologies. The nodes of Intra and Inter organization communication are indicated with different colors.

Thus, mobile information communication is an essential component of construction projects' information management in the construction and operation phases of built facilities as it facilitates communication between large physical distances. It would also help the construction industry in facing pressures to decrease costs, improve productivity and have a competitive edge in terms of quality of service and customer satisfaction (Venkatraman and Yoong, 2009). But, use of ICT and specifically mobile communication technologies in the construction industry is still at a nascent stage (Singh, 2006, Koseoglu and Bouchlaghem, 2006, Rebolj and Menzel, 2004). Some of the identified reasons for this scenario are that IT has not been adopted effectively at construction sites and also most construction IT solutions, like integrated building models (including complex product and process models) require highly organized and standardized project environments, which are not found in real-life con-

Table 1. Inter and intra organization information and data communication structure matrix

Org. 1	**Head Office**	Telecommunication	Telecommunication/ Mobile	Mobile	Telecommunication/ Mobile	Telecommunication
		Internet	VPN/Intranet	VPN/Intranet	VPN/Intranet	LAN
	Site 1 - Project Site Office	Telecommunication/ Mobile	Telecommunication/ Mobile	Mobile	Telecommunication/ Mobile	
		Internet	VPN/Intranet	WLAN	LAN	
	Site 1 - Site Mobile staff	Mobile		Mobile		
		Mobile Internet		WLAN		
	Site n - Project Site Office		Telecommunication			
			LAN			
	Org. 2 – Head Office	Telecommunication				
		LAN				
		Org. 2 – Head Office	**Site n - Project Site Office**	**Site 1 - Site Mobile staff**	**Site 1 - Project Site Office**	**Head Office**
			Org. 1			

struction projects (Rebolj et al., 2000). Thus, currently, the transfer of information between the construction site and office-based IT systems is tedious, time-consuming, and error prone (Reinhardt et al., 2005), which emphasizes the requirement to have an integrated communication system between project sites and head offices. Research has also indicated that use of mobile or wireless technologies should be integrated with central project repository (Kondratova, 2004) and information management structure of the projects (Rebolj and Menzel, 2004) and should not be adopted only for automation of existing processes.

Construction industry is dominated by small and medium enterprises (SMEs) (Katranuschkov et al., 2001) and they cannot afford the complicated conversion from paper to electronic processes requiring expensive information exchange technology (Ribeiro and Lopes, 2002). Due to the perceived budget constraints, they are also risk averse regarding IT investment, (Peansupap and Walker, 2005) as for most of the project team agencies, standard devices and services have to be available on the market at affordable prices in order to be attractive (Rebolj et al., 2000). Thus, investment in IT is slowly increasing, still many construction organizations are waiting for concrete proof that this investment would add value and improve their "bottom line" (Singh, 2006).

Construction industry is project oriented (Kondratova, 2004), where project teams are temporary and include organizations of different sizes and variable IT use. Moreover at any one point of time a construction organization is a part of multiple project teams. Thus, use of mobile communication technologies can be enhanced if it is adopted at the industry level.

This chapter introduces the topic and based on the research background discusses issues affecting adoption of mobile communications by the construction industry. Further, adoption of mobile communications with respect to developing countries is studied through the case study of a project being executed in India with a brief description of the Indian construction industry and its status of adoption of mobile communications. The chapter concludes with summarized discussion and a statement of future research directions.

MOBILE COMMUNICATIONS AND CONSTRUCTION INDUSTRY

Construction industry has benefited from the telecom revolution and widespread use of mobile phones. Another promising mobile tool is handheld personal computer or personal digital assistant (PDA). If PDA is used in an integrated and systematic way, which means an organized combination of PDAs, mobile phones providing wireless data transmission, and network-based document management, then its speed, memory capacity, communication possibilities, reliability, small size and long power independence, as well as its level of hardware and software standardization, gives the PDA a powerful potential in the information chain of a construction project (Rebolj et al., 2000).

Some of the studied requirements for adoption of mobile communications at construction sites are (Ward et al., 2004): the need to reduce defective work caused by incomplete, ambiguous, duplicate, missing or incorrect information; unsuitability of the site for bulky paper documentation; lack of verification on the data collected; errors in data translation between the site and office; lack of data re-use throughout the company; and the ability to meet the reactive nature of the construction process and sequence of work. Thus, mobile devices addressing these requirements are suitable for the construction industry. As per Huang et al. (2002), Broyd (1999) claims that the industrial adoption of high technologies in construction sector can be described as a technology-process-culture (TPC) triangle. Similarly effective adoption of ICT by construction organizations requires utilization of this triangular paradigm and equal consideration of associated technological and non-technological aspects (Ahuja et al., 2006). Thus, mobile communication system for construction sites is required to be studied with respect to the use of appropriate technology as well as the management issues which need consideration.

Technology Issues

Construction sites have rough working conditions and following factors have been found critical in hardware selection: screen size, outdoor readability, battery power, physical unit size and robustness (McCullough, 1997; Elzarka & Bell, 1997; Alexander et al, 1997 cited in Ward et al., 2004). Though, now these factors are being considered by hardware manufacturers, still number of suitable mobile devices for construction sites are limited.

Due to multi enterprise and multi functional characteristic of construction projects, for effective data interface, the software utilized in mobile devices should be compatible with other IT applications being utilized by the project team organizations. Also, through wireless networks (WLAN), mobile communication is required to be integrated with overall project and organization information structure (Menzel et al., 2004; Meissner et al., 2003), so that the project data is stored in a central database and can be accessed by the site as well as office team in real time. Such database would allow for a single point of data entry with all users having access to all of the construction data captured by other personnel (Ward et al., 2004).

Kimoto et al. (2005) have discussed that with respect to the construction industry, the necessary functions for the mobile systems are: mobility and durability of hardware, compatibility of hardware and operating systems, compatibility of data between mobile and PC, expressivity of display, stability of system, operability of user interface, and processing speed.

In the field of construction, drawings are among the most important types of documents, and therefore software for managing and accessing them is a necessary requirement for mobile computing in construction and Geographic Information Systems (GIS) are already available for some PDAs (Rebolj et al., 2000).

At construction sites where working conditions are unsafe, workers require mobile devices which are easy to use and preferably wearable devices with voice data input facility using Voice XML (Extensible Markup Language) technology. Among the wearable devices there are head mounted displays including the ones based on the Light-Guide Optical Element (LOE) technology, belt worn computers of varying sizes ranging from a normal PDA to the size of a buckle of a belt (Kuladinithi et al., 2004). But, Kondratova (2004) reports that a feasibility study, that focussed on the use of wearable computers in construction (Fuller and Sattineni, 2002), showed that it is largely the relatively high cost of rugged, industrial, wearable computers that prevents the wide spread use of these devices on construction sites. Also, these devices need to be integrated with overall information system structure.

In construction projects with large sites or marine projects etc. where tracking the location of equipment and resources is important, handheld computers can be augmented with a GPS (Geographic positioning system) receiver to reference the location (Kondratova, 2004). For example, a handheld computer with a GPS receiver was used for construction damage assessment after the September 11 terrorist attack in USA (Bacheldor, 2002 as cited in Kondratova, 2004). Barcodes and RFID (Radio frequency identification device) tags (Jaselskis and Misalami, 2003) can be utilized for efficient material management. Mobile communication structure should support the site staff to set up video conferencing at site for resolving issues and seeking expert advice.

Efficiency of mobile devices in the field is affected by the environmental conditions including data transmission speeds (Ward et al., 2004) and often environmental restrictions are decisive when choosing the mobile computing system, especially the device to be used (Menzel et al., 2004).

Management Issues

System implementation requires cost-benefit analysis where cost would include the initial cost, training cost and operations and maintenance cost. The benefits would be tangible benefits like reduced cost of remedial work due to effective communication within the project team and successful completion of project as well as intangible benefits like satisfied clients and enhanced market reputation leading to more projects being assigned to the organization.

In multi enterprise scenario of construction projects, for effective use of mobile communications there is a requirement to have common use of IT systems between all the agencies and as per Ward et al. (2004) efforts should be made towards providing a standard set of interfaces or user profiles from which the project manager can tailor the data collection to meet such requirements at the start of the contract. Adequate contract clauses should define the project information structure and component of mobile communications to be adopted by all the agencies.

System is required to be periodically evaluated and some of the identified factors for measuring performance of wireless networks are (Ward et al., 2004): throughput (packets of data per second); losses (lost packets); link quality (quality of wireless link); signal strength; and delivery rates (number of successfully completed packets).

Mobile communication technologies are changing and developing fast and becoming an integral component of our life. But, with equal consideration of the above discussed management issues, we can adopt mobile communications effectively for construction projects.

INDIAN CONTRUCTION INDUSTRY

The evolution of Indian Construction Industry was almost similar to the construction industry evolution in other countries: founded by Government and slowly taken over by enterprises (wikepedia.org, 2007). The industry ranked 12th in the world, employing more than 32 million people, and contributing 5 percent to the GDP (Gross domestic product) and moving towards 8 percent growth rate has become one of the key drivers of the booming Indian economy (Singh, 2006). Similar to the global trend, majority of the construction organizations are SMEs and a survey has reported that it is not just the big organizations writing the new Indian growth story but entrepreneurs leading the SMEs are also contributing in good measure (Indiapressagency.com, 2006).

Country has witnessed a telecom revolution due to which use of mobile phones is a norm in the country and by 2013, India is expected to have the third largest online population in the world (The Times of India, 2009). Government has proposed national mega infrastructure projects like the River Linking Projects, the Garland Port Linking Projects, Hydropower, Thermal and Nuclear power projects for electricity generation etc. which need the highly skilled technical manpower (Kumar et al., 2006) and use of modern management and communication techniques.

A recent survey was conducted across India to assess status of ICT adoption in the Indian construction industry. Data analysis results with reference to wireless and mobile communications showed that (Ahuja et al., 2009):

- Use of mobile phones is a norm for communication in the country and in the construction industry. But for construction project management, mobile internet had been used only by 15% of the surveyed organizations.
- RFID and barcoding had not been used by any of the surveyed organizations for materials management. Also, no evidence of use of these technologies was found through Literature Review and through discussions with experts in the industry.
- In majority of the surveyed organizations, more than 80% of the office staff had access to computers, but in only about 20% organizations more than 80% site staff had access to computers.
- Organizations with higher turnover have higher adoption of ICT.

This paper presents case study of ICT and mobile communications adoption at 'Renovation and Expansion Project of the Indira Gandhi International Airport' at the capital city of India, New Delhi. The information about the project description presented in the case study has been referred from public domain web sites and media reports. The description of adopted information structure is presented after discussions with the concerned personnel.

CASE STUDY: RENOVATION AND EXPANSION OF INDIRA GANDHI INTERNATIONAL AIRPORT, NEW DELHI, INDIA

Indira Gandhi International Airport (IGI Airport) is in New Delhi, the capital city of India. It is named after a former prime minister of the country.

Project Need

IGI Airport is the busiest airport in India in terms of number of daily flights and the second-busiest in India in terms of the passenger volume (The Times of India, 2008). At the start of the expansion project, the airport had five terminals including: two domestic terminals 1A and 1B, the international terminal T2, the Haj terminal used for Haj pilgrimage and international cargo terminal. It covered 1900 acres site and the main runway was one of the few runways in Asia that was equipped with CAT IIIB instrument landing system allowing landing in visibility as low as 50 m (airport-technology.com).

Owing to the booming Indian Aviation industry and the entry of numerous low-cost private carriers, the airport saw a huge jump in the passenger traffic and failed to cope with the demand. The capacity of Terminal 1 and Terminal 2 together was

12.5 mppa, but in 2008, total passenger count at the airport reached 23.97 million (wikepedia.org, 2009). At the beginning of the project, the existing runway and terminal facilities were severely overstressed by more than 150% of their capacity utilization at peak hours resulting in tremendous discomfort to all (Shankar Narayanan, 2008). Thus, the booming economy of India leading to increasing passenger and cargo traffic growth and inadequate airport infrastructure led to the major expansion of the airport.

Project Details

The expansion project is being carried out under Public Private Partnership (PPP) Initiative of the Government of India (airport-technology.com) and is being executed by DIAL, a joint venture consortium of a leading Indian developer the GMR Group (54%), Airports Authority of India (AAI) (26%), and Fraport & Eraman Malaysia (10% each). GMR is the lead member of the consortium; Fraport AG is the airport operator, Eraman Malaysia - the retail advisors and IDF - the financial investors (newdelhiairport.in/about-dial.asp).

Following an international competitive bidding process, in January 2006, the consortium was awarded the concession to operate, manage and develop the IGI Airport and on April 4, 2006 DIAL entered into Operations, Management and Development Agreement (OMDA) with the AAI. The initial term of the concession is 30 years extendable by a further 30 years (newdelhiairport.in/about-dial.asp). DIAL has awarded the Design and Construction Contract of the major component of Phase I work to Larsen & Toubro Ltd. (L&T) (newdelhiairport.in/milestones.asp), a leading contracting organization of India.

L&T's construction division has vast experience in building a variety of major airport structures in India and abroad and is the largest airport builder in India when it comes to design and construction of aviation infrastructure. In the ongoing drive of the Indian government for upgrading airport infrastructure in the country, L&T has been a part of construction of the first greenfield airport at Bengaluru on PPP basis followed by construction of Hyderabad airport and is now a part of modernization and expansion of Delhi and Mumbai international airports (all these cities are major metropolitan cities of the country) (ECC Concord, 2008).

Phase I of the expansion project is underway. The expansion and renovation work of Phase I is expected to cost around US $2bn (airport-technology.com) and the work components of the masterplan for Phase I and their status as on July, 2009 is as follows (newdelhiairport.in/about-dial.asp):

- Constructing a new runway - DIAL has already commissioned a new runway 11-29 at IGI Airport on September 25, 2008 (newdelhiairport.in/about-dial.asp). The 4.43-km runway strip is the longest in Asia and is equipped with a state-of-the art instrument landing system at both ends that will allow landing even in low visibility times (projectsmonitor.com/index.asp). It is capable of handling A380 sized aircrafts (economictimes.indiatimes.com).
- Constructing a new departure terminal - DIAL has inaugurated a new state-of-the art domestic departure terminal 1D (T1D) on 26th February, 2009. T1D is expected to increase the capacity of domestic departures to 10 mppa (newdelhiairport.in/about-dial.asp).
- Constructing a new integrated passenger terminal (Terminal 3) - Its construction is underway (newdelhiairport.in/about-dial.asp).

Additional works of Phase I also included upgrading the existing terminals (newdelhiairport.in/about-dial.asp) and the facilities at the existing terminals have already been upgraded.

The first phase of the airport is designed to handle 60 mppa. This phase is scheduled to be completed by 31st March 2010 and will be fully operational before the Commonwealth Games (newdelhiairport.in/about-dial.asp) to be hosted by New Delhi in 2010. In 2010, all international and full service domestic carriers will operate from Terminal 3, while Terminal 1 will be developed as an exclusive terminal for low cost carriers (wikipedia.org, 2009).

Terminal 3

Terminal 3 would be a state-of-the-art 520,000m^2 (aconex.com), two tier complex featuring Common Use Terminal Equipment (CUTE), an advanced 5 level in-line baggage handling system with explosive detection technology for greater efficiency and security, more than 160 check-in counters, more than 90 immigration desks for international passengers, and 76 aerobridges (newdelhiairport.in/about-dial.asp).

At the time of commissioning, Terminal three (T3) will handle 100% of international traffic and 50% of domestic traffic and would have an integrated 4,300 space multi-storey car park at the terminal (airport-technology.com). Access to the new terminal would be via a 6 lane approach road. The Airport would also be connected via dedicated high speed Metro line connecting the terminal T3 with the city centre (newdelhiairport.in/about-dial.asp).

Later Phases

In subsequent stages, the airport will be further developed with the increase in passenger demand and more terminals and runways would be added in a modular manner to form a U shaped complex with an ultimate design capacity of 100 mppa (newdelhiairport.in/about-dial.asp).

The low cost carriers would also move to the new terminal complex. Two new terminals, Terminal 4 and 5 would be built at a later stage, and once completed, all international flights would move to these two new terminals, while Terminal 3 would then solely be used for handling domestic air traffic. A new cargo handling building is also planned. (wikipedia.org, 2009)

Project Information and Data Communication Structure for Phase I

This section discusses the project information and data communication structure adopted by L&T, the prime contracting agency of Phase I. To execute this fast track project in a very tight schedule of 39 months, the project organization comprises three different Strategic Business Units (SBUs) working in complete synergy as there is continuous interaction for interfacing works, where Inter BU resource sharing to achieve optimization at higher levels in the project is a unique feature of this project (Shankar Narayanan, 2008). The project team in detail includes about 70 consulting organizations, 45 of which are outside India (aconex.com), over 16000 workmen and 1200 design and construction staff with the peak workmen strength likely to exceed 23000 (Shankar Narayanan, 2008). L&T has a Project Head Office (PHO) at site and small offices at 20 strategic remote site locations for effective work execution and coordination as the project area is spread over a very large area. For successful completion of this complex project, L&T had to adopt an efficient information management system. System architecture of information and data communication structure is shown in Figure 1.

Intra Site Communication

Site staff primarily relies on mobile phones for telecommunication. For better connectivity of mobile phones, at various locations at the site, the local telecom service providers have installed 'cell on wheels', usually referred to as 'COW'. It is a mobile cell site that consists of a cellular antenna tower and electronic radio transceiver

Figure 1.

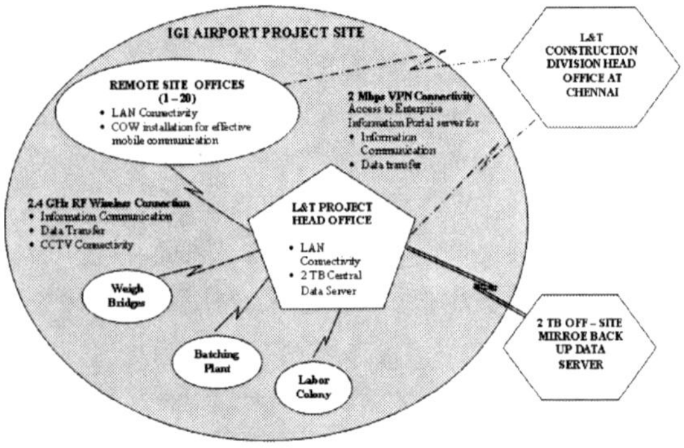

equipment on a truck or a trailer and designed to be a part of the cellular network.

All site staff have Internet access through 6Mbps wireless Internet connectivity. But, the connectivity is only for the work related websites and only between 1PM – 2PM and between 7PM -8PM, staff have access to personal networking websites like yahoo, gmail etc. Senior executives have access to mobile Internet cards for anytime and anywhere Internet access required for effective decision making and coordination. Mobile Internet is not preferred at site because of low speed connectivity. Thus, the remote site offices are positioned strategically so that for drawings and documents reference and for Internet access, no staff has to travel more than 10 mins.

Intra office data communication at site head office and at each remote location office is through dedicated LAN of each particular office. Inter office data communication at site is through wireless on 2.4 GHz RF. Local LAN of each office is created using the traditional wired infrastructure but the wireless network is the backbone of the project communication network (Network Computing, 2009). The Alvarion NLOS radio frequency antennae access is not only limited to line of sight communication, but even provides communication facility in the under construction -10 M lvl. basement.

Wireless is also used for video surveillance, which facilitates real time site inspections. At strategic locations at site, including at batching plant and labor colony, web cameras are placed and are connected through a CCTV system. Due to the connection between the wireless network and LAN, the authorized staff can zoom in on remote locations under web camera surveillance, which helps in monitoring the status of the construction activity in real time and helps the senior staff to respond quickly to emergency situations, when immediate decisions need to be taken (Network Computing, 2009).

The whole site data is stored at a central site server at the project head office as all the computers at site including those at the batching plant and weigh bridges are connected with the central site server through wireless network. The data server has a capacity of 20 TB. As a part of the 'Data Disaster Recovery Plan', every day the server data is mirrored off to an off site location server of similar capacity.

The information system provides following facilities to the project site staff:

- Domain server
- Windows Software update system for automatic updates of the required software
- Antivirus update system for data security
- Proxy server or gateway system used by policy managers for providing conditional access to the users

Staff access to site is monitored and controlled through the biometric system. It provides timesheet information in the MIS for calculating working hours of the site staff and provides site security data.

Intra Organization Communication

Head office of the construction division of L&T is at Chennai, another metropolitan city of India. It hosts an 'Enterprise Information Portal' (EIP) server for total intra organization communication. EIP is a web enabled Project Management system, developed, optimized internally on a 24x7 VPN network infrastructure (lntecc.com). IGI project site offices including project head office and remote site offices are connected with this EIP server through 2 Mbps VPN connectivity. This provides real time access of head office data server to the whole site staff.

Inter Organization Communication

Aconex, the online document management and collaboration system that securely and efficiently manages project information in construction, government and infrastructure and energy and resources projects is being utilized for document management (aconex.com). It provides a central repository for all project documents and also all project drawings and documents are communicated through the Aconex server. It is customized through a work flow system, which informs the users about the drawings and documents sent to them by other users. Every document and mail distribution is searchable and accessible from remote locations (aconex.com). This system also provides authorized documents access to users of different agencies working on the project. As reported in (aconex.com), in first 15 months of its implementation, the system had been used by over 970 users to store and manage more than 350,000 documents and distribute over 1.1 million correspondence items. For this project, Aconex has enabled 70 organizations to share information and collaborate, 45 of which are located outside of India. The system has a user friendly structure as new participants are quickly comfortable using the system and there is also a helpline system available (aconex.com).

The project site head office has a facility for videoconferencing that facilitates real time meetings between geographically separated project team agencies. It helps in efficient coordination, collaboration and decision making.

Problems Faced in System Implementation

L&T faced following problems in implementing the above discussed information system.

- The 2.4GHz free band used for Radio Frequency is the only permissible Radio band and this required permission from Airport Authority of India (AAI) as the project site is within an operating airport and any undesired interference with Flight navigation system can have disastrous consequences.
- Due to the fast pace of construction, the link location as well as the heights have to be frequently changed.
- Erection of glass façade in the building caused reflection problem of the Radio signals which also required relocation & height alterations of the links.
- The sub-contractors are using the same band frequency for their Internet connection, which also required permission from AAI.
- The usage of Aconex Document Management system is implemented through a web based system. Its implementation had its own difficulties as it required a project wide standard procedure to be adopted for document exchange. Some of these procedures were at conflict with the own internal organization procedures, but this difficulty was overcome by prioritizing the Project needs to all other local constraints.

Case Study Discussion

A mega project of the scale of IGI airport expansion project requires real time effective information and data flow between all the agencies and in this project, the information structure is contributing to the successful completion of project. Some of the reasons of the success of the adopted information system are:

- A sustainable system configuration was developed after due consideration of the project information requirements, site conditions, IT capabilities of the project site staff and other sub-contracting agencies and other associated technical as well as administrative issues.
- The adopted systems are user friendly including the Aconex document management system. Thus, the staff does not have a resistance to the adoption of developed systems.
- Also, the staff is provided adequate training before implementation of a new system and are supported by helpline systems.
- Adoption of the information system by all the sub-contracting and associate agencies is regulated through required contract clauses. Thus, all the project team agencies have adopted Aconex and are working on a common platform.
- For effective intra organization communication, L&T has integrated their site information systems with the EIP of the head office of their construction division. Thus, the intra organization administrative issues are also dealt in real time.

L&T did face some technical problems in implementing the system, but these were overcome as it has already constructed airport projects of similar scale and had previous experience of such projects, also the project needs were prioritized to all other local constraints.

As discussed earlier, use of barcoding and RFID for material management is not a norm in the country and has not been reported in the literature. But, in this project, since the computers at weigh bridges and the batching plant are connected with the head office data server, it facilitates real time accounting for receipt of bulk material and material consumption at the batching plant. Also, the biometric site access system facilitates manpower resource management.

Adoption of a document management system across all project team agencies and a videoconferencing facility at the project head office facilitating real time discussions and decision making, are very important for this project where the project team agencies are spread over different continents and the project is to be executed at international standards with fast track construction.

DISCUSSION

Mobile computing and communication is an intrinsic requirement of construction projects. Innovations in mobile communication devices and communication techniques are happening and these technologies are being adopted in different sectors. But, adoption of these technologies in construction projects is limited because of heterogeneous structure of construction project teams due to the variable size and variable IT capabilities of associating project team agencies, due to harsh and remote site locations, and because at present requirement of mobile communications is limited to large, remote and hazardous project sites. But, in such projects also to adopt new technology, bigger organizations need to take a proactive approach in designing, implementing and facilitating sub-contracting and associating agencies in implementing such a system as was observed in the discussed case study.

Construction sites have harsh working conditions and remote sites sometimes do not even have telecommunication links. Also, there is a requirement to study use of wearable computers with voice data input facility. So, more research should be conducted in adapting the mobile devices for construction sites after understanding of the work requirements of site personnel. Construction projects have multi enterprise and multi functional structure. Thus, more research is required in management and technology issues required to integrate widely available mobile communication devices with project information structures and organization information structures.

Majority of the construction organizations are SMEs with limited IT capabilities and large capital required for adoption of new technology, electronic conversion of data and documents and training required for the project staff. Initiatives for adoption of new technology like mobile communications has to be taken by large organizations and government bodies. Large organizations should invest in the required infrastructure and regulate mobile communications adoption by their sub contracting agencies who are primarily SMEs. This was also studied in the case study of IGI airport expansion project. L&T the prime contracting agency has implemented the WLAN at site for effective intra organization communication and an electronic document management system which is being utilized by all the sub contracting and associate agencies. As discussed above, they faced problems in implementing these systems due to site conditions and due to the requirement of having similar communication procedures across all agencies working on the project. These problems were resolved with the help of technical experts, by inclusion of appropriate clauses in the contract documents drawn with sub-contracting agencies and by facilitating these agencies in adopting the new technology. Government should also provide incentives to SMEs for adopting mobile communication. Since most of these organizations are waiting to see the impact of usage of these technologies in the industry, the government should compile and report the success of utilization of these technologies in different projects across the country. This would facilitate modifying perception of organizations specifically SMEs with respect to benefits of adoption of mobile communications in construction projects.

Consideration of all the above discussed issues would lead to higher adoption of mobile communication in the industry and a pragmatic shift in the industry.

FUTURE RESEARCH DIRECTIONS

Future research should focus on integration and further modification of widely available mobile devices for communication by the construction industry. The research should also study the issues which require attention by the government bodies and large organizations.

CONCLUSION

Mobile communications are required to be adopted for effective construction project management. But, research is required to study the construction site work processes, specific requirements of site staff, and management and information structure of project teams to enhance mobile communications adoption by the construction industry. Majority of the construction organizations are SMEs and an initiative would be required by large organizations and government bodies to enhance IT capabilities of SMEs and provide them incentives so that mobile communication can be adopted at the industry level. Case study of 'Indira Gandhi Airport expansion and renovation project' being executed in India shows that a sustainable mobile communication structure adopted for project team and integrated with overall organization communication structure helps the project team in successful construction project management.

REFERENCES

Aconex (2009, June 17). Aconex selected for world's second largest terminal. Retrieved July 2009 from http://www.aconex.com/Corporate/News/Latest-news/Aconex-selected-for-DIAL.html

Ahuja, V., Yang, J., & Shankar, R. (2006, November). Strategic use of ICT for construction organisations - Requirements and implementation issues. In *Proceedings of INCITE / ITCSED 2006 Conference* (Vol. 1, pp. 235-250). New Delhi, India.

Ahuja, V., Yang, J., & Shankar, R. (2009). Study of ICT Adoption for Building Project Management in the Indian Construction Industry. *Automation in Construction*, *18*(4), 415–423. doi:10.1016/j.autcon.2008.10.009

Airport-technology.com (n.d.). *Indira Gandhi International Airport Delhi, India*. Retrieved July 2009 from airport-technology.com/projects/delhi/

Alexander, J., Coble, R., & Elliott, B. (1997). Hand-held communication for construction supervision. In [Minneapolis, MN.]. *Proceedings of ASCE Construction Congress*, *V*, 972–979.

Bacheldor B. (2002, March 18). Handheld system assesses damage to see how buildings survived, *Information Week*.

Broyd, T. (1999). What Impact can IT Realistically have on Construction? In *Proceedings of Berkeley-Stanford CE&M Workshop – Defining a Research Agenda*. Stanford, Berkeley, USA: University of Stanford and University of California.

ConcordECC, *31*(1), 3.

Delhi Indira Gandhi International Airport. (n.d.). Retrieved July 2009 from www.newdelhiairport.in/about-dial.asp

Delhi Indira Gandhi International Airport. (n.d.). Retrieved July 2009 from www.newdelhiairport.in/milestones.asp

Elzarka, H., & Bell, L. (1997). Development of pen-based computer field applications. *Journal of Computing in Civil Engineering*, *11*(2), 140–143. doi:10.1061/(ASCE)0887-3801(1997)11:2(140)

Filos, E. (2002). European collaborative R&D projects related to the "Smart organization". A first evaluation of activities and implications for construction. In Turk & Scherer (Eds.), *Proceedings of the Conference on eWork and eBusiness in AEC* (pp. 27-32). Swets and Zietilinger, Lisse.

Fuller, S., & Sattineni, A. (2002). Investigating the use of wearable computers in construction. In *Proceedings of the First International Conference on Construction in the 21st Century Challenges and Opportunities in Management and Technology* (pp. 583-588), Miami, Florida.

Huang, C., Fisher, N., & Broyd, T. (2002, September). Development of a triangular TPC model to support adoption of construction integrated system. In D. Greenwood (Ed.), *Proceedings of ARCOM 18th Annual Conference* (Vol. 2, pp. 427-438). University of Northumbria, UK

Indiapressagency.com. (2006, July 18). *Indian SMEs shining bright: Survey*. Retrieved July 2009 from http://indiabusinessweek.com/News/Economy/sme.html

Jaselskis, E. J., & El-Misalami, T. (2003). Implementing radio frequency identification in the construction process. *Journal of Construction Engineering and Management*, *129*(6), 680–688. doi:10.1061/(ASCE)0733-9364(2003)129:6(680)

Katranuschkov, P., Scherer, R. J., & Turk, Z. (2001). Intelligent services and tools for concurrent engineering – An approach towards the next generation of collaboration platforms. *ITcon*, *6*, 111–128.

Kimoto, K., Endo, K., Iwashita, S., & Fujiwara, M. (2005). The application of PDA as mobile computing system on construction management. *Automation in Construction*, *14*, 500–511. doi:10.1016/j.autcon.2004.09.003

Kondratova, I. (2004). Voice and multimodal technology for the mobile worker. *ITcon*, *9*, 345–353.

Koseoglu, O. O., & Bouchlaghem, D. (2006, November). Design and construction collaboration using mobile visualization. Vol. II, *Proceedings of INCITE / ITCSED 2006 Conference* (Vol. 2, pp. 201-218) New Delhi, India.

Kuladinithi, K., Timm-Giel, A., & Gorg, C. (2004). Mobile ad-hoc communications in AEC industry. *ITcon*, *9*, 313–323.

Kumar, R. R., De, J. B., & Radhakrishna, K. (2006, November). Human resource development in construction industry through information technology (IT). In *Proceedings of INCITE / ITCSED 2006 Conference* (Vol. 3, pp. 303-326), New Delhi, India.

Larsen & Toubro Limited (n.d.). *Information Systems.* Retrieved October 2009 from lntecc.com/html/information_systems.htm

Magdic, A., Rebolj, D., & Suman, N. (2004). Effective control of unanticipated on-site events: A pragmatic, human-oriented problem solving approach. *ITcon*, *9*, 409–418.

McCullough, B. (1997). Automating field data collection in construction organizations. In [Minneapolis, MN.]. *Proceedings of ASCE Construction Congress*, *V*, 957–963.

Meissner, A., Mathes, I., Baxevanaki, L., Dore, G., & Branki, C. (2003). The cosmos integrated IT solution at railway and motorway construction sites - Two Case Studies. *ITcon*, *8*, 283–291.

Menzel, K., Keller, M., & Eisenblatter, K. (2004). Context sensitive mobile devices in architecture, engineering and construction. *ITcon*, *9*, 389–407.

NetworkComputing (2009). *A wireless umbrella for Delhi airport*. Retrieved July 2009 from http://www.networkcomputing.in/LT-ECC-Division-EDGE-001Jun009.aspx

Peansupap, V., & Walker, D. (2005). Factors affecting ICT diffusion: A case study of three large Australian construction contractors. *Engineering, Construction, and Architectural Management*, *12*(1), 21–37. doi:10.1108/09699980510576871

Rebolj, D., Magdic, A., & Cus-Babic, N. (2000). *Mobile computing in construction.* Slovenia: Maribor.

Rebolj, D., & Menzel, K. (2004). Mobile computing in construction. *ITcon*, *9*, 281–283.

Reinhardt, J., Garrett, J. H. Jr, & Akinci, B. (2005). Framework for providing customized data representations for effective and efficient interaction with mobile computing solutions on construction sites. *Journal of Computing in Civil Engineering*, *19*(2), 109–118. doi:10.1061/(ASCE)0887-3801(2005)19:2(109)

Ribeiro, F. L., & Lopes, J. (2002, September). An approach to e-business in construction. In D. Greenwood (Ed.), *Proceedings of ARCOM 18th Annual Conference* (Vol. 2, pp. 475-484) University of Northumbria, UK.

Shankar Narayanan, R. (2008). Delhi International Airport – Moving on fast track. *ECC Concord*, *31*(1), 31–34.

Singh, A. (2006, November). Training and workforce planning in construction industry through e-learning and IT management systems. In *Proceedings of INCITE / ITCSED 2006 Conference* (Vol. 3, pp. 341-360). New Delhi, India.

The Times of India (2008, July 10). Delhi beats Mumbai to become busiest airport. Retrieved July 2009 from http://timesofindia.indiatimes.com/India/Delhi_is_countrys_busiest_airport/articleshow/3216435.cms July 10, 2008

The Times of India (2009, July 27). India will be third largest net user by 2013: Study. Retrieved July 2009 from http://lite.epaper.timesofindia.com

Venkatraman, S., & Yoong, P. (2009). Role of mobile technology in the construction industry – A case study. *International Journal of Business Information Systems*, *4*(2), 195–209. doi:10.1504/IJBIS.2009.022823

Ward, M., Thorpe, T., Price, A., & Wren, C. (2004). Implementation and control of wireless data collection on construction sites. *ITcon*, *9*, 297–311.

wikipedia.org. (n.d.). Indian Construction Industry. Retrieved January 2009 from http://en.wikipedia.org/wiki/Indian_Construction_Industry

wikipedia.org. (n.d.). Indira Gandhi International Airport. Retrieved August 2009 from http://en.wikipedia.org/wiki/Indira_Gandhi_International_Airport#cite_note-8

ADDITIONAL READING

Abbott, K. R. (2002). *Voice enabling Web applications: VoiceXML and beyond.* New York: APress, Springer-Verlag.

ARUP Study. (2003). *Appropriate use of Information Technology on a Construction Site,* Retrieved from http://www.cite.org.uk/PII/bowden/constructionsitecomms.pdf

Baxevanaki, L., Bozios, T., & Mathes, I. (2001). Mobile user support for the construction industry. *Journal of the Institution of Telecommunication Engineers*, *2*(3), 123–129.

Björk, B.C. (2001). Document Management – A key technology for the construction industry. *Construction Informatics Digital Library*, Series:ecce2001

Boehling, H. (2002). Satellite-based remote multi-project reporting and controlling in construction industry applying IFCs and XML standardization, In Turk & Scherer (Eds.) *Proceedings of the Conference on eWork and eBusiness in Architecture, Engineering and Construction* (pp. 635-636). Swets & Zietilinger, Lisse.

Chen, C., Singh, K. D., Kuester, F., & Carsten, M. (2004). Portable input devices for human computer interaction. In *Proceedings of 1st International Forum on Applied Wearable Computing, IFAWC 2004* (pp. 107-120).

COSMOS. (n.d.). Construction Site Mobile Operations Support. EU Project EP-27021. Retrieved from http://evora.omega.it/~projects/cosmos and http://www.ipsi.fraunhofer.de/mobile/projects

Crabtree, I. B., & Rhodes, I. B. (1998). Wearable computing and the remembrance agent. *BT Technology*, *16*, 118–124. doi:10.1023/A:1009642301754

Das, S. R., Perkins, C., & Royer, E. M. (2000). Performance comparison of two on-demand routing protocols for ad hoc networks. In *Proceedings of the IEEE Conference on Computer Communications (INFOCOM)* (pp. 3-12).

Egbu, C. O., & Boterill, K. (2002). Information technologies for knowledge management: their usage and effectiveness. *ITcon*, *7*, 125–136.

Eisenblätter, K., Menzel, K., & Scherer, R. J. (2004). About the implantation process of mobile computing in AEC. In K. Beucke (Ed.), *ICCCBE 2004 – Proceedings of the Xth International Conference on Computing in Construction and Building Engineering*, Weimar.

Flood, I., Issa, R. R. A., & Caglasin, G. (2002). Assessment of e-business implementation in the construction industry, In Turk & Scherer (Eds.), *Proceedings of the Conference on eWork and eBusiness in AEC*, Swets & Zietilinger, Lisse, (pp. 27-32).

Hernandez, T. (2000). Mobile CAD goes onsite. Computer applications – mobile computing for construction. *Building Design & Construction*, *41*(9), 19.

Issa, R. R. A., Flood, I., & Caglasin, G. (2003). A survey of e-business implementation in the US construction industry. *ITcon*, *8*, 15–28.

Kuladinithi, K., Timm-Giel, A., & Görg, C. (2003). Potential productivity gains in the field through wireless technologies. In *Proceedings of the 10th ISPE International Conference on Concurrent Engineering* (pp. 725-731).

Lee, S. J., Royer, E. M., & Perkins, C. E. (2003). Scalability study of the ad hoc on-demand distance vector routing. *ACM/Wiley . International Journal of Network Management, 13*, 97–114. doi:10.1002/nem.463

Navarette, G. (1999). In the palm of your hand: Digital assistants aid in data collection. *Journal of Management Engineering, 15*(4), 43–45. doi:10.1061/(ASCE)0742-597X(1999)15:4(43)

Phair, M. (2000, September 25). Handheld device gets a second wind. *Engineering News Record.*

Rivard H. (2000). A survey on the impact of information technology on the Canadian architecture, engineering and construction industry. *ITcon, 5.*

Ruikar, K., Anumba, C. J., & Carrillo, P. M. (2005). End-user perspectives on use of project extranets in construction organizations. *Engineering, Construction, and Architectural Management, 12*(3), 222–235. doi:10.1108/09699980510600099

Shiratuddin, M. F., Perdomo, J. L., & Thabet, W. (2002). 3D visualization using the Pocket PC. In Turk & Scherer (Eds.), *Proceedings of the Conference on eWork and eBusiness in AEC,* Swets and Zietilinger, Lisse (pp. 639-644).

Sun, M., & Howard, R. (2004). *Understanding IT in Construction.* Spon Press.

Ward, M., Thorpe, A., Price, A., & Wren, C. (2003). SHERPA: Mobile wireless data capture for piling works. *Journal of Computer Aided Civil and Infrastructure Engineering, 18*(4), 200–220.

Zurbes, S. (2000). Considerations on link and system throughput of Bluetooth networks. In *Proceedings of the 11th IEEE International Symposium on Personal, Indoor and Mobile Radio Communications* (pp. 1315–1319).

KEY TERMS AND DEFINITIONS

Barcoding: Technology used for material tracking and management.

Construction Project Management: Management of construction projects.

India: An Asian developing country experiencing an economic boom.

Mobile Communication: Communication between two parties, out of which one or both are mobile.

Small and Medium Enterprises: Organizations which are not very large. Categorization based on size is specific to each country and industry.

Videoconferencing: Meeting or conference between geographically separated parties and through web media.

Wearable Computers: Computers which could be a part of the clothing.

Chapter 4
Requirements Engineering in the ICT4D Domain

Kristina Pitula
Concordia University, Canada

Daniel Sinnig
Concordia University, Canada

Thiruvengadam Radhakrishnan
Concordia University, Canada

ABSTRACT

Requirements engineering is an important stage in any software development. It is more so in the case of software development for social development projects in rural areas of the developing countries. ICT4D which stands for "Information and Communication Technologies for Development" is gaining more and more attention as computing is more widely affordable. This article is concerned with requirements engineering in the ICT4D domain. In many developing countries, a significant effort is being put into providing people in rural areas with access to digital content and services by using Information and Communication Technologies. Unfortunately most ICT4D projects pursue a top-down development model which is driven by the technology available and not by the very needs and social problems of the people living in rural communities (Frohlich et al., 2009). Existing technologies are often applied in a non-inclusive manner with respect to the local population, without sufficient adaptation or re-invention, and often without regard for user's needs and their social contexts.

INTRODUCTION

It has been shown that current (traditional) requirements elicitation techniques require that end users are able to understand and to articulate their problems and that they are able to express their needs. Expressing oneself, with the right problem descriptions and the right needs, however, are all learnt skills, which because of the socio-economic situation, people in rural communities may not have had the opportunity to develop (Pitula & Radhakrishnan, 2008).

In order to overcome these problems we propose a requirements management process especially suited for ICT4D projects. The process supports both bottom-up and top-down develop-

DOI: 10.4018/978-1-61692-818-6.ch004

ment. Needs and problems are elicited from end users using *structured digital storytelling*; an elicitation technique suitable for stakeholders with limited literacy to express themselves and their needs. In a bottom-up manner, the various "stories" are processed, analysed and abstracted into sets of "goals", "needs", and "objects". This abstracted information is then used in a top-down manner to identify the potential areas of technology intervention and will eventually lead to the production of the requirements specification.

The remainder of this article is structured as follows: In sections 2 and 3 we provide relevant background information on the topic of requirements engineering and detail the ICT4D context. Then, in sections 4 and 5, we propose structured digital storytelling as a requirements elicitation technique and contrast it with conventional elicitation techniques. Section 6 provides a description of our requirements management process for ICT4D projects. Section 7 discusses the validation of our approach while section 8 provides an outlook to next generation tool support. Finally we review relevant related work (section 9), provide an outlook to future avenues and conclude (section 10).

REQUIREMENTS ENGINEERING

Among the various software engineering disciplines, understanding the requirements of a project is one of the most difficult tasks faced by a software engineer (Pressman, 2005). In fact, a recent survey shows that 71% of software projects are late, go over budget or even fail completely (Standish-Group 2004). The most significant factors causing projects to be challenged or fail is related to the mismanagement of requirements (Taylor, 2000). Incomplete or incorrect requirements inevitably propagate into the later stages of software development and hence lead to the implementation of a product that does not meet user expectations. Such failures can be avoided, if end users and stakeholders are involved earlier in the software development process and if appropriate elicitation techniques are employed.

In order to facilitate end-user involvement, the requirements engineering community has proposed different methods (e.g., contextual inquiry, interviewing, requirements workshops, focus groups) to elicit software requirements and to facilitate the communication between users and analysts. In this paper we propose a novel elicitation technique, called *structured digital storytelling* (SDS), which can be used in a complementary fashion and, as we will demonstrate in this paper, is particular suited for eliciting requirements within an ICT4D context.

During requirements production, the elicited information is systematically transformed into a set of software requirements. Throughout this process it is important to establish *traceability links* that clearly relate requirements to their sources; the originally elicited information. In general terms, requirements traceability is defined as the ability to describe and follow the life of a requirement in both forward (post-requirements traceability) and backward direction (pre-requirements traceability) (Orlena & Finkelstein, 1994). Within an ICT4D context, requirements traceability is particularity important, if we take into account the fact that requirements come from different sources, such as regulatory and governmental bodies, international and national development agencies, product managers, and of course, the actual end users (Leite, 2003). These stakeholders all have different requirements on the product. Using requirements traceability an implemented feature can be traced back to the person, or organization, that requested it during requirements elicitation. Well established traceability links allow showing that every requirement is related to one or more stakeholder's needs and that all needs have been addressed by the specified requirements.

ICT4D CONTEXT

Eliciting software requirements in an ICT4D context poses a number of challenges due to the nature of the projects. Such projects target marginalized communities with the goal of assisting them in improving their socio-economic situation. The projects are driven by high-level social and economic development goals that almost invariably are initiated from outside the targeted community. Many involve multiple stakeholders such as social workers, agronomists, government and business representatives, etc. coming from the public, private, and non-profit sectors, and these need to work in partnership to be the most effective. The intended beneficiaries typically have limited schooling, low literacy levels, and low disposable incomes while the developing regions and countries where the projects take place are characterized by inadequate infrastructures, intermittent power and connectivity, underdeveloped economic markets and distribution and support networks, and a lack of trained personnel. Extreme climatic conditions such as heat, cold, dust, or humidity introduce additional operating constraints. All these factors give rise to the following challenges when eliciting software requirements in an ICT4D context:

1. Understanding and incorporating the diverse goals of the different stakeholders to converge on a single, agreed upon set of achievable software project goals
2. Identifying all the environmental constraints that will impact project goals
3. Getting input from the targeted end-users with respect to project goals

In the case of the stakeholders, their diverse backgrounds and areas of expertise often lead to a set of disparate high-level goals. If these are incomplete or vaguely stated, they can easily be misinterpreted. These goals, along with their underlying assumptions, must be elaborated in detail so that they can be mapped into operational project goals and constraints, and any conflicts identified and addressed. This is essential in order to converge on a single set of operational goals that drive the software requirements and according to which the project's success will be measured. Although end-users are crucial stakeholders in any project, often they are not consulted when ICT4D project goals are set (Tongia & Subrahmanian 2006), particularly when stakeholders from outside the community drive the project.

With regards to the environment, the technical, economic and cultural conditions which characterise the context-of-use introduce novel constraints that might compromise a project's success if not addressed. Devices must be appropriate for the prevailing operating conditions (climate, infrastructure, available support), and provide relevant services that are affordable and accessible to their intended users. Here economic and social factors come into play. For example, to keep costs low, shared facilities may be used. However this may introduce confidentiality and privacy concerns, and other social considerations, such as access being restricted because of age, gender or social standing (e.g. women may not be able to visit sites frequented by men or people from lower castes may be denied access) (Garai & Shadrach, 2006). And even if services are relevant and accessible, other factors such as personal obligations, public opinion or local customs may prevent users from being able to benefit from them. Determining what is relevant, accessible, and applicable requires the input of end-users.

The early involvement of end-users is recognised as critical in developing high-quality requirements. However, because of social and literacy conditions, people are likely to have problems articulating their information needs. Hence traditional requirements elicitation techniques such as interviewing or elicitation through questionnaires are inappropriate. Instead, we propose using *structured digital storytelling* (STD) techniques, supported by a suitably designed interactive multi-media software tool, to elicit their information needs. In a non-literate society stories are the primary mode of communication. Using a storytelling approach, we can involve end-users early in the elicitation process and thus put their problems and needs foremost in driving requirements.

STORYTELLING

Storytelling as a technique is applied in a wide range of domains. Stories constitute an art form, a form of entertainment and a fundamental mode of communication. In the social sciences, oral histories are used to provide alternative views on historical events based on first-hand experience, to capture cultural information, and to explore social issues. In business, storytelling is viewed as an integral part of organizational knowledge management (Snowden, 1999) while in software, stories are used in the design process to communicate among stakeholders. More recently, storytelling has been put forward as a means of eliciting requirements in domains such as healthcare, where access to end-users and the actual context-of-use is restricted (Gausepohl, 2008).

Storytelling has been proven useful for people with limited literacy to express themselves and their needs. A recent study by Kerr (2003) demonstrated the potential of digital storytelling for expressing community information, issues and frustrations; the very sources that will shed light on the *real* needs of users in rural areas. In this paper, we modify the digital storytelling concept with a goal-oriented specialization, which leads to our notion of *structured digital storytelling* (SDS).

At its simplest, storytelling consists of someone telling their personal story on some topic and the narration being recorded. An interviewer may guide the narrative to ensure that items of interest are clarified and expanded. Recent digital technologies support the authoring of sophisticated multimedia stories that can be made accessible to a broad audience. *Interactive Voice Response* (IVR) systems are an alternative approach for collecting user input. IVR applications use structured dialogues to ask a sequence of questions when eliciting information. Our approach combines the two. We propose adding a multimedia, structured dialogue interface onto digital storytelling technology to assist people in expressing their information needs through stories which can then be shared in the community. Instead of asking direct questions about their information needs, the villagers' needs can be elicited through a series of questions, short stories, 'what if scenarios', or by hearing their neighbours' views regarding the problems they face and their potential solutions. The structured dialogue ensures that relevant themes are covered, while hearing stories told by their neighbours will inspire people to tell their own stories.

STORYTELLING VS. CONVENTIONAL ELICITATION TECHNIQUES

Commonly used techniques for eliciting requirements from end-users include *interviews*, *focus-groups* and *ethnographic* studies (Leffingwell & Widrig, 2003). These are briefly described below along with their advantages and disadvantages.

Interviews consist of meeting with end-users individually to ask them questions regarding their problems and needs. Interviews can be structured, semi-structured, or unstructured. With a structured interview, a set of questions is pre-defined and all users are asked the same questions. With unstructured interviews, the interviewer asks questions on-the-fly. A semi-structured interview combines the two, with the interviewer asking questions from a pre-defined list, and follow-up questions as needed. Interviews have the advantages that both verbal and non-verbal responses can be observed, and users can be probed in-depth with follow-up questions. Among the disadvantages is that it is difficult to elicit tacit knowledge. Also, information on the context-of-use is not readily observable. Given the novel nature of the ICT4D context, there is no way of ensuring that all relevant aspects are covered by the questions. Furthermore, because of the socio-economic differences, end-users may be intimidated by the interviewer, or have difficulty answering direct questions. In addition, conducting individual interviews is time-consuming.

Focus groups are similar to interviews, with the difference that users participate in a group rather than individually. In a focus group, a facilitator presents the group a series of pre-defined questions or topics which the participants then discuss. Along with the advantages and disadvantages of interviews, focus groups have the advantage that the discussion among participants may reveal more requirements and they are less time-consuming than individual interviews. The disadvantages are that participants may feel uncomfortable stating opinions that differ from those of the group, leading to "groupthink". Also certain participants may dominate the discussion, leaving other valid viewpoints unexplored.

An **ethnographic study** is a technique borrowed from the domain of anthropology. Such studies consist of going onsite to talk with users and observe their activities and behaviours with respect to the proposed intervention. This includes observing what supporting artefacts they use as well as the environment in which the activities take place. The advantage of such studies is the insights they provide on the context-of-use. Their disadvantage is that thorough studies are extremely time-consuming. This can be addressed by 'rapid ethnography" techniques. However the risk there is that the study is too superficial, devolving into what Dourish refers to as "scenic fieldwork" (i.e. "we went there and this is what we saw") (Dourish, 2006). With superficial studies there is a risk that aspects which are not directly observed are not considered.

Structured digital storytelling (SDS) differs from the other forms of elicitation mentioned above in that the narrator is largely left on his or her own to tell their story. Among the advantages we foresee in this approach is that storytelling capitalises on the villagers' primary mode of communication. Although the questions provide some guidance, an interviewer is not present to influence the narration. Thus it is possible to identify problems and needs not initially envisaged and contextual factors that might otherwise be overlooked. Among the challenges to address, storytelling is by nature a social exchange that is enriched by the presence of an audience. One disadvantage with respect to the other elicitation techniques is that there is no one present to provide clarifications, guidance or immediate follow-up on items of interest. Also, participants may focus their story on one aspect, leaving other equally relevant aspects unmentioned. Confidentiality and self-censure are also considerations, particularly when dealing with sensitive subjects.

INCORPORATING 'SDS' IN REQUIREMENTS GATHERING PROCESS

In this section we detail our requirements management process especially designed to support ICT4D projects. The process is based on the ideas and principles defined in the previous section. That is, the process promotes the application of *structured digital storytelling* in an expanded context to elicit requirements from users. The process also applies well known software engineering principles such as, goal-oriented development, traceability and user-driven requirements elicitation and validation.

An overview of the proposed process is given in Figure 1. Its various phases of the process (viz. *elicitation*, *abstraction*, *specification*, and *validation*) are described next, with a particular focus on requirements elicitation and requirements abstraction.

Figure 1. Traceable Requirements Specification Process

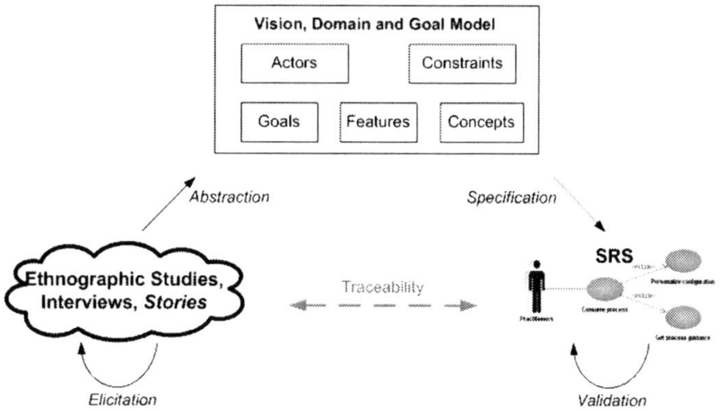

Requirements Elicitation

Given the high-level goal of delivering information of value, requirements elicitation generally starts with answering the question: "What categories of information are vital to this rural society in some prioritized order". These categories identify the potential areas of intervention for a project, and invariably relate to a community's economic activity, and the health and welfare of its families. As mentioned in the previous section, because of social and literacy conditions, people living in rural areas are likely to have problems articulating their information needs. Hence traditional requirements elicitation techniques such as interviewing or elicitation through questionnaires are inappropriate. Instead, we propose using SDS techniques, supported by a suitably designed interactive multi-media software tool, to elicit their information needs.

SDS can be applied at any stage of requirements elicitation, to assist in identifying areas of intervention, to identify and validate high-level goals and constraints, or to elaborate and validate operational goals and conditions. The use of an SDS approach does not exclude the use of other elicitation techniques. Indeed we see it as complementary, assisting stakeholders in validating that they are focusing on the right problem, and that the problem is thoroughly understood in the context in which it occurs.

Once the area of intervention and topics of interest have been identified and a set of corresponding questions formulated, the collection of stories is relatively straightforward. An application (such as the E-Tool presented in the next section) that plays the questions and records responses can be made available on a suitable device (e.g. a laptop or mobile phone) in a location such as a community center, and villagers invited to try it. Participation can be encouraged by enlisting the support of respected members of the community and having them record their own stories to serve as examples in the story library. The time involved corresponds to the time it takes to tell the story, allowing a reasonable number of stories to be recorded in a few days. Because resources such as interviewers and facilitators are not involved, the application can readily be deployed in a number of villages, increasing the number of the stories collected and coverage of issues. Different topics can be addressed by changing the application prompts while different linguistic regions are easily supported by simply translating and rerecording the application prompts, making this elicitation method highly adaptable.

Requirements Abstraction

During the *abstraction* phase the information elicited is processed, analysed and categorised. Since stories are told in people's local language, the first step after data collection consists of translating the stories into English. This is projected to be a manual activity, performed by local individuals (ideally from the same rural community), who possess a high degree of literacy in both the local language and English. Although this is a relatively labor intensive process, it is straightforward and does not require highly skilled labor.

Following this the transcriptions are analysed to determine what issues the participants mention, their importance, and the various factors that play a role therein. It is here that a skilled analysis is required, to identify the various issues and related factors or conditions. For example, farmers complain of unfair pricing at the markets where they sell their produce. At the same time, farmers who have taken out loans are compelled to sell to the lender, thus bypassing the open market system. Any attempt to address the issue of unfair pricing must also consider how many farmers are constrained by loans and how they might benefit from a possible solution. Such issues are prioritised based on how critical they are and the frequency with which they are mentioned, and these along with the other factors are related back to the high-level goals. Additionally needs, domain concepts and desired features are abstracted as well. Due to the often large amount of information, we plan to employ manual as well as automated or semi-automated natural language processing to extract the information stated above. The applicability of corpus linguistics to document processing in requirements engineering has been shown in several problem domains and at different levels. E.g., Rolland and Proix (1992) provide a general background for the applicability of natural language processing to requirements engineering.

The primary artefact resulting from the abstraction phase is a detailed goal model, supplemented by a domain model and detailed *software* vision document. Starting with high-level goals that identify the purpose of the system, these are successively decomposed until a set of requirements by which these goals can be met is attained. According to (Van Lamsweerde, 2001) a goal is achieved by satisfying a set of subgoals. Constraints are simply properties and conditions that must be satisfied when achieving those goals. Goals may support each other or be in conflict, and constraints will apply to certain goals. A goal structure can be used to represent the relationships among the goals and constraints, linking high-level goals at the top of the structure to operational requirements in the leaves. The links indicate the AND/OR relationships among the subgoals, and positive and negative support among the goals and with respect to the constraints. The information gathered during the elicitation phase can assist in this, by prioritising goals, and identifying additional goals and constraints to satisfy, based on end-users' input.

Requirements Specification and Validation

During *specification*, the goals, needs, features, and concepts are transformed into a complete set of software requirements. For the purpose of requirements production, the augmented goal model of the previous phase serves as primary input for a traditional requirements engineering process. Depending on the technology available, this information is then used in a top-down manner to identify the areas of "technology intervention or adaptations", which will eventually lead to the production of the requirements specification. Throughout the process of requirements production and refinement, statements from the different sources are integrated into a single set of requirements in the final *software requirements specification* (SRS). This specification unambiguously depicts the functional and non-functional requirements of the envisioned software product.

Finally the SRS obtained is *validated* to ensure that it reflects stakeholders' needs and is sound and complete. The former is achieved by establishing traceability links to the user stories gathered during elicitation. It is ensured that each user need is addressed by at least one requirements statement

Figure 2. E-Tool Interface for Listen to Stories (left) and Tell your own Story (right)

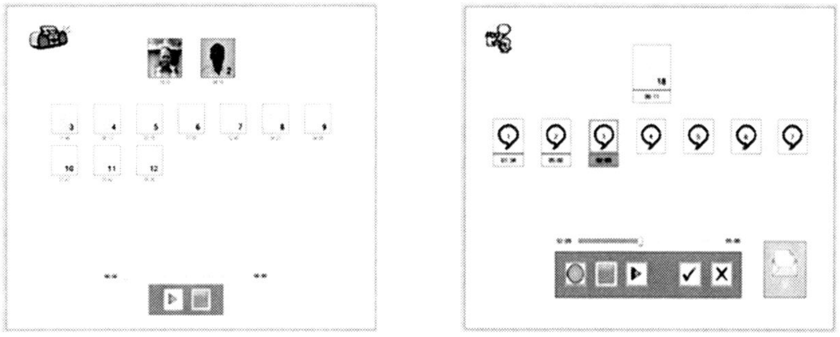

and any unnecessary system features that do not reflect any user needs, are excluded. It is important that the requirements engineers remain in close contact to domain experts, who will be able to reconcile differing goals and needs. In addition to the traceability links, the SRS is also analyzed using informal validation techniques (e.g., walkthroughs, reviews, or checklists) and formal methods (e.g., static analysis or model-checking).

VALIDATION OF THE SDS APPROACH – A PRELIMINARY EXPERIMENT

In order to test the viability of our storytelling approach for requirements elicitation, we conducted three field studies in two rural areas of India using a prototype elicitation tool (called E-Tool). Our objective was to determine if SDS can effectively be used as a requirements elicitation tool within an ICT4D context. Towards this end we elicited stories on two different topics in two different rural regions with distinct regional languages. Below we first provide a brief description of the E-Tool and the rural Indian context before describing the experiment.

The Original E-Tool

In order to facilitate SDS we have the developed a prototypical tool called E-Tool. It is designed to be a self-contained application for collecting narrations from villagers with minimal intervention from outside support staff. It runs on a laptop computer equipped with a microphone. The application provides (1) an introductory video explaining the reason for collecting the stories as well as an overview of how to operate the application, (2) a story library where the stories are stored and villagers can listen to them, and (3) an interview feature which guides users through a series of questions, letting them tell their own story. The application is designed to be easy to use by a non-literate population. Consequently, drawing on the experience of (Medhi et al. 2007), instead of text, navigation aids are provided using graphical icons, buttons with distinct colors and audio prompts to identify and guide users through the various options. The user interface of the E-Tool application for the "listen to stories" and "tell my own story" functionalities are illustrated in Figure 2. E-Tool Interface for Listen to Stories (left) and Tell your own Story (right). A 'recording session' elicits a sequence of recordings, where each recording focused on a simple theme relevant to the end user. An example would be: "Tell us about what crops do you cultivate." A small sequence (no more than five plus or minus two) of thematic recordings in a session is pre-planned based on a predefined course-grained task model developed by an analyst.

Rural Indian Context

In India there are wide disparities in socio-economic conditions. Whereas the new economy centered on urban areas is giving rise to a comfortable middle class (and aspirations to those below), the traditional economy of the rural areas, largely based on farming, has left much of the rural population in extreme poverty with few options out (Robinson, 2007; Rezwan, 2009). Over 70% of the Indian population still lives in rural areas, many barely surviving on subsistence farming, seasonal work and occupations dictated by caste, with incomes at or below the international poverty line. Small plots, land depletion, poor yields, usurious lending practices and corrupt agents all conspire to put many farming households into a downward spiral of debt. A crop failure under such conditions has disastrous effects, pushing many farmers into despair, and in certain states suicide among farmers has reached crisis proportions. Although the government tries to regulate many aspects of farming, unregulated and corrupt practices persist. Villagers, many of whom have no schooling themselves, view higher education as the only way for their children to aspire to a better future. However, while primary and secondary education is government funded and locally accessible, higher education is not, and there are many private institutions of questionable merit whose primary goal is profit. Word-of-mouth is the only source of information for many parents and children regarding career choices and available education programs. It is for this reason that we focus our study on these two problem areas, namely:

1. The problems faced by farmers, and
2. The problems faced by parents and children when considering higher education.

Field Studies

As stated earlier, a total of three studies were conducted. Two of these were conducted near Chennai in Tamil Nadu state and one was conducted near Bangalore in Karnataka state. All three took place in farming villages that are typical of the rural Indian context. In Tami Nadu, stories on farming and higher education were elicited while in Karnataka only farming stories were elicited. All together 30 stories were collected, 17 on farming and 13 concerning higher education. These were told by both male and female participants representing a broad age range, from children to the elderly, and a cross section of financial situations, from the very poor to those who considered themselves relatively well off (by local standards). The farming stories were collected in local homes, on the main street of the village, or in the case of Karnataka, in the office of a local NGO[1] involved with farmers while the stories pertaining to education were collected at two local schools.

With regards to the farmers, participants were primarily male (only 3 were female). They were aged 30 years and over, including two over 60 years old. While one participant had never attended school, the majority had completed some schooling, but only 4 had a higher education (mainly agronomy) and spoke English. The farmers' plot size varied from 1 to 10 acres and farming was the primary activity off all but two. Their financial situation varied from very poor to relatively well-off. However, only two did not need to borrow money to finance their farming activities.

In the case of higher education, both the parent and child participated in telling a story. Girls and boys were equally represented, all attended school, and were among the top in their class. A higher number of mothers participated, likely due to the time and location of the study (which took place at local schools in the middle of the day). None of the parents had a higher education themselves, all were employed in typical traditional activities, and many were poor, with some among the poorest in the village.

The studies required a certain amount of preparatory work. Once the area of intervention was selected (higher education and farming) it was necessary to identify the specific topics within each that participants would be asked to talk about. For this we drew on local experts in the respective areas to identify a list of relevant subtopics. These were formulated into open-ended questions which were then translated and recorded in the local language using a laptop computer and inexpensive microphone. A respected individual from the community was then asked to record their story to serve as an example.

In all three studies, a local person assisted in organizing the study and recruiting participants. The sessions were conducted informally and group participation was encouraged. On arriving participants were either shown the video, or given a brief explanation of the purpose of the study in the local language. They were then shown how to operate the application and invited to record their own story. Each participant recorded his or her story in turn while the others listened. Late arrivals quickly picked up what was going on by observing the others. After finishing their story, participants were offered a small gift (monetary or a box of candy) for their participation.

The results of our study exceeded our expectations. All the participants were able to tell their own story, and were enthusiastic about doing so. Villagers participated readily and quickly picked up the operation of the application. Once they began talking, they became involved in telling their story and were not distracted by the mechanics of recording. Although in almost all cases they participated in groups, their stories were highly personal and did not show any signs of "groupthink". At the same time, the group provided an audience for the teller, making the narrations a natural communicative exchange. A preliminary analysis of the data indicates that the stories are useful in identifying the participants' primary concerns and reveal many environmental factors that might otherwise be overlooked. Presenting the results of this analysis is beyond the scope of this paper. Below we list some practical findings with regards to applying the SDS technique:

- Participants do not answer the questions directly when a sub-topic is brought up; instead information is spread across the entire narrative. Consequently, any analysis pertaining to a sub-topic must consider the entire narrative.
- The narrations tend to be high-context, omitting contextual information that participants share in common and assume to be known. They also show a wide range of style, from very effusive to sparse. This appears to be correlated to the socio-economic situation of the person telling the story. For this reason it is recommended that participants be recruited from all socio-economic strata present in the community in order to provide sufficient background information on the issues.

This study clearly demonstrated that storytelling is an effective and acceptable way for people to express their concerns and needs with regards to some problem regardless of their literacy level. By conducting our experiment on different topics in different region and languages we also show that our approach is readily adaptable to new contexts in terms of population group, language, and topic addressed. Now that we have established this, we are working on making this elicitation technique accessible to a wider audience from a technical perspective.

THE NEXT GENERATION E-TOOL

We are currently developing a mobile version of the E-Tool in which structured audio-stories can be created using a handheld cell phone. The motivation behind this is the high penetration and ubiquitous availability of cell phones both in urban and rural contexts, especially in develop-

Figure 3. Non-Textual Interface of the E-Tool Mobile

ing countries. For example, mobile phones have a 65% market penetration rate in India[2]. Using a handheld device, completed stories are uploaded to a central server. Based on a pre-analysed task model, the UI will prompt the user to express their information needs in a storytelling format. It is assumed that the end user will have minimal cognitive abilities to recognize iconic symbols and numbers from 0 to 9, and follow a voice command. Similar to the desktop version, navigation is made quite simple by using symbols that are native to the specific community of people. To facilitate human interactions, the phones are assumed to be equipped with 'speaker-phone-output' or an ear-phone plugged in for listening to the speech output while viewing the screen. Screenshots of the prototypical mobile UI are portrayed in Figure .

An ongoing decision that needs to be made is how we plan on delivering voice clips to the server site. We have narrowed our choices to two delivery methods: the multimedia messaging service (MMS) and the wireless application protocol (WAP). The rationale for using the MMS service would be that it is an already implemented service in most cell phones and it does not use cellular data which tends to be expensive and, in certain markets, is not readily available. The downsides of this choice are compatibility issues between different phones. WAP, on the other hand, is the industry standard for data transmission through cell phones. It allows for direct communication to the server and for automatic upload of audio clips. Its benefits include simplified programming, direct access to files and less user input. Its downside is that it relies on mobile data.

RELATED WORK

Over the past decade, although numerous ICT4D projects have been attempted, few have achieved long-term sustained success. A too great emphasis on technical success with inadequate consideration of end-users' needs and of the social development aspect of the projects are among the factors that have contributed to this lack of success (Heeks, 2008). The need to involve intended beneficiaries in determining project goals and constraints has long been recognized in social and economic development circles, and a number of participatory approaches such as *Participatory Rural Assessment* (PRA) based on the work of Chambers (1983) exist for accomplishing this. However, such approaches are primarily geared towards identifying and introducing social and economic interventions and would need to be adapted to assist in determining software requirements for suitable ICT tools. Furthermore, participatory approaches rely on techniques such as interviews, focus groups and community meetings, and thus encounter the same disadvantages as previously described when used for eliciting requirements in an ICT4D context.

In sociology, storytelling has proven itself as a means for developing a collaborative analysis from the bottom-up. In a recent study by Kerr (2003) it was applied to identify the problems and constraints of homelessness as perceived by the homeless themselves. The resulting analysis revealed a number of significant issues that do not emerge from conventional top-down analyses where input is solicited from people such as social service providers, public officials and academic experts. There was a comparable divergence in the nature of potential solutions and associated issues as viewed from the top-down versus bottom-up. Furthermore, the research process of telling and listening to stories served as a catalyst to the homeless to become active in changing their situation.

In the ICT4D context, digital storytelling has been proposed as a means for sharing information among semi-literate people in rural villages. A study by Frohlich et al. (2009) has shown the viability of storytelling as a means of communication in rural India. The study showed that villagers were enthusiastic about creating and listening to stories. Interestingly, the study also revealed a certain tension between those interested in creating and disseminating serious "development" content, and others more interested in creating personal and cultural content for entertainment purposes.

More recently, storytelling has been proposed as a means for eliciting requirements in domains where access to end-users and the actual context-of-use is restricted. In (Gausepohl, 2008), the author describes a field study in the healthcare domain in which the requirements elicited using focus groups and interviews are compared to those elicited using focus groups and stories. The focus of this study was to determine any differences in the number, breadth and depth of themes addressed and the amount of time required by participants. The study concluded that there was no significant difference with respect to the number and breadth of themes addressed. However, storytelling and focus groups were more effective in eliciting more diverse context-of-use and social information, and storytelling required less time than interviews.

With regards to a goal-oriented approach to requirements engineering, this has been proposed as a means of dealing with complex systems in the software engineering domain (Cheng & Atlee, 2007). However, to date this paradigm has not been applied to elaborate requirements in an ICT4D context.

SUMMARY

In this chapter we discussed the particularities of requirements engineering within an ICT4D context. We proposed structured digital storytelling (SDS) as a complementary elicitation technique for software requirements. We contrasted SDS with traditional elicitation techniques (such as contextual inquiry, focus groups, and ethnographic studies) and established that SDS is not only a complimentary elicitation technique but also superior when dealing with semi and non-literate users.

We also presented a requirements management process that is especially suited for ICT4D projects. Our process adopts a range of techniques (i.e. storytelling, goal-oriented requirements engineering, and requirements traceability) each well established in its own domain, and combines them to arrive at an approach that addresses the specific challenges encountered when eliciting and specifying software requirements in the ICT4D domain. Our process supports both bottom-up and top-down development. It starts off with the analysis of the information gathered during the elicitation phase. In a bottom-up manner, the various stories, interviews and ethnographic studies are processed, analysed and abstracted into a detailed goal model.

This abstracted information is then used in a top-down manner to identify the potential areas of technology intervention and will eventually lead to the production of the requirements specification. In order to ensure that each requirement addresses a particular user need, we establish, for each step of the requirements process, traceability links between the various artefacts. Using these traceability links we are able to trace requirements back to their original statements (stories) at the later stages of development and testing. A study (Orlena & Finkelstein, 1994) has shown that the majority of problems attributed to poor requirements traceability are due to inadequate "pre-requirements traceability management".

SDS is supported by our ETool prototype which allows for recording and playback of elicited stories. The tool is designed to be easy to use by a non-literate population and thus equipped with a minimalistic, text-free user interface, which instead of text, uses graphical icons, buttons with distinct colors and audio prompts to identify and guide users through the various options. We are currently developing a mobile version of the E-Tool in which structured audio-stories can be created using a handheld cell phone. The motivation behind this is the high penetration and ubiquitous availability of cell phones both in urban and rural contexts, especially in developing countries. Other future avenues deal with the validation of the proposed process and the development of additional tools, such as a Goal-Task-Modeler or a Natural-Language-Analyser.

REFERENCES

Chambers, R. (1983). *Rural Development: Putting the Last First*. Prentice Hall.

Cheng, B. H. C., & Atlee, J. M. (2007). Research Directions in Requirements Engineering. In *Proceedings of Workshop on the Future of Software Engineering held at ICSE 2007* (pp. 285-303).

Dourish, P. (2006). Implications for design. In *Proceedings of SIGCHI Conference on Human Factors in Computing Systems* (pp. 541-550).

Frohlich, D. M., Rachovides, D., Riga, K., Bhat, R., Frank, M., Edirisinghe, E., et al. (2009). StoryBank: mobile digital storytelling in a development context. In *Proceedings of the 27th international conference on Human factors in computing systems* (pp. 1761-1770). Boston: ACM.

Garai, A., & Shadrach, B. (2006). Opportunities and Challenges. In *One World South Asia*. Taking ICT to Every Indian Village.

Gausepohl, K. A. (2008). *Investigation of Storytelling as a Requirements Elicitation Method for Medical Devices*. Masters Thesis in Industrial and Systems Engineering, Virginia Polytechnic Institute.

Heeks, R. (2008). CT4D 2.0: The Next Phase of Applying ICT for International Development. *IEEE Computer*, *41*(6), 26–33.

Kerr, D. (2003). "We Know What the Problem Is": Using Oral History to Develop a Collaborative Analysis of Homelessness from the Bottom Up. *The Oral History Review*, *30*(1), 27–45. doi:10.1525/ohr.2003.30.1.27

Leffingwell, D., & Widrig, D. (2003). *Managing software requirements: a use case approach*. Boston: Addison-Wesley.

Leite, J. C. S. P. (2003). *Perspectives on Software Requirements*. Springer.

Medhi, I., Sagar, A., & Toyama, K. (2007). Text-Free User Interfaces for Illiterate and Semi-Literate Users. *Information Technologies and International Development*, *4*(1), 37–50. doi:10.1162/itid.2007.4.1.37

Orlena, C. Z., & Finkelstein, A. C. W. (1994). An Analysis of the Requirements Traceability Problem. In *Proceedings of 1st International Conference on Requirements Engineering* (pp. 94-101). IEEE Computer Society Press.

Pitula, K., & Radhakrishnan, T. (2008). A Multimedia Tool to elicit Information Needs in Rural Communities. In *Proceedings of HCI for Community and International Development Workshop at CHI 2008*, Florence, Italy.

Pressman, R. S. (2005). *Software engineering: A practitioner's approach*. Boston: McGraw-Hill.

Rezwan (2009, August 25). India: A wave of suicides among farmers. *Global Voices Online.* Retrieved September 12, 2009 from http://globalvoicesonline.org/2009/08/25/india-a-wave-of-suicides-among-farmers/

Robinson/Sunna. S. (2007, May 17). Seeds of Despair. *Time.com.* Retrieved September 12, 2009 from http://www.time.com/time/magazine/article/0,9171,1622180,00.html

Rolland, C., & Proix, C. (1992). A Natural Language Approach for Requirements Engineering. In *Proceedings of CAiSE 1992 on Advanced Information Systems Engineering*, Manchester, UK.

Snowden, D. (1999). Story telling: an old skill in a new context. *Business Information Review, 16*(1), 30–37. doi:10.1177/0266382994237045

Standish-Group (2004). *CHAOS Third Quarter Research Report* (Technical Report).

Taylor, A. (2000). IT Projects: Sink or Swim? *The Computer Bulletin, 41*(1), 24–26. doi:10.1093/combul/42.1.24

Tongia, R., & Subrahmanian, E. (2006). Information and Communications Technology for Development (ICT4D) - A Design Challenge? In *Proceedings of International Conference on Information and Communication Technologies and Development* (pp. 243-255).

Van Lamsweerde, A. (2001). Goal-Oriented Requirements Engineering: A Guided Tour. In *Proceedings of Fifth IEEE International Symposium on Requirements Engineering* (pp. 149-163).

ENDNOTES

1. Non-government organization
2. Telecom Regulatory Authority of India

Section 2
M-Technology Applications

Chapter 5
In Defence of Mobile Technologies:
Exploring the Socio-Technological Dimensions of M-Learning

Ayse Kok
University of Oxford, UK

ABSTRACT

There has been a rapid growth in the research concerning mobile phones and the delivery of the learning experience in developing countries in recent years. The aim of this chapter is to improve understanding of this expanding research area and in so doing consider the potential for mobile phone applications for the delivery of educational services for the poor. The current state of knowledge is assessed by reviewing the existing research articles drawn from both peer-reviewed academic journals and non-peer reviewed studies and other practitioner-orientated sources. Issues relating to educational needs and the measurement of impacts have been comparatively neglected, whilst application design and adoption have received greater attention. Emphasis tends to be on devices and new ways to deliver services, but ignores the broader context of educational services for the poor and tends to be technology-led (Duncombe, 2006). In order to correct this imbalance in research, the paper identifies key points relating to concepts, methodologies, issues addressed and evidence presented and provides pointers to future research directions.

INTRODUCTION

There is an increasing and unprecedented adoption of mobile technologies in developed and developing countries alike (Keegan 2003). Mobile technologies are revolutionising education, transforming the traditional ways of learning and teaching into 'anytime' and particularly, 'anyplace' education. Empirical studies have reported the advantages of using mobile technologies in learning environments, including supporting group work on projects, engaging learners in learning-related activities in diverse physical locations, and enhancing communication and collaborative learning in the classroom (Liu et al., 2003). This paper seeks to address this question by assessing the current state of knowledge concerning the potential of mobile phones as a delivery mechanism for learn-

DOI: 10.4018/978-1-61692-818-6.ch005

ing services for the poor in developing countries. While the more technical aspects of wireless access and a detailed technical analysis of the latest mobile technologies are not included here, it is acknowledged that the technical capabilities of mobile devices are a crucial factor in developing and sustaining feasible m-learning projects. A number of handheld computers offer telephone functionality as well as e-mail and Internet access, alongside conventional personal digital assistant (PDA) features. At the same time, a number of mobile telephones offer some computing or data functionality, making the distinction between phone- and data-centric handheld devices ever more difficult to draw (Becta 2004). Therefore, it is assumed that both handheld computing devices (for example, a PDA) and mobile phones are in fact handheld devices, and will be referred to as such in this paper. Limited research has been conducted on the potential of mobile technology for educational use in developing countries. Wood (2003) states that using mobile technologies in education may contribute to combating the digital divide in developing countries, as this technology is generally cheaper than desktop computers, particularly mobile phones and PDAs. In order to develop a model that could be used for m-learning adoption, it is essential to gain a greater understanding of the considerations associated with the use of mobile technologies. Therefore, after providing a brief overview of the history of mobile technologies this paper will delve into the theoretical frameworks regarding the use of these technologies in education in the less developed world.

A Brief Background and Definition of M-Learning

Due to the enormous growth and development of the Internet over the past decades and the experimental use of the WWW and e-mail in education, e-learning emerged as an educational concept during the 1990s and has grown into a globally accepted, even necessary mode of delivery in most educational institutions. Web-based Learning Management Systems such WebCT, Blackboard and others are already widely used across the globe.

Further Internet developments over the past decade brought about a greater need for wireless connections and the development thereof. Wireless communication received enormous boosts when mobile phones reached the market. By 2000, landline telephones and also wired computers were beginning to be replaced by wireless technologies. The whole world was literally going mobile as the turn of the millennium approached. Apart from mobile phones, other wireless and mobile computational devices such as Laptops, Palmtops, PDAs (Personal Digital Assistants) and Tablets also rapidly entered the market – some devices of course with more success than others for particular markets. It is only since the turn of the millennium that educational institutions started to experiment with wireless and mobile technologies and that the concept of m-learning started to emerge. Figure 1 depicts the development of mobile technologies in relation to the Internet.

Given these historical developments, m-learning is a term coined to cover a complex array of possibilities opened up by the convergence of new mobile technologies, wireless infrastructure and e-learning developments. As with any emerging paradigm, there are many attempts to define its essence. It is worth quoting some of these definitions in order to capture the common threads inherent in the term m-learning:

1. "M-Learning is the intersection of mobile computing and e-learning: accessible resources wherever you are, strong search capabilities, rich interaction, powerful support for effective learning, and performance-based assessment. E-Learning independent of location, time or space" (Quinn, 2000).
2. "A new m-learning architecture will support creation, brokerage, delivery and tracking of learning and information contents, using

Figure 1: A general overview of the development of mobile technologies

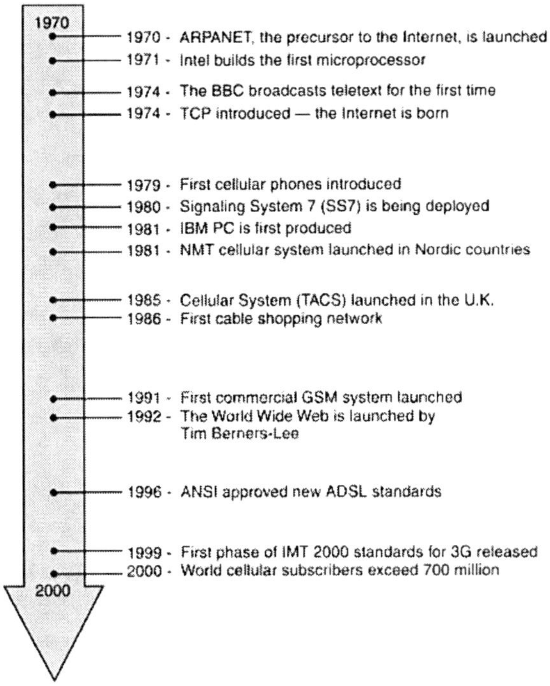

ambient intelligence, location-dependence, personalisation, multi-media, instant messaging (text, video) and distributed databases" (Mobilearn, 2003).

3. "Three ways learning can be considered mobile "learning" is mobile in terms of space; it is mobile in different areas of life; it is mobile with respect of time" (Vavoula and Sharples, 2002).

Throughout this chapter, the author will refer to the first definition when using the word 'm-learning'.

RATIONALE

This chapter builds upon a succession of reports from donor and industry sources that highlight the potential of mobile phone applications to address the learning needs of those currently excluded from formal educational services in developing countries (Becta, 2004; Belt, 2001; Berger, 2004) Belief in the potential of mobile phones to help meet the educational needs of the poor has been driven by rapid expansion of networks into previously un-served regions and communities of developing countries during the last decade. The impact has been most noticeable in the least developed countries (LDCs) of sub-Saharan Africa and South Asia where existing fixed-line infrastructure was particularly weak and underdeveloped. Mobile communications are now the fastest growing technology in the developing world and research has already demonstrated that use of mobile phones has had significant socio-economic impact in poor communities.

Because mobile phones are increasingly becoming part of the everyday lives of the poor, it is argued that they have potential to become a low cost accessible delivery channel for learning services, thus facilitating innovations including m-learning. Existing initiatives (Keegan, 2003) are already demonstrating the viability of such services in developing country environments. Research also suggests that the poor majority are in need of, and are increasingly demanding, a broader range of micro-learning services that could potentially be delivered via mobile phones or via mobile phone operators. These have been specified as low cost solutions that can underpin regular savings, facilitate informal learning, reduce the digital divide and deliver knowledge. However, studies also caution that the learning needs of the poor are fungible and embody a complex set of interactions across a broad portfolio of predominantly informal learning settings (Jensen, 2002; Juniu, 2003). There is a primary requirement, therefore, to more fully understand the inter-relationship between the suggested *potential* for mobile phone applications and the *reality* of the educational preferences and behaviours that the poor majority exhibit.

This chapter seeks to address this question by assessing the current state of knowledge concerning the potential of mobile phones as a delivery mechanism for learning services for the poor in developing countries. Thus, the overall aims are as follows:

1. Provide a literature review and analysis of research concerning mobile phones and learning (m-learning) in developing countries – with particular reference to those currently underserved and excluded from formal financial systems.
2. Identify key research trends and gaps relating to: a) concepts; b) methodologies; c) issues addressed and questions raised; d) evidence presented; and e) future research directions in the area.

This review is opportune given there is growing global interest in the role of mobile phone technologies in developing countries from donors, governments, regulators, and the educational sector. At present the field of research is disparate. A number of studies have been conducted some of which are academic, but with a greater preponderance of ground level surveys carried out for specific purposes by practitioners. Consequently, academic research and conceptual understanding of mobile phones in the development of learning services is lagging behind the rapid pace of change on the ground. Lack of primary research and lack of conceptualisation of how mobile phones can interact with existing learning networks should be a major concern. Besides, gaps in research related to social perspectives concerning trust, cultural norms, and other links to measures of social should also be emphasized. Interestingly, literature reviews of m-learning undertaken in the industrialised countries come to similar conclusions – reporting the application of business-driven models in preference to analysis of socio-cultural factors and comparison between mobile, traditional or other electronic learning services (Liu et al., 2003).

METHODOLOGY FOR LITERATURE REVIEW

Because the initial developments within this research field have been heavily influenced by the activities of practitioners, this literature review incorporates both 'academic' sources (peer reviewed – journals, working papers, conference papers) and 'practitioner' sources (exclusively non-peer reviewed – consultant's reports, official reports, journal articles, conference papers and other occasional and published papers). A number of criteria were set for the identification, selection and classification of literature sources (Figure 2).

- First, it was decided that the review would be time limited – identifying research that has been reported in the literature since the year 2000 with a cut off point of June 2009. Given m-learning in developing countries is a new area of practice and research, this timeframe allowed for the identification of some influential early studies as well as the bulk of recent studies that have appeared in the past three to four years.
- Second, the scope of the review would need to cross academic boundaries. Articles were drawn from a spectrum of disciplines incorporating social impact assessment at one end to learning models and technical system design at the other. Thus, research from a broad range of disciplines is a feature of the topic area with articles drawn from educational studies, technology and innovation, management and information systems (M&IS) and information & communication technology for development (ICT4D).
- Third, the type of content surveyed was limited in terms of the extent to which the research article addressed m-learning as a defined research area in a developing country context. Only articles dealing with m-

Figure 2. Summary of Literature Sources

	2000	2001	2002	2003	2004	2005	2006	2007	2008	Total
Agricultural Economics		1								1
Contemporary South Asia				1						1
Computer Fraud and Security								1		1
Computers and Security					1					1
Electronic Commerce Research and Applications									1	1
Information Technology for Development								1		1
Innovations: Technology, Governance, Globalization								2		2
International Journal of Information Management				1						1
Journal of Economic Studies								1		1
Managerial Finance									1	1
Small Enterprise Development					1					1
Technology in Society			1							1
Other peer reviewed articles				1			2		1	4
Peer reviewed articles (total)		**1**	**1**	**3**	**2**	**0**	**2**	**5**	**3**	**17**
Consultative Group to Assist the Poor (CGAP)							1		2	3
UK Department for International Development (DFID)							3	1		4
The World Bank							2			2
Vodaphone Research								2		2
United Nations Conference on Trade and Development (UNCTAD)								1		1
Swedish International Development Agency (SIDA)								1		1
Other non-peer reviewed articles				1	1	2		5	4	13
Non-peer reviewed articles (total)				**1**	**1**	**2**	**6**	**10**	**6**	**26**
Total articles reviewed		**1**	**1**	**4**	**3**	**2**	**8**	**15**	**9**	**43**

learning as a core issue – research studies assessing the application of mobile technologies to educational services in developing countries – were included in the review. Thus, search criteria cross referenced key words linked to: a) mobile technologies (*e.g., mobile networks, cell phones, mobile, and mobile phones*) with b) those linked to education (*e.g., educational services, m-learning*) and c) those linked with the developing country contexts (*e.g., developing countries, poor, uneducated*).

The spread of research articles suggests a peak of interest in m-learning for development from 2003. This is largely the result of donor, mobile and telecommunications industry funded researchers and other independent consultants progressing research in the area as well as a smaller number of academic studies. Because the majority of the articles were non-peer reviewed, there was a need to separate out the more deeply researched studies from the anecdotal evidence and other literature based on opinion or policy prescriptions. It was useful, therefore, to identify articles where primary data had been collected,

Figure 3. Framework for Review of Analysis

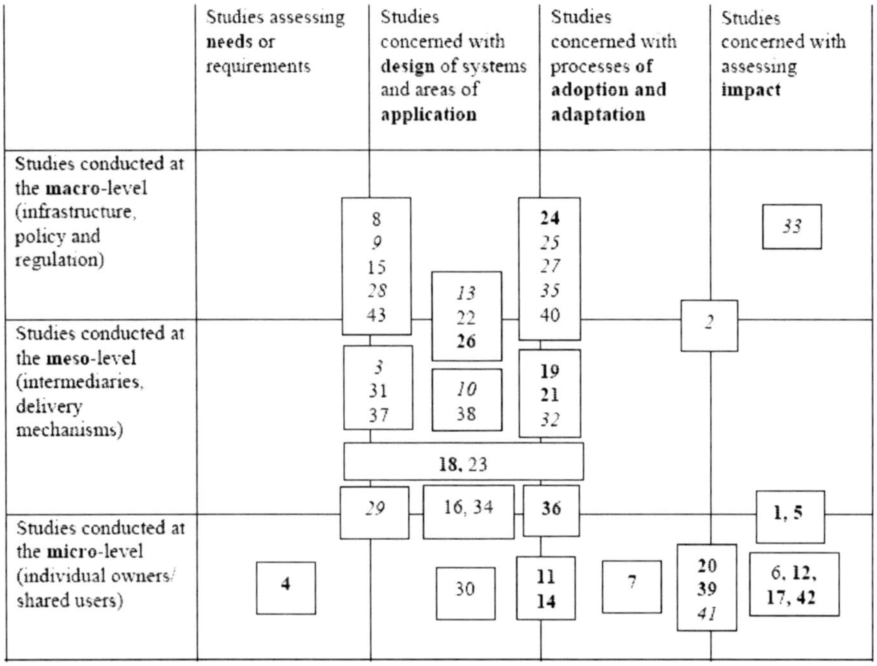

Bold: analysis of data collected from primary sources (17 articles)
Italic: analysis of data collected from secondary sources (13 articles)
Normal: no data collected or analysed (13 articles)

and to critically assess the findings of such studies in the light of the theoretical and methodological approaches adopted.

Methodological approaches were classified along qualitative—quantitative. This distinguishes between: a) quantitative studies which tend to be more representative in terms of sampling, but possibly contribute less to theoretical understanding; b) qualitative studies which largely focus on analysis of individual case studies, which make no claims to the general population, but tend to provide more in-depth data concerning processes and contribute more to theory. Also added to this list were: c) mixed methods studies; and d) studies that were purely descriptive.

The following section further reviews the literature according to the approach to framing m-learning research (Figure 3).

Design Considerations

In terms of *technical design*, the need to take greater account of user perceptions and capabilities is emphasised by Hughes & Lonie (2007) and Vaughan (2007). Similarly, Infodev (2006) note that GLOBE Telecommunication's m-learning initiatives in the Philippines were founded on existing SMS technology, using SIM-based menus over GSM networks, and as such presented no serious technical design constraints. In common with other studies, a study of 215 low income users in South Africa (Ivatury & Pickens, 2006) found that factors concerning the utility and social context of the m-learning system were far more important in determining usage patterns than technical design considerations. Williams & Torma (2007) further articulate the complexity of organisational and institutional factors confronting application design in a comparative study of the three most well publicised m-learning initiatives that are extending their reach to poor users.

Adoption Patterns

The overall level and pace of adoption of m-learning services in developing countries is relatively low and confined to more affluent users. Brown et.al (2003) demonstrate this in a study of adoption of m-learning in South Africa using a sample of respondents located in urban areas. This target group may have been expected to be early adopters. Whilst 91% of the 162 respondents owned a cell phone, only 6% had used m-learning, but 35% indicated an intention to use it. At the time of the study, the use of cell m-learning was found to be very low in comparison with e-learning. This may be expected in a relatively developed urban environment, and indeed the slowness in take up of m-learning has also been a feature of the industrialised countries due to the advantages conferred by alternative and pre-existing delivery channels.

METHODOLOGICAL ISSUES IN M-LEARNING RESEARCH

The conceptual approaches are differentiated according to a schema adapted from Heeks and Bailur (2006) which creates a hierarchy moving from 'shallower' conceptualisation to 'deeper' theoretically-based approaches – as follows:

A. *Theoretically-based approaches* which make clear use of an identifiable theory that can be applied or tested.
B. *Framework-based approaches* that make use of a framework for analysis that is derived from a body of theoretical work.
C. *Model-based approaches*: models that are applied, but without reference to a deeper body of knowledge.
D. *Concept-based approaches* that make use of a defined concept such as 'information poverty', but which is not theoretically grounded.
E. *Category based approaches* that make use of a prescribed set of factors to carry out analysis.

Studies that draw upon deeper *theoretically-based approaches* tend to be located in the field of economics, and are quite varied depending upon the level of analysis employed. At the micro-level, Williams & Torma (2007) draw on theories of (i) consumer choice and demand; (ii) network externalities; (iii) switching costs; (iv) information technology value; and (v) adoption and diffusion, creating a multi-stranded approach for researching m-learning, focusing on testing hypotheses linked to processes of transition from pre-existing transaction networks to adoption and use of m-learning.

Recourse to deep theory, however, is at its early stage of development and only isolated examples of research are identifiable. Much more prevalent is shallow theory which takes the form of *framework- or model-based approaches* that have been applied to researching applications, design and adoption, in the main through approaches borrowed from information systems (IS) and innovation research and applied to m-learning. They include the technology acceptance model (TAM) (Davis et al., 1989); diffusion of innovation (Rogers, 1983); and the decomposed theory of planned behaviour (Taylor & Todd, 1995). The TAM has been used particularly to provide conceptual underpinning for studies of m-learning adoption such as Brown, et.al (2003) who define and test a range of factors (e.g., relative advantage, compatibility, complexity and trialability) in relation to a study of early adopters undertaken in South Africa.

Emerging *model-based approaches* are drawn from educational research, and they have been applied principally to understand design and adoption (Williams & Torma (2007) but also to specify and analyse needs and consider impacts (Williams & Torma, 2007).

The importance of developing effective business-models is emphasised in most industry-led studies – unsurprising given that most of m-learning initiatives are essentially market-led and market-facing.

At the technical end of the spectrum, an interesting model has been developed by Hughes & Lonie (2007) which suggests an m-learning architecture suited to rural applications. The proposed system which is being tested in an Indian school purports to be user-friendly in design and documentation, tied to paper forms, localised and can be used offline. Other *concept-based approaches* are used to explore the transformational impact of m-learning and include the ripple-effect of technologies and information poverty conceptualising impacts of m-learning in terms of empowerment of individuals economically and socially.

Methodological Issues

The number of studies employing a rigorous approach to methodology for the collection and analysis of primary data are a small proportion of the reviewed articles as a whole, and geographical coverage is biased towards Africa and South and East Asia. Another feature of the research area is that all the surveys are snapshots and as yet there is no availability of time series data or 'before and after' studies concerning adoption or impact.

Detailed qualitative research is thin on the ground. Case studies drawing on primary evidence are provided by Duncombe (2006). These provide valuable insights, but are representative of exploratory or work in progress, and overall, there is a noticeable lack of in-depth qualitative case studies that could provide a basis for theorising.

Another issue linked to methodology concerns gauging the impartiality of the studies that are non-peer reviewed. Many of the key studies have been funded by industry research bodies and the affiliation of the authors is often to agencies or to consultant organisations that are partially funded by the mobile industry. The orientation towards industry needs is reflected in the use of business and market survey techniques in many of the reports concerned. A number of studies, however, were carried out by external consultants that were independently funded by donors, and these seem to provide more scope for impartiality, a greater focus on social research indicators, and a greater degree of critical analysis.

RESEARCH GAPS AND FUTURE RESEARCH DIRECTIONS

There is an indication, with some exceptions, that knowledge and understanding of the learning needs of poor communities have not been sufficiently linked to existing m-learning research or to the development and implementation of initiatives. This may be due to the fact that most, if not all, m-learning initiatives are commercially driven by the mobile phone industry as a value added service that is primarily designed to expand market share and generate revenue. Consequently research has been informed by business models that emphasise market development rather than social models that may put greater emphasise on community needs assessment. This orientation is followed by the current actors in the field of research, whose primary focus is system functionality and business value, focusing on technical and organisational aspects, rather than on the disruptive or ripple-effects the initiatives may have. Thus, emphasis tends to be on devices and new ways to deliver services – this is important, but ignores the broader context of financial services for the poor and tends to be technology-led.

Gaps in Conceptual Approach

Greater emphasis on researching social and cultural contexts may serve to highlight informal mechanisms to a greater extent and identify the potential disruptive effects of m-learning initiatives. Such negative impacts may arise because mobile phones are primarily a demand (market) driven innovation, so their use is likely to be distributed unevenly according to a range of non-dependent variables. This represents another significant gap in conceptualisation (and methodology). The factors that are likely to be significant include gender, empowerment, social inclusion or exclusion and the manner in which use of mobile phones (for m-learning or otherwise) impacts upon social structure and cultural norms. This is a nascent research area, but an important one given the increasing level of mobile penetration in poor communities.

Gaps in Methodological Approach

The most apparent gap in the use of methods was the lack of in-depth qualitative studies analysing primary data – in contrast to the loosely-positivist mixed-method approaches which tend to dominate the area of study. Lack of depth of qualitative data may explain lack of conceptualisation as such studies have played an important role in other avenues of research concerning the application of ICTs to development (Duncombe, 2006). As indicated earlier, action research was evident in the review, particularly in the descriptive studies but with lack of overt reference to methodology.

Most studies investigating m-learning in a poverty context used a mixed-method approach. The best examples of these make use of multi-method approaches and multiple sources of data and a quasi-experimental research design that compares users and non-users of m-learning initiatives. Most studies, however, exhibit gaps in methodology which raise questions of reliability and validity of the findings. Three main issues stand out. The first relates to whether the results of surveys are representative (in terms of the wider population, but also the extent to which they address the potential for m-learning initiatives to meet the learning needs of the poor). Some methodological problems encountered include the issue of causality – the extent to which the utility or wider social impact of m-learning initiatives can be attributed (i.e., how to disentangle the effects of the mobile phone from other contributory factors). This is a key constraint for studies that seek to measure broader socio-economic impact. This is also addressed in some of the studies through use of comparisons with non-user groups. However, such counterfactual assessment is of limited value unless it can be demonstrated that the user group and the control group are directly comparable in terms of a wide range of non-dependent variables. The *bi-directionality of influence* should also be taken into account. In other words, does, for example, the use of m-learning lead to a reduction in poverty, or is it a reduction in poverty that leads to (or facilitates) the use of an m-learning service? Another issue concerns the extent that lessons learned can be transferred from one context to another. Because most studies are exploratory, their geographical and locational scope is limited – often to urban or peri-urban areas – or focusing on groups that are culturally distinct. As yet, due to the small number of studies, there is little indication of how findings may differ when the studies are replicated in different cultural, locational or country contexts.

CONCLUSION AND POINTERS FOR FUTURE RESEARCH

This review indicates a rapid expansion of research into mobile phones and learning applied to developing countries, and as yet there has been no systematic attempt to review how this research has progressed both conceptually and methodologically.

Overall, this literature review suggests the following practical implications:

- A high level of practitioner involvement and the high level of positive interaction between the donor community and the mobile phone industry. This has set in motion a research agenda that seeks to seriously address the potential of new technologies to serve the needs of poor communities via m-learning.
- Specific attempts to develop theoretical models, and create a deeper understanding of m-learning applications, most noticeable in the area of application development and technology adoption.
- A small number of primary research studies that have developed rigorous methodologies for data collection and analysis, and where those approaches and lessons learned have been documented and shared.

On the other hand, the following remarks should also be taken into account for the progress of m-learning:

- That the research area has become too overtly 'technology-led' and driven by a mobile industry-donor nexus which (and in the absence of a strong evidence base) has tended to over-'hype' the potential of m-learning applications for poor users.
- There has been lack of focus on assessing learning needs *a priori* to specifying m-learning solutions, and by and large importance has not been attached to analysing the relationships between the technical and systemic aspects of m-learning services and the behaviours and preferences of poor users.
- In this respect there has been lack of focus on methodologies that emphasise user involvement such as participatory methods and action research, or where these approaches have been used they remain undocumented.
- The research area overall is unbalanced in terms of methodological approach, with use of surveys and quasi-experimental techniques outweighing qualitative approaches – which could be used to build in-depth case studies that can form the basis for theorising. With a few notable exceptions, the research reviewed tends to lack depth of evidence and analysis. Issues of validity of findings and attribution of causality have yet to be addressed sufficiently.
- Overall, there is lack of geographical diversity, with the state of current knowledge based upon a relatively narrow evidence base. There is also lack of published research being conducted by developing country institutions and researchers.

In order that such research can constitute a reliable source of data and evidence it is important that research methods are enhanced. This can be achieved by building on the current quantitative and mixed-method approaches (which are essentially quasi-experimental) to improve their reliability through use of larger and more carefully constructed samples, and repeat studies, but also through use of extensive and clearer statements of the methodology which will aid the transferability lessons learned. Outside the confines of current research, there is also an urgent need to broaden out the research methods used. This should involve the construction of more detailed qualitative case studies (possibly within a mixed-method approach) with more effective use of triangulation of research methods and sources of data. More detailed qualitative data may help map the complexity of causal chains of impact, with greater emphasis placed on understanding and theorising the micro-processes that cause the poor user to interact with a mobile device or to make use of a service delivered via mobile networks.

REFERENCES

Becta (2004). *Becta Home Page*. Retrieved April 27, 2009 from http://www.becta.org.uk/subsections/foi/documents/technology_and_education_research/ handheld_computers.doc

Belt, H. (2001). Wireless Devices in the Classroom. *Learning and Training Innovations*. Retrieved April 3, 2009 from http://www.ltimagazine.com/ltimagazine/article

Berger, G. (2004). Putting perspective on cellphone euphoria. *Mail & Guardian*. Retrieved April 27, 2009 from http://archive.mg.co.za

Brown, I., Cajee, Z., Davies, D., & Stroebel, S. (2003). Cell phone banking: predictors of adoption in South Africa – an exploratory study. *International Journal of Information Management*, *23*, 381–394. doi:10.1016/S0268-4012(03)00065-3

Duncombe, R. A. (2006) Analysing ICT applications for poverty reduction via Micro-enterprise using the livelihoods framework. Development Informatics Working Paper Series, No.27, Institute for Development Policy and Management, University of Manchester. Retrieved June 2, 2009 from http://www.sed.manchester.ac.uk/idpm/publications/wp/di/index.htm

Heeks, R. B., & Bailur, S. (2006) Analysing eGovernment research: perspectives, philosophies, theories, methods and practice. *iGovernment Working Paper Series*, Paper No.16, Institute for Development Policy and Management (IDPM), the University of Manchester. Retrieved May 2, 2009 from http://www.sed.manchester.ac.uk/idpm/publications/wp/igov/educigov.htm

Hughes, N., & Lonie, S. (2007). M-PESA: mobile learning for the "uneducated" in Kenya . *Innovations: Technology, Governance, Globalization*, *2*(1-2), 63–81. doi:10.1162/itgg.2007.2.1-2.63

Ivatury, G., & Pickens, M. (2006) Mobile phone learning: evidence from South Africa, Consultative Group to Assist the Poor (CGAP). The World Bank and United Nations Foundation, Washington, D.C. Retrieved May 14, 2009 from http://www.cgap.org

Jensen, M. (2002). The African Internet: A Status Report. Retrieved June 1, 2009 from http://www3.sn.apc.org/Africa/afstat.htm

Juniu, S. (2003). Implementing wireless technology in the classroom: the iPAQ project. *College Planning and Management*, *6*(10), 38–40.

Keegan, D. (2003). *The future of learning: From eLearning to mLearning*. Hagen: Femstudienforchung, Germany. Retrieved May 11, 2009 from http://learning.ericsson.net/mlearning2/old_sites/book.html

Liu, T., Wang, H., Liang, J., Chan, T., Ko, H., & Yang, J. (2003). Wireless and mobile technologies to enhance teaching and learning. *Journal of Computer Assisted Learning*, *19*(3), 371–382. doi:10.1046/j.0266-4909.2003.00038.x

Mobilearn (2003). The Mobilearn Project Vision. Retrieved May 3, 2009 from http://www.mobilearn.org/vision/visiton.htm

Quinn, C. (2000). M-Learning. Mobile, Wireless, In-Your-Pocket Learning. *Linezine*. Retrieved May 14, 2009 from http://www.linezine.com/2.1/features/cqmmwiyp.htm

Vaughan, P. (2007). Early lessons from the deployment of M-PESA, Vodafone's own mobile transactions service. In D. Coyle (Ed.), *The Transformational Potential of m-Transactions,* Policy Paper Series, No.6. Vodaphone, London. Retrieved April 4, 2009 from http://www.vodaphone.com/m-tranactions

Vavoula, G. N., & Sharples, M. (2002). KleOS: A Personal, Mobile, Knowledge and Learning Organisation System. In M. Milrad, H.U. Hoppe & Kinshuk (Eds.), *IEEE International Workshop on Wireless and Mobile Technologies in Education* (pp. 152-156). Los Alimatos, USA: IEEE Computer Society.

Williams, H., & Torma, M. (2007). Trust and fidelity: from under the mattress to the mobile phone. In D. Coyle (Ed.), *The Transformational Potential of m-Transactions*, Policy Paper Series, No.6, Vodaphone, London. Retrieved April 4, 2008 from Available at: http://www.vodaphone.com/m-transactions

Wood, K. (2003). Introduction to Mobile Learning (M Learning). Retrieved April 4, 2009 from http://ferl.becta.org.uk/display.cfm?page=65&catid=192&resid=5194&printable=1

Chapter 6
Mobile Learning in China

Yuqin Yang
Nanyang Technological University, Singapore

Qiyun Wang
Nanyang Technological University, Singapore

ABSTRACT

Mobile learning (m-learning in short) has received increasing attention in the new millennium. Considerable interest in exploiting the unique capability of mobile technologies for teaching and learning arises from educators and technical developers in recent years. In this chapter, we present an overview of the development of m-learning in China, including the construction of m-learning infrastructure, and the research projects conducted by universities and companies. Also, suggestions for future development of m-learning are provided.

INTRODUCTION

M-learning is nowadays a common term, not a buzzword. There is considerable interest from educators and technical developers in exploiting the unique capabilities of mobile technologies for teaching and learning. The future is becoming more wireless-oriented.

A nation looking for a lasting economic success must raise the literacy level of its citizens. The importance and centrality of education as a fundamental human right have been well documented in the literature. Ezeomah (1982) and Aleyidieno (1985) state that making education as a fundamental human right provides a viable springboard for transforming social and economic policies. Moreover, providing an equal educational opportunity of for all is an important indicator of social justice. However, achieving an equal right to be educated for all is a challenge, especially in China, as she is less developed, has a large population, and has huge economic and cultural gaps between different regions.

The China's Ministry of Industry and Information Technology (MIIT, 2008) reports that the number of mobile phone users is 688,340,000 in November 2008 in China. It is about 47.3% of the country's population. Moreover, the number

DOI: 10.4018/978-1-61692-818-6.ch006

is increasing at a rate of 1.01% per month and 17.51% per year, and the number will be up to 738,000,000 in 2010. The mobile users sent more than 56.87 billion short messages (SMS), 3.01 per day per person in November 2008.

In addition, the China Internet Network Information Center (CNNIC) reports that, there are 155, 000,000 mobile phone users who use wireless technologies to access the Internet in 2009 in China. Of all these users, 38.8% are between 10-19 years old, 37.6% between 20-29 years old, 15.5% between 30-39 years old, 5.4% between 40-49 years old, 1.4% between 50-59 years old, 0.8% under 10 years old, and 0.5% above 60 years old. The majority of these users are senior school graduates (43.3%) whose ages range from 12 to 14 and junior school graduates (28.5%) whose ages vary between 15 and 17. 6.1% are primary school graduates, 11.3% are senior secondary school graduates and 10.2% are bachelors (CNNIC, 2009). The statistics results show that educating Chinese population via m-learning has great potential in China, as m-learning may help to provide a more equal access and brighter opportunities for all people regardless of races, colors, ages and living places.

The aim of this chapter is to present an overview of the use of different mobile technologies and applications in China, including the construction of m-learning infrastructure, and research projects conducted by universities and companies. Issues and suggestions for future development of m-learning are also provided. This chapter will help readers to better understand the development of m-learning in China.

LITERATURE REVIEW: THE DEFINITION OF M-LEARNING

M-learning is defined rather ambiguously even though it has been popularly used in literature (Traxler, 2007). Different researchers have given various definitions from different perspectives. The following summarizes the definitions in the literature.

Define M-Learning Based on the Mobile Devices

M-learning is often defined as learning conducted via the use of mobile devices. In general, mobile devices refer to any electronic devices which are small enough to accompany us in every moment of our daily life (Caudill, 2007). The broad categories of mobile devices include PDAs, iPods, Plamtop Computers, mobile phones, and MP3 players. The majority of authors view m-learning as learning connected to a mobile device (Aderinoye, Ojokheta, & Olojede, 2007; Ally, 2004; Cui, et al., 2001; Keegan, 2002; QIA, 2008; Sharma & Kitchens, 2004; Trifonova, Georgieva, & Ronchetti, 2006; Wexler, et al, 2007). This definition focuses on the facet of technology in m-learning (Traxler, 2007).

Define M-Learning in the Context of Learning Experiences and Learning Environments

Learners do not learn in a vacuum. They learn while competing and collaborating, they learn by doing and by correcting misconceptions they have from previous learning activities (Laouris & Eteokleous, 2005). Some researchers define m-learning from this prospective. For instance, Sharples (2005) defines m-learning as a learning process, in which learners collaborate with their peers and teachers, construct the meaning of knowledge. Colazzo, Molinari, Ronchetti & Trifonova (2003) state that a m-learning process can be viewed as any learning and teaching activity which is facilitated by mobile technologies or in settings where mobile tools are available. Liu, Li & Liu (2007) and Yu (2007) define m-learning as a novel educational approach delivered, enhanced and supported by wireless and mobile devices and technologies. It enables students to obtain educational contents, resources and services at anytime anywhere.

Define M-Learning in the Context of E-Learning

M-learning is closely related to e-learning (Cavus & Ibrahim, 2009). Many researchers and educators view m-learning as the immediate descendant of e-learning. Quinn (2000) and Pinkwart, et al. (2003) define m-learning as e-learning that uses mobile devices. Some researchers argue that m-leaning is a unique kind of e-learning (Caudill, 2007). Others think m-learning is an extension of e-learning (Shrivastava, Yadav, & Shrivastava. 2005); or m-learning is a new and unique component of distance learning. Georgiev and colleagues (2006) define it as a new stage of the development of e-learning.

In addition, Nyiri (2002) views m-learning from a more general perspective, who defines it as learning that arises in the course of person-to-person mobile communication. Also, Laouris and Eteokleous (2005) develop a model of m-learning that includes the full spectrum of actors (such as time, space, learning environment, content, technology and method) in a systematically correct and complete way.

A closer analysis of the above definitions indicates that no precise definition of m-learning is attainable. It is a challenge for educators and researchers to come up with a commonly acceptable definition of m-learning which is able to guide research, development and implementation of m-learning (Ally, 2007). In general, m-learning is a combination of 'mobile' and 'learning'. It refers to the use of mobile devices or technologies; the mobility of learners; and the learning experiences or learning environments that support anytime anywhere learning. Many definitions in the past simply overemphasized the technologies and ignored the key component of learning. Recently educators highlight that the focus of m-learning should be shifted to learning rather than the form of mobility. In this chapter, m-learning is broadly defined as any type of learning that takes place in informal or formal learning settings, where individual inquiry and collaboration are supported by the use of mobile technologies.

EDUCATIONAL POLICIES AND INITIATIVES IN CHINA

Educational Policies and Initiatives

Education is one of China's fundamental national policies. As one of the priorities of China's economic and social development, education is a matter of great concern to the government. In recent years, a lot of education policies and programs involving the construction of information infrastructure and educational informatization have been undertaken. These policies and programs include the Development Outline for Educational Informatization, the National Educational Technology Standards, the Program of Modern Telecommunication Education (Modern Distance Education) in Universities, the Program of Modern Telecommunication Education in Primary and Secondary Schools in western and central areas of China, the Program of Digital Campus of Colleges and Universities, the Program of Connection among all Primary and Secondary Schools of Our Whole Country.

Meanwhile, a series of educational initiatives have been undertaken. The 2003-2007 Action Plan for Invigorating Education, approved and disseminated by the State Council, is one of the national principal initiatives. This action plan lists six major projects which China will implement in the coming five years, including quality-oriented education, innovation in vocational education and training, teaching quality and teaching reform in institutions of higher education, employment promotion for college graduates, educational informatization, and building up a competent teaching and administration system. According to the plan, the government's investment in education will expand faster than the growth of the nation's fiscal income. Also, the 9th (1996-2000) and10th (2001-2005) Five-Year Plans for Educational Development have made significant contribution to the development of education.

Furthermore, China has invested a large amount of money in education, especially in educational informatization. Ministry of Education of the People's Republic of China (MOE in short) reported that China's investment in education accounted for 3.32% (1214.807 billion yuan) of its gross domestic product (GDP) in 2007, increased by 0.32% compared to 2006, and China aims to reach 4% of its GDP in a relatively short time period (MOE, 2008). And the increased investment will mainly be put into rural education, particularly the compulsory education in countryside. For educational informatization, China invested 27.2 billion yuan in 2005, and reached 30 billion in 2006.

Telecommunication Policies and Reforms

In recent years, China has achieved significant achievements in building the information society and educational informatization. However, with more than 700 million residents living in rural areas, the infrastructure, such as Internet and computers, is still far from adequateness for popularizing education, eliminating illiteracy, and providing equal educational opportunities to all.

On the other hand, in recent years, there has been a steady growth in the infrastructure development and use of mobile phones in China. An increasing rate of accessibility, especially with the arrival of new 3G services, encourages more and more people to purchase and use mobile phones.

So under its unique macro-economic and political environment, China has adopted a special telecommunication policy and strategy for telecommunication reform, which includes m-learning as well. As a fundamental and strategic industry, the Ministry of Posts and Telecommunications (MPT) of China has monopolized telecommunication operations for more than four decades. With the goal of completely separating the functions of the government and the enterprise, the Ministry of Information Industry (MII) was established in 1998. MII became a neutral regulator by taking over the regulatory functions of MPT, while ceding the functions of enterprise.

In parallel with the reform and reorganization, China also gives a high priority to the legislature work and industrial supervision of telecommunications. In 2000, Regulations of Telecommunications of China and the Administrative Methods of Internet Services were promulgated, and a centralized telecommunication regulatory body was formed.

To further accelerate the ICT development, the national 9th Five-Year Plan established an information and telecommunication industry as one of the crucial industry. With the reform as a driving force, China's telecommunications developed steadily and rapidly throughout the 9th Five-Year Plan period. The 10th Five-Year Plan continued to make it as a basic national policy to further facilitate the nation's development through science and technology; to push industrialization by virtue of the IT development; and to explore a path for IT development suited to the national conditions. The 11th Five-Year Plan highlights great importance to modern distance education.

Challenges in Adopting M-learning

Though some achievements have been made in telecommunication education, many issues and challenges are still involved in m-learning. The following presents some of the issues and challenges.

Illiteracy. Since 2000, the number of illiterate people has decreased by an average of 2 million per year through the promotion of the Nine-Year Compulsory Education policy. And China has lifted 46.5 million people out of illiteracy (Yang, 2007). However, illiteracy is still a critical problem in China. According to the recent Fifth National Census done by National Bureau of Statistics of China (2000), there are more than 80 million illiterates at or above 15 years old, and 72.7% of them are women. The adult illiteracy rate in China is 9.08 percent. What's worse, imbalanced development of literacy education exists. The 23 million population of illiteracy among youths and adults disperses in 7 western provinces, where the level of socio-economic development is relatively low. The illiteracy rate among youths and adults in the above-mentioned regions ranges from 10% to 5%, and illiterate population mainly lives in remote areas where minorities inhabit. China's illiterate population is large, and hence eradicating illiteracy is a challenge.

Poor infrastructure. Insufficient infrastructure is a critical problem for achieving connectivity in rural areas, especially in the western regions of China. There is currently a lack of incentive for telecommunication providers to invest in broadening their western networks, mainly due to a lack of purchasing power and low population densities in these areas. Although the government has invested a large amount of money into educational informatization, the infrastructure is inadequate to adopt m-learning or e-learning. What's even worse, there is still disparity of investment in education across China, though the proportion of educational investment between urban and rural areas lowers to 1.2:1.

Digital divide. A severe digital divide exists not only between China and developed countries, but also among its own regions and social groups. The term digital divide refers to the gap between those with regular, effective access to digital and information technology, and those without this access. It generally encompasses both physical access to technology hardware and, more broadly, skills and resources which allow for its use.

The digital divide between urban and rural regions is obvious in China. According to the survey report done by CNNIC (2007), by the end of June 2007, the number of the rural Internet users had reached 37.41 million. The penetration rate of the Internet for all 737 million rural residents was only 5.1%. Also, at the end of 2006, every 100 households in rural areas possesd only 2.7 computers, which was far behind the figure of urban areas (47.2 computers). Certainly, the main reason for not using the Internet is the lack of facility.

Political/social climate. The current top-down planning process is a double-edged sword. The positive side is that the nation can coordinate all the industries to accomplish a project efficiently. However, the negative side is that the subordinate group may be the passive recipients rather than proactive planners for the ongoing project. So the effect and efficiency may drop down significantly. What's even worse, the substantial developments are often neglected. For instance, in the educational informatization process, significant attention is paid to the hardware and infrastructure while less emphasis is paid on the effect of its' actual use. Little efforts are made to discover what is happening in the educational context and what needs to do. Corresponding instructional and infrastructural support and training are seldom provided. Although the Chinese government has spent millions of yuan on its education programs, achievement remains low. Furthermore, people may not be willing to embrace m-learning, as changing mindset is a gradual process.

M-LEARNING PROJECTS IN CHINA

The domain of m-learning includes a wide variety of applications and new teaching and learning techniques. In this section, some ongoing research projects are described.

Construction of M-Learning Infrastructure

The first m-learning pilot project in China is called Mobile Learning: Theories and Practice. It was initiated by the Educational Lab of Modern

Figure 1. The Framework of Mobile Campus Program of Shanghai TV University

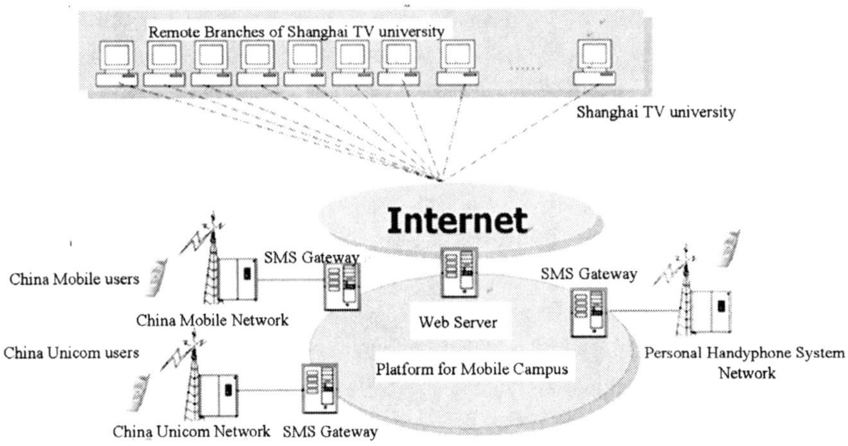

Educational Center in Peking University. From 2002 to 2005, four m-learning platforms were developed based on (Liu, 2007):

i. the GSM network and mobile devices. This platform uses a Short Messaging System (SMS) to exchange information so that the mobile network and Internet can share the same resource.
ii. GPRS. This platform supports a variety of GPRS-enabled equipments, such as mobile phones, PDAs or computers.
iii. ontology for resource production and delivery. This platform improves the developmental norms of educational resources and educational services for dynamic expansion and customization, and lays a sound foundation for the Semantic Web.
iv. the Semantic network. This platform aims to improve the intelligence of the educational service platform and the construction of the multi-purpose educational service platform though the use of the Semantic Web and ontology technology.

However, the implementation of this project was not that satisfactory, due to three major reasons: the wireless infrastructure in colleges and universities was inadequate at that time; the ownership rate of handheld devices was low; and the technical support was weak.

Another m-learning project on infrastructure improvement was initiated in 2001 and supported by the MOE. This project is collaboratively carried out by educational institutions with high reputations in instruction and use of technology for teaching and learning: Tsinghua University, Peking University, and Beijing Normal University. This project aims to: i) build an mobile educational information network via the SMS and GPRS platforms of the China Mobile Communications Corporation; and ii) build an m-learning service system for leading universities' students. This project has formed a backbone system for m-learning in China (Liu, 2007).

M-Learning in Universities

In recent years, many universities in China have begun to experiment m-learning. A noteworthy pilot study of m-learning to date is the Mobile Campus Program of Shanghai TV University (see Figure 1). It enables more than 100,000 students in 41 remote branches to use smart phones to access the SMS System. This system can rapidly deliver educational and administrative information to students (Sun, 2008).

Shanghai TV University and the Lanzhuo Information Technology Corporation jointly developed a Mobile Learning System (MLS) in 2007. This system is connected to Shanghai TV University's Online Classroom (OC) and provides the following services:

- Academic SMSs. The system can automatically generate English test items and deliver the items to students through SMSs. It sends answers and explanations in the following day. The faculty can edit test questions, answers and explanations, determine learning scope, schedule the delivery time, and check students' responses. Teachers can also provide corresponding support via BBS in OC.
- Academic services. Students can use smart phones to communicate with lecturers in real time. The lecturers are also able to monitor the class progress through laptop computers. The teachers and students can discuss through mobile devices or SMS messages. Facilitators can also send information or feedback to the learners.
- Academic summaries. The system can deliver summaries of content, examination questions, or guidelines to students when needed.
- Administrative SMSs. Administrators can send SMSs to students on matters like timetable changes, examination deadlines, or assignment results.

The Mobile Campus program has been commonly used in the university. By January 2007, the enrollment of distance students had reached 83,892, and a large number of students used "Mobile Campus" to learn (Sun, 2008). The students' passing rate of English II was 93.3% in 2006. A survey result showed that most participants agreed that the Mobile Campus: (a) contributed to their learning; (b) helped them to understand lecture's key points; (c) helped them to identify key areas; and (d) increased their access to college services. However, some issues were found in the project such as low learning efficacy, limits of mobile devices, limited m-learning recourses, and little guidance is given in the learning process (Sun & Chen, 2007).

The Mobile Campus program of Shanghai TV University is still ongoing. The final report on the evaluation of the project and other issues (such as reliability, security etc) affecting the m-learning project in the university is not available at the moment.

Nanjing University, Panasonic Communication Industrial Company and Software Consultant Corporation have started another program called Computer Aided Learning (CAL) Unite Multimedia Education Technology (MET) since 1999 (Sui et al., 2004). This system integrates the advanced multimedia technology, mobile communication technology and network technology into the program design. It has been implemented at Nanjing University since 2000 to support flexible learning at anytime and anywhere.

Lastly, the Android project started in 2007 embraces a consortium of 35 partners, including Google and other 33 companies (such as China Mobile Communications Corporation, NTTDoCoMo, Motorola, SprintNextel, South Korea's Samsung, Intel, eBay, and LG). The consortium is responsible for the development of a free software program called Android, which is to be installed in smart phones. Peking University is in charge of multimedia material development for mobile devices. Android makes smart phones more convenient and portable and break the barrier of fee charge by various service providers.

M-Learning in Companies

As a distinguished training company in China, the New Oriental Education & Technology Corporation starts a learning plan, which allows users to access the learning content on the websites of http://www.mobiledu.cn/web/web/index.jsp and http://www.koolearn.com/ (Liu et al., 2007). Users can download English learning materials or other learning content as long as they have their mobile phones installed certain software.

The ETS (Educational Testing Service) and Nokia initiate jointly the New Ways of Learning English project. The ETS provides customized solutions for teacher certification, English-language learning, and elementary, secondary and post-secondary education. Also, it conducts educational research, analysis and policy studies. Nokia in China helps to develop English learning courses especially for Nokia phone users. The courses developed include practice questions, business English, survival English, academic English, and American English phonology. Nokia phone users can download the courses through Nokia's Mobiledu.

Another company which has made great contribution to m-learning is Beijing Whaty Technology Development Co., Ltd (Liao, 2006). This company develops a m-learning platform, which consists of an administration platform, a course development and management platform, and an interaction center. The m-learning administration platform provides access to the learning content and administration, and facilitates to track report and manage the learning process. Also, it caters to, and focuses on different educational and administrative requirements, such as learner account creation, management and transmission of content. The course development and management platform has the functions of folder creation, file uploading, and content creation and editing. It also provides tools for content authoring and editing. Teachers can easily convert existing e-learning courses for m-learning and distribute them over the Internet. The interaction center provides a place where students and peers can share information and interact with one another.

Research on Pedagogies for M-Learning

A number of research studies have been conducted to investigate the cognitive and pedagogical aspects of using mobile technologies in education in China. Some of the studies propose approaches for m-learning design (Lun, 2007; Xiong, 2005), while others suggest directions for application designers (Jiang, 2008; Liu, 2007). Most of the research studies claim that mobile devices are more likely to be an extension, rather than a new tool to replace existing tools (Zhu, 2008; Sun & Chen 2007). Also, it seems that not all kinds of learning contents and activities are appropriate for m-learning. Here is one example of the research studies on m-learning in China.

The Application of Handheld Networked Learning Systems in Subject Teaching is a collaborative research project done by Beijing Normal University and Innovation Noah Electronic (Shenzhen) Co., Ltd. This study explores how to use mobile devices and systems to promote teaching and learning across all subjects, and to enable a rich, interactive learning environment.

This project has multiple purposes. It aims to develop personal mobile technologies, wearable learning technologies and supporting systems for learning based on a deep understanding of how people learn in multiple contexts. It intends to produce creative learning and teaching methods and technologies for m-learning, and to design meaningful activities using mobile technologies to assist students in information and numerical literacy development. Also, it supports knowledge sharing and conversation among learners.

Furthermore, this project explores what factors and requirements are crucial to the design of m-learning environments and activities, and suggests how applications can be effectively designed with an understanding of these factors and requirements. This project is implemented based on the premise that m-learning should not replace traditional teaching and learning, but support students and teachers by providing just-in-time services (Yu, 2007).

This project has produced a number of mobile devices and resources, such as Mobile Mind English, Noah Learning Machine, and Noah learning machine-Graphing Calculator. More information about the project can be found on the web site http://aplei.noahedu.com/Tutorsite/index.htm.

FUTURE RESEARCH DIRECTIONS AND CHALLENGES

Despite the tremendous growth and potential of the mobile technology, m-learning is still in an embryonic stage in China. From the macroscopic perspective, the lack of a mature social context hinders the rapid development of m-learning. Device manufacturers, resource providers, network operators and educators have not formed an integral open m-learning system. From the microscopic perspective, technology is an integral part of m-Learning. However, the mobile technology has limitations in designing virtual learning environments and promoting friendly interaction. For example, the current mobile devices have certain constraints of small screens, poor input, and limited battery lifespans. Much of the current m-learning research is driven by the technical capabilities of new devices. These new capabilities inspire new practices. However, applications of mobile devices by following certain learning theories are still lacking.

In the future, the success of learning and teaching with mobile technologies will be measured by how seamlessly it weaves itself into our daily lives (Naismith, Lonsdale, Vavoula & Sharples, 2004, p. 36). In order to find ways to ensure that m-learning is highly situated, personal, and collaborative, more research can be carried out in the areas of instructional support, individualized and collaborative learning, teaching and learning strategies, and the development of wireless learning management systems.

In terms of providing instructional support, the existing m-learning infrastructure is not yet ready for truly m-learning. The problems in course support, device availability, and stability remain. Future research should focus on what instructional support is crucial for mobile devices, and what enabling technology affects the wide diffusion of m-learning; and how to match programs' interface design with learners' cognitive and social abilities.

With respect to individualized and collaborative learning, m-learning provides just-in-time help and just-for-me features. More research can be done on how learning activities and learning environments using the mobile technology are designed to support innovative educational practices, and how instructional materials are customized for users with special needs and to enhance collaboration between mobile learners.

Regarding the teaching and learning strategies, future research can study what models can be used to better support the learning process when communication is mediated by the mobile technology. Further research studies can focus on the impact of mobile devices on the learning process. Additionally, more research should investigate what key requirements a wireless learning management system needs, and how to design and develop such a system.

CONCLUSION

China is the biggest developing country in the world. In the recent years, China has achieved enormous economic development. Also, she has made great progress in the use of the mobile technology in education. A number of companies and universities have started their research and developmental projects in mobile learning. The mobile technology has also shown great potential for teaching and learning. Characteristics such as accessibility and availability make the mobile technology an effective tool for teaching and learning. In the near future, it is expected that learning will move gradually from the classroom to the learning environment both real and virtual (Naismiths, et al, 2004). For this to happen, the usability of mobile devices for educational purposes need to be further improved, and educational institutions would have to integrate more mobile and handheld devices into the formal learning environment. Also, more innovative learning design and delivery models suited for m-learning need to be further developed.

REFERENCES

Aderinoye, R. A., Ojokheta, K. O., & Olojede, A. A. (2007). Integrating mobile learning into nomadic education programmes in Nigeria: Issues and perspectives. *International Review of Research in Open and Distance Learning*, *8*(2), 1–17.

Aleyidieno, S. (1985). Education and occupational diversification among young learners: The problem of harmonising tradition practices with the lessons of our colonial heritage. In *Issues on Development: Proceedings of a Seminar held in Zaria*. January 12-13. Zaria, Nigeria: Ahmadu Bello University Press.

Ally, M. (2004). Using learning theories to design instruction for mobile learning devices . In Attewell, J., & Savill-Smith, C. (Eds.), *Mobile learning anytime everywhere: A book of papers from MLEARN* (pp. 5–9). London: Learning and Skills Development Agency.

Ally, M. (2007). Mobile learning. *International Review of Research in Open and Distance Learning*, *8*(2), 1492–3831.

Caudill, J. G. (2007). The growth of m-learning and the growth of mobile computing: Parallel developments. *International Review of Research in Open and Distance Learning*, *8*(2), 1–12.

Cavus, N., & Ibrahim, D. (2009). M-Learning: An experiment in using SMS to support learning new English language words. *British Journal of Educational Technology*, *40*(1), 78–91. doi:10.1111/j.1467-8535.2007.00801.x

CNNIC. (2007). *Survey report on the Internet development in China*. Retrieved October 5, 2009 from http://www.cnnic.net.cn/uploadfiles/pdf/2008/1/17/104156.pdf

CNNIC. (2009). *Survey report on the Internet development in China*. Retrieved October 5, 2009 from http://www.cnnic.net.cn/uploadfiles/pdf/2009/1/13/92458.pdf

Colazzo, L., Molinari, A., Ronchetti, M., & Trifonova, A. (2003). *Towards a multi-vendor mobile learning management system*. Retrieved April 24, 2009 from: http://trifonova.net/docs/Towards%20a%20multi-vendor%20Mobile%20LMS%20(long).pdf

Cui, G. Z. (2001). M-learning-a new direction of modern educational technology. *Journal of Modern Educational Technology*, *6*, 15–18.

Ezeomah, C. (1982). Movements and demography of Fulani Nomads and their implications for education development. In *Proceedings of the 1st Annual Conference on the Education of Nomads in Nigeria, Jos, Nigeria*.

Georgiev, T., Georgieva, E., & Trajovski, G. (2006). *Transitioning from e-Learning to m-Learning: Present issues and future challenges*. Retrieved April 28, 2009 from http://ieeexplore.ieee.org/iel5/10923/34372/01640716.pdf?tp=&isnumber=34372&arnumber=1640716.

Jiang, Y. C. (2008). *Design and development of mobile game in learning model*. Unpublished master's thesis, East China Normal University, Shanghai, China.

Keegan, D. (2002). *The future of learning: From e-learning to m-learning*. Retrieved June 28, 2009 from http://www.eric.ed.gov/ERICDocs/data/ericdocs2sql/content_storage_01/0000019b/80/1a/b6/8e.pdf

Laouris, Y., & Eteokleous, N. (2005). *We need an educationally relevant definition of mobile learning*. Paper presented at mLearn 2005, the 4th World Conference on Mobile Learning. Cape Town, South Africa. Retrieved June 6, 2009, from http://www.mlearn.org.za/CD/papers/Laouris%20&%20Eteokleous.pdf

Liao, C. (2006). *A comprehensive m-learning platform*. Retrieved April 28, 2009 from http://all.zcom.com/mag2/shehuikexue/wenkejiaoti/30982/200608X/15388980

Liu, H. Y. (2007). *The design and development of web-based SMS system for course answering*. Unpublished master's thesis, South China Normal University, Guangzhou, Guangdong, China.

Liu, J. S., Li, Q., & Liu, J. M. (2007). Review of mobile learning. *Journal of E-education Research*, *7*, 21–25.

Lun, M. H. (2007). *Theory study and application design of mobile learning content*. Unpublished master's thesis, Beijing Jiaotong University, Beijing, China.

MIIT. (2008). *The increasing numbers of mobile phone users*. Retrieved October 5, 2009 from http://it.people.com.cn/GB/1068/42899/8137255.html

MOE. (2008). *Report on the investment in education*. Retrieved October 4, 2009 from http://www.moe.edu.cn/edoas/website18/84/info1229326947658184.htm

Naismith, L., Lonsdale, P., Vavoula, G., & Sharples, M. (2004). *Literature review in mobile technologies and learning*. Bristol, UK: NESTA Futurelab. Retrieved March 3, 2009, from http://elearning.typepad.com/thelearnedman/mobile_learning/reports/futurelab_review_11.pdf

National Bureau of Statistics of China. (2000). *The Fifth National Census*. Retrieved October 4, 2009 from http://www.stats.gov.cn/tjgb/rkpcgb/qgrkpcgb/t20020331_15434.htm

Nyiri, K. (2002). *Towards a philosophy of m-learning*. IEEE International Workshop on Wireless and Mobile Technologies in Education, August 29-30, 2002, Teleborg Campus.

Pinkwart, N., Hoppe, H. U., Milrad, M., & Perez, J. (2003). Educational scenarios for the cooperative use of Personal Digital Assistants. *Journal of Computer Assisted Learning*, *19*(3), 383–391. doi:10.1046/j.0266-4909.2003.00039.x

QIA. (2008). *Mobile learning*. Retrieved June 28, 2009, from http://www.excellencegateway.org.uk/page.aspx?o=ferl.aclearn.page.id958

Quinn, C. (2000). *mLearning: Mobile, wireless, in-your-pocket learning*. LiNE Zine. Fall.

Sharma, S.K. & Kitchens, F.L. (2004). Web services architecture for m-learning. *Electronic Journal on e-Learning, 2*(1), 203-216.

Sharples, M. (2005). Learning as conversation: Transforming education in the mobile age. In *Proceedings "Seeing Understanding, Learning in the Mobile Age",* Budapest, April 28–30, 2005, 147-152.

Shrivastava, M., Yadav, L. L., & Shrivastava, A. (2005). *From distance education to e-learning: Changing the roles of tutor and learner*. Retrieved June 28, 2009, from http://www.ignou.ac.in/ICDE2005/PDFs/theme2pdf/theme2_153.pdf

Sui, Q. J., Zhang, Y. P., & Zhang, J. B. (2004). Review of mobile learning. *Journal of Education Exploration*, *8*, 66–67.

Sun, Y. T. (2008). The practice and research of m-learning and mobile service. *Journal of Distance Education in China*, *8*, 68–70.

Sun, Y. T., & Chen, X. (2007). Research on the building of mobile campus in the open universities. *Journal of Educational Informatization in China*, *10*, 7–9.

Sun, Y. X. (2008). *Research on m-learning system architecture and application*. Unpublished master's thesis, Shanghai Jiaotong University, Shanghai, China.

Traxler, J. (2007). *Defining, discussing and evaluating mobile learning: the moving finger writes and having writ. . . .* Retrieved May 10, 2009, from http://www.irrodl.org/index.php/irrodl/article/view/346/875

Trifonova, A., Georgieva, E., & Ronchetti, M. (2006). 'Has the time for university's mobile learning come? Determining students' readiness'. *WSEAS Transactions on Advances in Engineering Education*, *3*, 1790–1979.

Wexler, S., Schlenker, B., Metcalf, D., Brown, J., Metcalf, D., Quinn, C., et al. (2007). *Mobile learning: What it is, why it matters, and how to incorporate it into your learning strategy*. Retrieved October 4, 2009 from http://www.elearningguild.com/assets/files/29/mobile_abstract.pdf

Xiong, Z. G. (2005). *Research on mobile learning & the design of mobile learning resources*. Unpublished master's thesis, East China Normal University, Shanghai, China.

Yang, J. (2007). *Illiteracy continues to decline in China*. Retrieved October 4, 2009 from http://news.xinhuanet.com/english/2007-08/01/content_6460066.htm

Yu, S. Q. (2007). From knowledge transfer to cognitive constructivism, then to situated cognition - the development and prospects of mobile learning. *Journal of China Educational Technology*, *6*, 7–18.

Yu, S. Q. (2007). *The implementation plan for 'the application of handheld networked learning systems in subject teaching (AHNLSST)'*. Retrieved April 5, 2009, from http://www.360doc.com/showWeb/0/0/211018.aspx

Zhu, W. P. (2008). *The research and development of title English learning software based on J2ME*. Unpublished master's thesis, Sichuan Normal University, Chendu, China.

Chapter 7
Blended Mobile Learning in Developing Nations and Environments with Variable Access:
Three Cases

Susan Smith Nash
AAPG (American Association of Petroleum Geologists), & University of Oklahoma, USA

ABSTRACT

This chapter presents an overview of the experience of implementing mobile technology solutions in developing nations in conditions of limited Internet access, challenging logistics, and informal learning settings. Specific cases include experiences in blended mobile learning in Paraguay, Iraq, Afghanistan, Colombia, Nigeria, and in remote forward operating bases with NATO and U.S. military professionals. The chapter discusses the factors in developing effective instructional strategies for diverse learning settings and non-traditional learners. Further, it details the unique advantages of mobile learning versus e-learning, why the attributes of pervasiveness, ubiquity, and spontaneity can contribute to learner success, particularly when coupled with informal social networking and support groups.

INTRODUCTION

Mobile learning offers advantages to learners, primarily centered on the fact that m-learning can be spontaneous, ubiquitous, portable, personal, contextual, and personal. However, limited access, hardware and software issues, and size and space limitations, can impede effectiveness. Further, individual learning style differences and the nature of the subject matter may require a flexible approach. While m-learning can refer to any portable device, including digital media players, and handheld computers (personal data assistants / PDAs) some authors and researchers, such as ULTRALAB (mlearning.org, 2009) focus on mobile phones that have the capability of transmitting and receiving text, audio, and video.

E-learning using laptops or desktop computers connected to the Internet offers robust, web-based solutions that can be multimedia rich, and effective synchronous and asynchronous collaboration.

DOI: 10.4018/978-1-61692-818-6.ch007

However, desktop based e-learning is not portable, nor is it ubiquitous. It is difficult to create the kinds of interactive situations that are possible with mobile learning, such as those where learners are gathering data from their environment and are sharing them with other learners. It is also difficult to foster an atmosphere of spontaneous learning, where the student is able to, for example, listen to a lecture or a debate while commuting to work.

Blending old and new technologies is often the most effective approach to incorporating mobile technologies in teaching, training, and learning programs. Ideally, a blending of mobile and desktop-based e-learning can result in a program that accommodates learners' differences, whether they consist of learning styles and preferences, or whether the differences have more to do with delivery of the materials, including access, software, hardware, instructional materials, and assessment. The key is to align instructional strategies with learning goals, while accommodating the learners' individual situations and the learning environment.

Creating the ideal blend of e-learning and m-learning requires careful planning, which begins with an examination of learning objectives, and then continues with a realistic evaluation of the conditions of learning (access, software, hardware), the logistics of the learners, and their likely learning preferences. After the factors are considered, then instructional strategies can be developed and implemented.

Even when all factors are brought to bear, successful implementation often depends on motivation – not just on the part of the learner, but also with the instructor and the instructional support team. The instructor and the online support team must respond in a positive, solution-centered way to unexpected issues that may emerge, and to help reduce frustration and learner anxiety, and must keep students focused and on task. Many of the success strategies must be developed based on in-the-field experience.

The goal of this chapter is to present, analyze, and discuss cases of mobile learning in developing countries, with an end to making recommendations with respect to appropriate and effective learning strategies, ideal blends of technologies (mobile, web-based, traditional books and paper, F2F, high-tech materials, digital pens, all-weather paper), the development of learning communities, the adoption of creative uses of podcasts and GIS to engage learners, reinforce experiential learning, the development of methods to address the affective domain: motivation, attitude, self-efficacy, self-determination to boost learner engagement and course completion, and the effective uses of mobile technologies for meaningful assessment and evaluation.

Specifically, the following cases are examined:

Mobile devices; almost no connectivity. The primary challenge consisted in limited connectivity. Handheld devices and portable media devices were used to provide e-book content, lesson plans, audio lectures by the subject matter experts, review materials. Some cases included informal collaborations as individuals formed study groups or friendships. This approach was used in Paraguay with the University of Oklahoma's College of Continuing Education and the Paraguayan American Cultural Center.

Mobile devices in remote locations with limited connectivity. The primary challenge consisted in limited connectivity, along with expensive books and materials. iPods were used to provide instructional podcasts (videos and audio), interactive grammar activities, instructions for activities, guides for collaborative / peer activities. This approach was used for military students deployed to Iraq, Afghanistan, and other locations.

Mobile devices used for content and interaction. Skype and web-conferencing software was used for synchronous elements. Email and blogs were used for asynchronous interaction. One major challenge was to meet the needs for portability while keeping the ability to communicate synchronously. Mobile devices used to download and store instructional content (ebook, video podcasts, lesson plans, audio lectures), and also to provide interactivity and to share timely information (field information from markets, photos, etc.). This approach was used for students in Afghanistan, Nigeria, and Colombia, with courses from the American Association of Petroleum Geologists.

BACKGROUND

Definitions of mobile learning differ, which can problematize the formation of a blended solution, since it may be difficult to separate the elements that make m-learning unique. In general, researchers either emphasize the technology used, or the informal nature of the learning experience. For example, Geddes (2004) looks at m-learning in a general sense, and suggests that it is a measurable change in behavior that results from the application of an anytime, anywhere mobile technology. Other authors (Quinn [2000], Harris [2001], Wood [2003]). look at the environment of learning, and focus on the fact that the learner is in an informal learning environment and not in their regular, more formal learning environment (home, workplace, educational institution).

Researchers (Kukulska-Hulme & Traxler [2005], Lee & Chan [2007], McConatha, Praul & Lynch [2008]) have generally pointed to the fact that mobile learning is spontaneous, ubiquitous, and pervasive (in the sense that it is integrated tightly and unobtrusively with daily activities) as the key elements of successful learning.

E-learning seemed to offer the same "any time, any place" access, which causes confusion in the definition of m-learning. However, an emphasis on the fact that m-learning allows ubiquitous engagement with the learning environment. Further, the pervasive nature allows learners to be spontaneous. These are the main attributes that make m-learning truly a sub-set of e-learning, and not simply the same thing.

Theoretical differentiations can be a bit ambiguous. However, when one looks at case studies where m-learning has been employed, the differences are clearer. For example, mobile learning that facilitates "learning by doing" (Dewey, 1938) by incorporating experiential learning in an interactive setting (lab, field, hospital, etc.), is clearly different than desktop-based e-learning, which becomes more formal (Lai etal, 2007). Alternatively, m-learning that allows individuals to have lifestyle-integrated learning events at home, on the way to work, while waiting for a family member, etc. (Lee & Chan, 2007) also clearly differentiates itself from the formal environment of desktop e-learning.

Differentiation between e-learning and m-learning can also become clearer when looking at the nature of the instructional materials. Because of the small size and limited memory, m-learning has tended to focus around the delivery of learning objects, and the objects typically are of small size. The range of learning objects includes text, images, audio, animations, and video. Likewise, interaction and collaboration are, by necessity, on the scale of microblogging, which is to way that there are limits to the words that can be shared via text, the size of images emailed to each other, and the complexity of the GPS or other information gathered in a field location and posted to a common location. Similarly, assessments and interactive quizzes have been small (McConatha, Praul, & Lynch, 2008).

Many authors (Thornton & Sharples, 2005; Gomez, 2007) see the future of m-learning as tied closely to the further evolution of social networks, and the comfort that learners have in sharing information via RSS feeds, "tweets" or graphics and media-rich sites (Facebook, YouTube, Flickr, Bebo, Orkut, etc.).

In addition, authors focus on the experience of mobile learners in the U.S., Australia, and Europe (Lee & Chan, 2007; McConatha etal, 2008), and they look at the convenience of mobile devices in creating spontaneous, pervasive, and ubiquitous learning environments. However, in some countries, m-learning is more of a necessity than a luxury, due to the abundance of cell phones and the limited number of households with high-speed internet connections (Malaysian Communications and Multimedia Commission, 2009). Further, learners who are accustomed to using mobile phones for a wide variety of applications will tend to feel comfortable with the medium (Suki & Suki, 2007).

Authors look at learning theories and keep them in consideration in order to maximize the efficacy of the learning experience. Mobile learning is often viewed as engaging, and ideal for the conditions of learning described by Gagne and others. In addition, it is recognized as ideal for developing schemata as the material is "chunked" by necessity, due size and space limitations. Critical thinking skills are fostered, and, where the mobile device is interactive (as in the case of mobile phones), then the learning experience can be collaborative, with opportunities for role-playing. In the case of m-learning for experiential learning, active learning is foregrounded. Cognitive theories of learning as applied to interactive multimodal learning environments that require dialoguing, controlling, searching, and manipulating have been explored by Moreno and Mayer (2007). How learning takes place in multimedia learning environments has also been extensively explored, not only as it relates to formal, but also to informal environments (Mayer, 2003; Lorencova, 2009).

BLENDED MOBILE LEARNING CASES

Case One: Almost No Internet Access Away from Home Base or Center

Early successes with mobile learning did not rely on the coverage possible with cell phones. Instead, they focused more on providing ubiquitous coverage, in ways that accommodated highly informal learning environments.

Paraguay

In the late 1990s, the University of Oklahoma's College of Continuing Education made a commitment to provide online courses and instructional materials to students in Paraguay, with the goal of enabling learners to go to a central location, the Paraguayan – American Cultural Center in Asuncion, the country's capital, and then download content to a portable device, which could then be accessed in areas without internet access. Approximately 60 students participated in the courses from 1998 until late 1999, which focused on business-related topics such as marketing, consumer behavior, and Business English. Mobile content was delivered in the form of pdfs of texts, mp3 files of audio lectures and presentations, and presentations that could be downloaded and accessed via mobile device. Paraguayan students were able to read lessons, listen to recorded content, and then to complete journal entries or instructional activities, which they then emailed to their instructors.

The approach blended synchronous and asynchronous learning approaches. Since students traveled to a central location to download the content, it created an ideal situation for collaboration and support. Informal study groups and collaborations occurred as students reviewed the instructional materials and each other's work. Later, they could listen to the audio files while riding the bus home, or have access, even if their homes and offices, often located in the outskirts of the city, or in rural areas, did not have Internet access.

- Students with daily internet access: 35%
- Students with occasional internet access: 65%
- Students who used mobile devices for content every day: 82%
- Students who used a blend of mobile devices and computers: 100%

Course Completion Rates

The course completion rates varied by subject:

- Marketing 18%
- Management 33%
- Technical English 42%
- Business English 45%

The success rate for students was highest for students who formed support networks and studied together when at the Paraguayan-American Cultural Center. It was also best for students who purchased the textbook and who were able to learn well with text, images, and questions that asked them to situate their learning.

The low completion rate for marketing was felt to be due to the fact that many students signed up for it, and that the course requirements included market research and a rather involved case study analysis, for which there were no examples, and where the guidelines were determined (after the fact) to have been confusing in a distance environment.

High completion rates in the English courses was felt to be due to the fact that the students where were taking those courses were also taking face-to-face English courses at the Paraguayan-American Cultural Center, and that the students formed informal study groups.

Reasons for Lack of Persistence. When asked why they did not complete their courses, students had a variety of explanations. Many had to do with the fact that work and family obligations interfered with their studies. They pointed out it was easier to stay focused in a classroom.

Others said that they started out well, but then lost interest when they failed to receive feedback in a timely manner. They also said that without feedback from peers or professors, they started to feel very isolated. They did not feel they were a part of a learning community. However, for the students who happened to also be taking courses onsite at the Paraguayan-American Cultural Center, and who formed study groups on their own, the experience was much more positive. They were able to complete the courses because they were able to work together, share their projects, and to discuss issues they found to be of interest.

Although the instructional material was the same, the students who worked together in informal study groups reported that their courses were relevant and the course content "engaging" and "interesting." Conversely, the students who were not involved in study groups and who were reading texts and listening to audio files only found the material to be "tedious" and "not interesting."

Clearly, retention and successful completion were functions of being a part of a learning community. So, in order to improve retention, study groups were formed, as well as scholarships and incentives. Even with the benefits, the completion rates did not approach those of the face-to-face courses. The director o the Paraguayan-American Cultural Center discussed the problem often, and pointed out that culturally, the idea of independent study is fairly alien, and in order to assure results, it would be best to have at least some weekly meeting and personal contact, even if it took the form of a weekly teleconference. They pointed out that having weekly telephone calls with the instructor would also help students set goals and to meet deadlines.

The instructors at the Paraguayan-American Cultural Center suggested as well that the course content itself was not very dynamic. There were no movies, simulations, or games. The content was presented in the form of text and audio, which was not very engaging. The fact that the courses needed to be assembled quickly, and the mobile devices did not accommodate complex animations or technology, made it difficult for learners who may have preferred simulations and role play. The goal was to be able to deliver text, images, and audio in a compact way to facilitate the learning process. The courses needed to have a compact design in order to load as quickly as possible via a slow connection with laptop or mobile device.

Another problem was limited interaction with the instructors, and limited feedback. This was considered very demotivating for many learners, who needed frequent feedback. In fact, lack of interaction and feedback is the primary problem with many early mobile learning programs. In the case of the Paraguayan courses, professor–student interaction was possible, but had to be done sporadically, when connectivity was possible. At that time, students emailed questions and their work. The students took exams in proctored locations and the exams were emailed or send via regular mail to the instructors.

Case Two: Limited Access

This case involved courses delivered for deployed military students. The program launched in 2001, shortly after September 11, in order to accommodate the needs of active duty students who could no longer take week-long intensive short courses on base because as active duty military professionals, they were either deployed to remote locations or away from base due to temporary duty assignments. Rather than drop out of the program, they preferred to take online courses.

From 2001 until 2008, there were approximately 1,500 online / mobile learning enrollments per year in graduate-level courses offered through the Department of Human Relations at the University of Oklahoma.

Before successfully implementing the program, it was necessary to address a few issues that troubled students and administrators. First, many students had a fundamental discomfort with online or mobile courses and were reluctant adopters. They had developed a belief system that mobile learning would be difficult, even unengaging.

It was also necessary to convince individuals that the logistical and access barriers could be overcome, and that mobile learning would be sufficiently robust to effectively deliver the content. In most cases, the students would not have access to the Internet except in limited circumstances; with "morale" computers in tents in the forward operating bases. When away from the bases, the students would need to rely on their iPods and other mobile devices to access content.

One unusually challenge consisted of the fact that space was very limited. The students were deployed, and had to carry all their belongings with them. Thus, it was necessary to use e-books, and to be able to save quizzes, papers, and other assessments onto a flash drive or on the mobile device itself.

Mobile content included text, graphics, podcasts (audio and video), and journal / discussion board.

Instructional Approach and Strategies

Although students may have started the program with doubts and uncertainty, they were highly motivated to complete their courses. Often, completion of a degree program signified a raise or a promotion.

Most of the students were officers in the U.S. military, and they chose the Master of Human Relations degree program because it contained a leadership track that directly corresponded to their goals. That said, there was some risk because the same students who were highly motivated to study the courses were also very demanding. They wanted to be able to apply the knowledge they gained, and demanded relevancy. As a result, the instructors and the instructional design team adjusted the course design and the approach to make sure that the mobile / hybrid format could accommodate the content.

Required work for the mobile courses differed from the face to face courses in significant ways. First, there were fewer assignments, and each assignment was longer. The reason for the shorter, longer assignments was the fact that connectivity was limited. The assignments could be submitted through a learning management system or via email. Instead of requiring peer interaction via the discussion board or forums, students had the option of maintaining a journal and sending it in separately. Flexibility helped students develop a sense of self-determination and self-confidence. Giving students a choice of assignments, and encouraging them to incorporate prior learning allowed them to apply the concepts encountered in the courses, and to feel comfortable.

Findings and Lessons Learned

Over an 8-year period, the completion rate of the courses stayed at an astonishingly high 90 to 95% completed. Course satisfaction consistently achieved high rates, and the course was almost always full, with a waiting list.

Students reported that they liked the fact that the courses consisted of immediately relevant material that could be applied to their own lives and challenges. They liked the flexible approach to the course materials (choice of readings, choice of writing assignments) and the fact that writing assignments required analysis and synthesis of the core readings with current events, controversies, case studies, and personal experience. Other elements helped alleviate student anxiety, such as flexibility with respect to due dates.

RECOMMENDATIONS AND FUTURE APPLICABILITY

Deployment, high-stress environments, rapid changes of location, erratic schedules, and radical changes in time zones made it difficult to access the materials in a consistent manner. There were many distractions, and the fact that some learners were suffering from the aftermath of traumatic situations made it very important to do the following:

- keep the organization of the content as clear and straightforward as possible
- connect the content to the real world for a truly situated learning experience
- maintain a solution-centered attitude
- develop affirming and encouraging responses in order to help students develop and maintain a positive sense of self and self-efficacy

Some of the activities that exist on the internet are not possible to implement in an environment where students have limited access, firewalls, and limited bandwidth.

However, in the future, the kinds of students who benefit from leadership courses where they look to their coursework for guidance and tools for achieving immediate career, family, and professional objectives, it is possible that more media-rich instructional content would be very relevant. In the future, it would be ideal to include the following:

- Media (audio and video podcasts)
- Simulations and serious games (decision-making, or strategy games are very effective)
- Role-playing in virtual worlds (Second Life)
- Recorded debates or interviews, particularly on topics that involve controversies and ethical dilemmas

Students were highly motivated to persist, and for that reason, the completion rates were higher than they might have been. The requirement for military members to have a master's degree in order to achieve certain ranks made the group of students highly motivated, despite difficult conditions of instruction and learning.

Case Three: Good Urban or "Base Camp" Connectivity for Downloads; Good Cell Phone Coverage for sending basic text messages

In 2005 through 2008, learners located in Afghanistan, Nigeria, and Colombia wanted to take technical writing and business writing courses offered by the American Association of Petroleum Geologists (AAPG), but they lacked instructors, classrooms, and consistent access to the Internet. There were very basic mobile devices which could be used for content and interaction. Skype could be used for synchronous elements. Email and blogs could be used for asynchronous interaction.

Mobile devices were used to download and store instructional content (ebook, video podcasts, lesson plans, audio lectures), and also to provide interactivity and to share timely information (field information from markets, photos, etc.).

Courses: Technical Writing

Instructional Approaches

The most effective approach was to use a building block approach to writing papers, which involved breaking down the task of writing reports and other technical documents into short, easily accomplished tasks. Because of the problems of distractions, it was important to engage students by making sure to give them the flexibility to write on topics that interest them. In this way, the instructional strategy made full use of the attributes of ubiquity, pervasiveness, and spontaneity that characterize mobile learning.

Specifically, it was effective to use guiding materials, such as flow-charts and checklists, and provide examples in order help develop a sense of self-efficacy, and give freedom with respect to topics in order to reinforce a high level of self-determination.

Student Completion Rate

- Afghanistan: 60% completion
- Colombia: 30% completion
- Nigeria: 40% completion

The high completion rate in Afghanistan was due in part to the fact that several individuals met together and worked on assignments as a group. In Columbia and Nigeria, the students worked alone. One Nigerian student used the course productively with his job with a multinational oil company to help him prepare effective presentations.

It is worth noting that one of the students in Afghanistan was truly remarkable. She was able to apply the assignments to several projects she was encountering in her job in helping build resources and infrastructure in Afghanistan. She was highly motivated, and was able to communicate very effectively via e-mail. The fact that she had a well-defined purpose and reason for developing her documents made a great deal of difference.

Lessons Learned

Students were working professionals who were very busy in their jobs. They were often on the road in remote locations. Internet access was expensive and not always available. At times, online access would improve dramatically, depending on the student's location. Some students were very comfortable with social networking and Web 2.0 integrated applications, despite the challenges of connectivity. Because of the problem with lack of access, and the distractions posed by social networking, it was necessary to have deadlines, and yet they needed to be flexible. Content was more engaging if it had practical connections to the workplace, and if they could use "live" documents as their content. For example, one student in Afghanistan used a proposal she was writing in response to a request for proposals posted by the US Agency for International Development.

While more connectivity may seem to mean a better experience with mobile learning, that is not always the case. The experience of utilizing numerous Web 2.0 applications in order to enhance the experience of connectedness and to create a learning community revealed a number of weaknesses in the assumption that more connectivity is better.

It became clear that the quantity if interaction is not as important as the quality. In fact, too many Twitter messages, YouTube videos, and FaceBook posts can lead to chaos. Information can be distracting rather than instructive, and it can be difficult for students to sort out the "noise" from the information they will need in order to succeed with their assessments.

*Limit the number of Web 2.0 apps. Keep them focused in order to avoid "noise" and distractions
*Place more emphasis on direct instruction (structured overviews) and practices assessments
*Make sure that activities are learning outcome-centered

Quality of interaction can be measured in several ways: 1) the degree to which the interactions contribute to the development of a bona fide learning community; 2) the degree to which learners participate in instructional activities, share them, and then respond to each other in a cogent way; 3) the degree to which the activities advance the learners as a whole to the achievement of course outcomes and learning objectives.

CONCLUSION AND RECOMMENDATIONS

High rates of success occurred when mobile learning took full advantage of the learner's environment, and when the content directly related to their personal and professional goals, as well as having a concrete linkage to real-life events. Learner satisfaction was enhanced by informal and formal support networks and the existence of tangible penalties for dropping out.

Even when penalties for dropping out did not exist, the programs that blended mobile learning with informal support groups and learning communities resulted in higher completion rates. Students working along reported that their material was engaging, but when they became confused and needed more guidance, there was no way to obtain guidance in a timely fashion. Further, getting feedback for graded work was a difficult proposition when there was limited connectivity. Students felt frustrated when they could not obtain timely feedback.

Perhaps the most significant opportunity presented by mobile learning is also its most significant challenge. Taking learning into an informal learning, and situating the activities within new contexts is positive, but, the reality is that pervasive, ubiquitous learning is always accompanied by distractors. Some of the learners were in combat support operations, and others were in very demanding jobs that required long hours. They had difficulty finding time to focus, and when they did, it was difficult to dedicate long periods of time. Thus, cognitive overload was a problem, and mobile learning needed to incorporate an immersion experience in order to block out distracters. The best approach in this case was to use audio and multimedia to maintain focus.

- Deliver the content in smaller bite sizes, smaller "chunks"
- Use more media when appropriate.
- Develop self-guided worksheets and instructional elements that avoid ambiguity
- Find a mechanism for providing student support
- Develop relevant activities that pertain to one's real-life needs
- Create way to incorporate prior knowledge
- Incorporate experiential learning where possible

FUTURE APPLICABILITY

While some of the mobile learning implementations did not have a very high completion rate, there are numerous situations where delivering content via mobile devices remains the most cost-effective choice. The need is particularly acute in areas where it is difficult to obtain instruction, materials, or where either the learner is constantly moving or changing locations.

Hostile learning conditions may be very difficult to overcome, but the demand for educational and training materials continues to grow. It is important to make sure that the following elements are accommodated:

Equipment must be rugged, with long battery life and resistance to water, sand, wind, temperature fluctuations, and other environmental conditions.

Instructional strategy must focus on small, discrete learning units that can accommodate frequent interruptions and distractions.

Instructional strategy needs to privilege experiential learning, with frequent opportunities for the student to make connections between the lessons and real life, so that the learning experience continues to relevant and meaningful.

Instructional materials must be engaging and readings need to be easily broken into digestible chunks, with frequent summaries and connections back to learning objectives.

Finally, it is important to keep a human face firmly attached to what is often a technology-driven endeavor. The beneficiaries of the instruction are not just the individuals who take the courses, but also all the individuals and communities whom their education touches. For example, the female Afghani geoscientist who was able to take a technical writing class through a combination of online and mobile learning was able use the course to further her technical writing skills to the point that she was able to compete for a grant to bring new infrastructure and economic development projects to her community. In this case, the beneficiaries of mobile learning are numerous, and they are in many different places.

REFERENCES

Chen, Y. S., Kao, T. C., Sheu, J. P., & Chiang, C. Y. (2002). A mobile scaffolding-aid-based bird-watching learning system. In M. Milrad, H.U. Hoppe, & Y. Kinshuk (Eds.), *IEEE International Workshop on Wireless and Mobile Technologies in Education* (pp. 15-22). Los Alamitos, CA: IEEE Computer Society.

Corlett, D., Sharples, M., Bull, S., & Chan, T. (2005). Evaluation of a mobile learning organiser for university students. *Journal of Computer Assisted Learning*, *21*, 162–170. doi:10.1111/j.1365-2729.2005.00124.x

Devi, C. (2005). New hotlink services for youths. *New Straits Times . Computimes*, *15*, 3.

Dewey, J. (1938). *Experience and Education*. New York: Macmillan.

Geddes, S. (2004). Mobile learning in the 21st century: benefit for learners. *The Knowledge Tree: An e-Journal of Learning Innovation*, 6.

Gomez, S. (2007). Scroll to "e" for Education. *Times Higher Education Supplement*, *1780*, 13.

Kukulska-Hulme, A., Evans, D., & Traxler, J. (2005). *Landscape study in wireless and mobile learning in the post-16 sector*. Bristol: Joint Information Systems Committee.

Kukulska-Hulme, A., & Traxler, J. (2005). *Mobile learning: A handbook for educators and trainers*. New York: Taylor & Francis.

Kvavik, R. B. (2005). *Convenience, communications, and control: how students use technology*. Retrieved July 18, 2009 from www.educause.edu/ConvenienceCommunicationsandControl%3AHowStudentsUseTechnology/6070

LaBrasca, S. (2009, July). E-Learning without boarders. *Learning Circuits*. Retrieved July 20, 2009 from http://www.astd.org/LC/2009/0709_labrasca.htm

Lai, C.-H., Yan, J. C., Ho, C. W., & Chan, T. W. (2007). Affordances of mobile technologies for experiential learning: the interplay of technology and pedagogical practices. *Journal of Computer Assisted Learning*, *23*, 326–337. doi:10.1111/j.1365-2729.2007.00237.x

Lee, K. S. (2006). More youngsters using handphones. *New Straits Times*, *25*, 6.

Lee, M., & Chan, A. (2007). Pervasive, lifestyle-integrated mobile learning for distance learners: an analysis and unexpected results from a podcasting study. *Open Learning*, *22*(5), 201–218.

Lehner, F., & Nosekebel, H. (2002). The role of mobile devices in e-learning – first experience with a e-learning environment. In M. Milrad, H.U. Hoppe & Y. Kinshuk (Eds.), *IEEE International Workshop on Wireless and Mobile Technologies in Education* (pp. 103-106). Los Alamitos, CA: IEEE Computer Society.

Littlefield, L. (2004). *Psychosocial Aspects of Mobile Phone Use Among Adolescents*. Melbourne: The Australian Psychological Society.

Lorencova, V. (2009). YouTube dilemmas: The appropriation of user-generated online videos in teaching learning. *Current in Teaching and Learning*, *2*, 62–71.

m-Learning.org. (2009). *Ultralab*. Retrieved July 14, 2009 from http://www.naec.org.uk/ultralab/ww3/projects/mlearning/

Malaysian Communications and Multimedia Commission. (2009). Facts and figures. Retrieved July 14, 2009 from http://www.skmm.gov.my/facts_figures/stats/ViewStatistic.asp?cc=52036251&srid=50919742

Marriott, L. (2006). What services are driving mobile internet usage? Retrieved July 18, 2009 from www.clickz.com/experts/ad/mobile/article.php/3606606

Mayer, R. (2003). The promise of multimedia learning: Using the same instructional design methods across different media. *Learning and Instruction*, *13*, 125–139. doi:10.1016/S0959-4752(02)00016-6

McConatha, D., Praul, M., & Lynch, M. (2008). Mobile learning in higher education: an empirical assessment of a new educational tool. *The Turkish Online Journal of Educational Technology*, *7*(3), 15–23.

McNeely, B. (2005). Using technology as a learning tool, not just the cool new thing. Retrieved July 18, 2009 from www.educause.edu/UsingTechnologyasaLearningTool%2CNotJusttheCoolNewThing/6060

Moreno, R., & Mayer, R. (2007). Interactive multimodal learning environments. *Educational Psychology Review*, *19*, 309–326. doi:10.1007/s10648-007-9047-2

Oblinger, D., & Oblinger, J. (2005). *Is it age or IT: first steps toward understanding the net generation*. Retrieved July 18, 2009 from www.educause.edu/IsItAgeorIT%3AFirstStepsTowardUnderstandingtheNetGeneration/6058

Pullchino, J. (2006). Mobile Learning Research Report. *The E-Learning Guild*. Retrieved July 10, 2009 from http://www.elearningguild.com/

Ramaley, J., & Zia, L. (2005). The real versus the possible: closing the gaps in engagement and learning. Retrieved July 18, 2009 from www.educause.edu/TheRealVersusthePossible%3AClosingtheGapsinEngagementandLearning/6064

Rogers, E. M. (1995). *Diffusion of Innovations* (4th ed.). New York: The Free Press.

Rushby, N. (2005). Editorial. *British Journal of Educational Technology*, *36*(5), 709–710. doi:10.1111/j.1467-8535.2005.00556.x

Salz, P. A. (2006a). Learning to go mobile. Retrieved July 18, 2009 from www.econtentmag.com/Articles/ArticleReader.aspx?ArticleID=15299&AuthorID=157

Salz, P. A. (2006b). Mobile delivery discoveries. Retrieved July 18, 2009 from www.econtentmag.com/Articles/ArticleReader.aspx?ArticleID=15719&AuthorID=157

Seppala, P., & Alamaki, H. (2003). Mobile Learning in Teacher Training. *Journal of Computer Assisted Learning*, *19*, 330–335. doi:10.1046/j.0266-4909.2003.00034.x

Sharples, M. (2000). The Design of Personal Mobile Technologies for Lifelong Learning. *Computers & Education*, *34*, 177–193. doi:10.1016/S0360-1315(99)00044-5

Star InTech. (2006). Go mobile, get embedded. *Star InTech*, *13*, 19.

Suki, N., & Suki, N. (2007). Mobile phone usage for m-learning: comparing heavy and light mobile phone users. *Campus-Wide Information Systems*, *24*(5), 355–365. doi:10.1108/10650740710835779

Thornton, P., & Houser, C. (2005). Using mobile phones in English education in Japan. *Journal of Computer Assisted Learning*, *21*, 217–228. doi:10.1111/j.1365-2729.2005.00129.x

Wagner, E. (2005). Enabling Mobile Learning. *EDUCAUSE Review*, *40*(3), 40–53.

Chapter 8
Mobile Technologies and Rich Media:
Expanding Tertiary Education Opportunities in Developing Countries

Trish Andrews
The University of Queensland, Australia

Robyn Smyth
The University of New England, Australia

Belinda Tynan
The University of New England, Australia

Andrew Berriman
The University of New England, Australia

Deborah Vale
The University of New England, Australia

Richard Caladine
The University of Wollongong, Australia

ABSTRACT

This chapter focuses on the emerging possibilities and issues arising from the rapid adoption of mobile technologies for learning in tertiary and higher education contexts in developing countries. In particular, it explores the implications for developing nations of the rapid proliferation of mobile devices. Many opportunities are forecast along with some lessons learned from an Australian investigation into the use of rich media technologies in higher education.

DOI: 10.4018/978-1-61692-818-6.ch008

INTRODUCTION

Mobile technologies and rich media can enable greater opportunities for situated and personal learning in both real and virtual higher education contexts in developing countries, thus providing increased access and opportunities for higher education for larger numbers of people for whom this kind of education is often difficult to obtain. As Naismith et al (2004) state

Mobile technologies are becoming more embedded, ubiquitous and networked, with enhanced capabilities for rich social interactions, context awareness and internet connectivity. Such technologies can have a great impact on learning. Learning will move more and more outside of the classroom and into the learner's environments, both real and virtual, thus becoming more situated, personal, collaborative and lifelong (2004, p. 5).

For the purposes of this chapter, which concerns learning in tertiary education contexts, we define mobile technologies as devices which enable interactive learning to take place anywhere, anytime and at the pace of the learner who we assume will be technology savvy, 'nomadic and continually on the move' (Ally, 2007). Such a definition assumes connectivity via mobile phone broadband networks for devices with the capability for basic functionality expected of any personal computer as well as devices enabled to receive Internet based services including email, browsing, social networking and shortly video and audio conferencing.

BACKGROUND

The take-up of mobile technologies worldwide, particularly mobile phones, is expanding rapidly with more than half the world's 6.5 billion people owning mobile phones in 2007, a significant increase from the 2 billion people identified as owning mobile phones in 2005 (Miller, 2007). Of this number, two thirds of mobile phone owners are located in developing countries. Developing countries are described by the World Bank as those countries where the majority of people are low or low middle income earners (Donner, 2008). While in the developed world there are currently more mobile phones than there are people, in the developing world one mobile phone is often shared amongst several users, expanding access to these devices. Subscriber rates in many parts of the developing world are increasing at a minimum of 25% year and 50% a year in Africa (Miller, 2007). Mobile phone ownership in South Africa is now reported as 60% of all people over the age of 16 (Kreutzer, 2008). Amongst the 2.4 billion people living in the world's poorest nations including India, Pakistan, Bangladesh and Vietnam, a median concentration of 7.5 mobiles per100 people was reported in 2006 (Donner, 2008) and growing rapidly. In the more advanced low-middle income countries such as China, Egypt and the Philippines, median penetration of mobile devices in 2006 was already 30 per 100 (Donner 2008).

While much of this ownership is amongst the more prosperous members of these nations, mobile ownership is spreading rapidly across all levels of these societies.

As mobile services are pulled increasingly into rural and low-income communities, mobile phones are riding the strength of rapidly growing networks, low power and maintenance and increasingly affordable pricing (Lehr, 2007, p 3).

This rapid acquisition of mobile technologies amongst the worlds' poorest nations and people is creating considerable opportunities for economic development, improved health care and improved access to education for the people living in these countries. In the developed world, a high level of access to broadband internet has been accompanied by an expansion of access to rich media technologies. This uptake of rich media technology is being further fuelled by the integration of these tools into popular social networking applications (Caladine et al, in press).

Rich media technologies are those technologies that enable presentation of visual and vocal cues, greatly enhancing the communication aspects of information and communication technologies (Baecker, 2003). Combined with mobile technologies, rich media technologies offer opportunities for situated and personalised learning and the ability to link learning contexts in unique ways to provide a range of educational opportunities for people in the developing world who often experience great difficulty accessing education. These technologies also offer ways to address a number of cultural issues that may be of significance to people in the developing world (Caladine et al, in press; Andrews et al, 2008).

Unreliable and expensive fixed networks, lack of infrastructure and limited expertise in computer based learning applications make mobile technologies an attractive proposition for developing countries, although they also come with their own set of challenges.

Factors Influencing Adoption of Mobile Technologies

There are a number of core considerations which should be kept in mind when discussing mobile technologies for tertiary education in developing countries and these include factors within and beyond the control of the institutions attempting to move to this form of delivery.

Infrastructure

Infrastructure for fixed line technologies such as DSL and cable is usually expensive and this can be prohibitive in many developing countries. Even in many developed countries, in areas where there are low numbers of subscribers services can be limited outside urban areas (Smyth, 2006). In developing countries, access to these kind of services are often non-existent as there is often no such infrastructure of any kind, terrain is often inhospitable to such infrastructure and the economic feasibility of developing it makes such provision highly unlikely (Smyth, 2006). However, 'new forms of wireless protocols (Wi-Fi, WiMAX, etc) are overcoming challenges of terrain, infrastructure and finance', (Smyth 2006, p. 3) rapidly increasing access to phone and internet technologies across developing countries.

Affordability

As outlined above, the uptake of mobile phone adoption in the developing countries has been nothing short of extraordinary (Donner 2008). There are many possible explanations for this but affordability made possible through pre paid subscriptions and cheap handsets is part of the explanation. Studies by Hodge (2005) and Esselaar and Stork, (2005) found that users preferred prepaid mobiles to fixed line phones even though per minutes rates for mobile usage are higher than fixed line rates (Donner, 2008). Further to providing affordability, Lehr (2007) comments that:

prepaid cards that allow for the purchase of air time in small increments also closely matches the budgeting and payment needs of many low income customers (p 3).

Government Policy

Proactive government policy has also played a leading role in this rapid adoption of mobile technologies. For example, reforms of India's telecommunications industry in 1999 (Fraunholz and Unnithan, 2004) are considered to be a major factoring influencing the rapid adoption of mobile technologies in that country. Additionally, liberalization of the telecommunications markets is considered a factor in other countries such as Guatemala (Ibarguen, 2003) which has had a much higher level of adoption of mobiles compared with other South Americans countries (Donner 2008). As well as promoting adoption of mobile technologies, Government policy can also create barriers. Lack of liberalization of the telecommunications sector in a number of countries where there has been less enthusiastic adoption of mobile technologies, such as Iran, Syria and Lybia is considered a major factor limiting mobile technology adoption (Donner 2008).

Shared Usage Models

The development of shared usage of mobile services is one that is having a major impact on providing access to mobile phones and services in communities that previously have had little or no access to such services. One model is the 'village phone' (Donner, 2008; Lehr, 2007). Typically these services are provided by entrepreneurs who borrow money to buy handsets configured for multiple users and bulk minutes they then on sell to individual users in the village (Donner, 2008; Lehr, 2007). One of the first models of this approach was the Grammen model, using a microeconomic approach, which was established in Bangladesh and has since been rolled out to other countries such as Uganda and Rwanda (Donner 2008). This approach has been very powerful in putting very poor rural communities in touch with a wide range of online and other services and opportunities.

There is also some evidence that there is informal shared usage amongst youth in very poor communities in South Africa (Kreutzer, 2009) where lack of personal ownership of devices does not prevent young people from accessing a variety of internet services. It is quite possible that such patterns of use amongst young people are common across developing countries. They have certainly been personally evident to the authors during work and travel in countries such as Bhutan, and sub-Saharan Africa.

Regardless of how it is being achieved, evidence suggests access to mobile technologies amongst the world's poorest people is growing at a very rapid rate (Kreutzer, 2009; Donner, 2008). Nevertheless, there is an onus to ensure that mobile access is inclusive, catering for the most marginalised learners no matter what their circumstances. One program being trialled with the nomadic peoples of Nigeria provides a useful model for ensuring that this particular group of marginalised learners is not left out:

Procuring mobile phones for these nomadic groups of learners will not only motivate them and instil positive attitudes towards learning, it will also help to sustain their interest in gaining literacy skills, especially through the distance learning approach (Aderinoye, Ojokheta, & Olojede, 2007, p.15).

Consideration then, needs to be given by institutions and educators to ways to make use of this rapidly growing access to mobile technologies in developing countries in ways that will motivate and encourage broader participation in appropriate teaching and learning activities.

Key Issues Relating to Use of Mobile Technologies and Rich Media in Developing Countries

The comprehensive review of mobile use in the developing world by Donner (2008) identifies mobile technologies as having promise but that the patterns of use receive limited attention. The thematic grouping of the review serves as a useful point of discussion for informing where this chapter aims to arrive in regards to expanding tertiary learning opportunities. In fact much of what has preceded this point has had a focus on the themes suggested: determinants of mobile adoption, impacts of mobile use and the interrelationships between mobile technologies and users (Donner, 2008). The penetration of mobile technology is well reported however it may as it is pointed out fail to actually detail pre-paid or indeed shared use which was alluded to in how individuals may share mobile devices which expands the level penetration considerably. While Donner (2008) reports that there are studies with a focus on the impact on economies the authors note that there appear to be fewer with a concern for educational impact for the developing world. In other parts of this chapter many studies that are referred to come from developed countries.

In this comprehensive review, Donner (2008) also details how policy influences both connectivity and accessibility with the chief concern to reduce the digital divide. There is no doubt that governments and providers have interrelated roles to play and they significantly influence both penetration and availability. While these factors continue to influence the use of mobile technology they will also have an impact on how they are used for different sectors such as tertiary education. Donner (2008) notes that "the rationale for a variety of impact studies is clear: the mobile is an incredibly powerful tool for exchanging ideas at a distance and for managing daily life" (p.14). For educators this presents both opportunity and enormous challenge. There is scope to promote elearning as the new frontier for mobile learning that can mediate distance and enable the exchange of meaningful learning experiences. The questions that need to be asked however seem to be numerous and highly complex. What role for example do sociocultural factors play in designing educational experiences? What economic constraints are there? How do personal and educational uses overlap? What software or systems might be required for different contexts that take into account cultural and language differences? How do different cultures use mobile devices and should this impact on possible education applications?

The implications for further research are numerous. If Donner's (2008) suggestions are anything to go by those with an interest in increasing tertiary education participation should review applications and use of mobile technology broadly. There is an obvious need for further research on mobile technology that explores the interrelationship between development as economically driven and everyday need in relation to tertiary education specifically. Bridging that divide will possibly 'be nice' as it would make the current two-dimensional emphasis in the literature harder to duplicate. Furthermore there is a requirement to better understand the needs and motivations of all users whether well-off or poor before the potential of tertiary education can be fully understood in developing country contexts.

Considerations for Tertiary Education in Developing Countries

The use of mobile technologies for learning sounds wonderful and quite liberating in theory but as Traxler (2005) and others identify, there are many issues that might impede mobile phone use for learning. These include screen size, monochromatic colour, battery life etc. In addition, there are the larger issues of network capability, ICT infrastructure generally (Sife, Lwoga, $ Sanga, 2007) staff readiness and the availability of spaces which provide conducive environments for learning. We need to remember that the very poor in our world deserve opportunities for learning but that we should not set them up for failure because of unrealistic expectations of their life circumstances. That said, there is enormous potential for social change associated with mobile technologies because, anecdotally at least, mobile devices seem to be ubiquitously accepted. The use of social networking sites such as 'flicker' to alert the rest of the world to events unfolding in Iran following disputed election results there seems to support this notion. There is a sense in which internet enabled mobile devices have the potential for educators to achieve the type of social liberation envisaged by social theorists' such as Paulo Freire (Escobar, Fernandez, Guevara-Niebla, & Freire, 1994).

The Concept of Mobility

The notion of mobility in learning is unique to mobile devices and one that has considerable implications for education in both developed and developing countries. The concept of mobility is one where time, space, pace and place take on new and different qualities to those that characterize non-mobile environments (Katz & Aakhaus, 2002). Mobile devices theoretically enable people to learn anywhere, anytime. They also create unique opportunities for students to communicate within and outside the classroom. Thus, the notion of mobility in relation to mobile technology usage for teaching and learning underpins much of what makes mobile learning different from elearning Kushla-Hulme & Traxler 2007).

While the idea of 'anywhere, anytime' learning can be attractive for educators, the notion of mobility in relation to education can also challenge accepted norms and the conceptions that both teachers and students have about their role/s in teaching and learning (Smyth, in press).

Mobile technologies deliver knowledge and information in ways that challenge formal learning, its institutions and its professionals. (Traxler, 2008, p.5).

Naismith et al (2004) view the concept of mobility as a key issue and one that significantly changes the ways in which people learn and interact with each other and that:

the ability to link to activities in the outside world also provides students with the capability to 'escape' the classroom and engage in activities that do not correspond with either the teacher's agenda or the curriculum (p.4) .

The notion of backchannel communication discussed in the next section is one way in which this mobility and ability to 'escape the classroom' can have both positive and negative impacts on learning and significantly change classroom dynamics. While this observation relates to the developed world there can be little doubt that this also is an issue for education in developing world contexts.

To a large degree, the influence of mobility on education and society in general are still emerging and at this juncture are not necessarily well recognized or understood, However what is understood is that the concept of mobility will have considerable impact on teaching and learning activities (Traxler, 2009). In developing countries in particular, appropriate consideration needs to be given to understanding and managing mobility in a context where mobile technologies could play a major role in facilitating access to online learning for a range of educational requirements and purposes.

The increasing availability of rich media technologies for mobile devices also provides learning opportunities moving away from text reliance, which further extends the possible applications of mobile devices for teaching and learning. The challenge for educators relates to how learning will be facilitated in the various virtual contexts in which students may opt to engage (Smyth, in press).

In relation to staff skills a recent project in the Philippines has highlighted that lack of skill in using technology can have a negative impact on successful integration of technology into teaching and learning activities Natavidad (n.d.).

Back Channel Communications

As outlined previously, the use of mobile technologies for teaching and learning is creating new classroom dynamics generating challenges for educators to manage these in constructive ways. An example of this is back channel communications. Back channel communication is the practice of running an online conversation at the same time as participating in a live presentation.

The research on back channel communication is in its early days. However a recent study by Madrell (2008) into the impact of back channel communication on cognitive load indicates that back channel communication may have both positive and negative effects.

The potential for distraction, split attention, and redundancy effects may indicate back channel interactions place unnecessary extraneous cognitive load on learners. However, findings may also suggest that the back channel interactions directly facilitate learning through more effective and efficient processing of the to-be-learned material (p.18-19).

What is clear is the mobile technologies facilitate backchannel communications that can be either productive or disruptive and that educators need to consider this in designing their teaching and learning activities for mobile devices.

Usage Patterns

Frequent and growing use of mobile technologies to access the internet, social networking and rich media is not confined to the developed world. Work by Kreutzer (2009) in Cape Town in South Africa has found that disadvantaged youth are highly connected to the internet and are making intense use of social networking and other digital technologies available. In the main this is facilitated by mobile devices. This study of 500 year 11 students from very impoverished environments found that the vast majority of these students used mobile devices daily for personal communications and that they made extensive use of mobile internet applications. Interestingly, while almost all students reported accessing mobile devices, 25% did not own a personal handset. Usage amongst these two groups showed little variation in relation to internet access, but considerable variation in relation to personal communication (Kreutzer, 2009). A small group (4%) owned their own sim card and used another person's mobile. Of particular note, almost all respondents used handsets other than their own to access advanced technical features. Downloading songs videos and games, instant messaging and browsing or googling the internet were amongst the most popular applications of mobile phone usage identified by Kreutzer. Significantly this study showed that this group preferred mobiles over computers for a range of activities. While this does not indicate usage by disadvantaged youth across all developing countries it does identify that disadvantaged youth can and do find ways to access the internet for a wide variety of needs and activities.

Familiarity with Technology

There is great deal of variation in the skills that users in both the developing and developed worlds bring to technologically mediated learning environments, be they fixed or mobile technologies. In recent years many educators have taken the stance that the current generation of students entering higher education in the developed world are digital natives (Prensky, 2001) with high levels of access to the internet and heavy users of social networking and other digital technologies. Acceptance of this belief has led to misconceptions about students' skills and resulted in failure to successfully integrate social networking technologies such as blogs and wikis into teaching and learning activities. Work by Kennedy et al (2008, 2007) has disputed this assumption and highlights the variations in the ways in which the 'net gen' use social networking and other digital technologies and provides considerable insights in to how these might be integrated into teaching and learning activities.

Teaching and Learning Approaches for Mobile Learning

Much of the interest in use of mobile technologies for teaching and learning in developing countries lies in their perceived suitability for a wide range of teaching and learning approaches, as well as their increasingly wide ranging availability. Notwithstanding this interest, mobile learning has proven difficult to define with early attempts focusing on the technology and more recent definitions including theorisation of teaching and learning' (Traxler 2009). While less appears known of educational applications in developing countries there is some evidence that mobile technology can cross the digital and distance divide (Donner, 2008).

Characterizing Mobile Learning

Kushla-Hulme and Traxler (2007) outline several characterizations of mobile learning which have strong implications for mobile learning theory, learning design and teaching and learning practice. These characterizations include: ' personal', 'spontaneous', ' opportunistic', 'informal', 'pervasive', 'situated', 'private', 'context aware', 'bite sized' and 'portable' (p.181). These words highlight the distinctiveness of mobile technologies in supporting learning that has 'an emphasis on … informality, movement and context (Kushla-Hulme and Traxler, 2007, p.181)' Considering these characteristics (Kushla-Hulme and Traxler, (2007) suggest that mobile learning is well suited to support learning which is personalised, situated and authentic' (p.181). Rich media technologies also have similar capacities (Smyth, in press, Andrews et al, 2008). This combination offers many possible opportunities for learners in the developing world regardless of context. Such opportunities do, however, have impacts on how both students and teachers view learning and approach their task/s. These need to be considered in the development of new pedagogies intended to support mobile learning to achieve the worthy ideals implied here.

Personalised, Situated, Authentic Learning

Kushla-Hulme and Traxler, (2007) describe personalised learning as learning 'that recognizes diversity, difference and individuality" (p. 184). A popular example of this is museum studies where students record items of interest to them, send text messages containing keywords and construct personalized websites to outline their interests (e.g Mulholland et al 2005), Identifying items of interest in a clinical practice setting and using mobile devices for recording them for later discussion and reflection as part of online and face-to-face activities is another approach (Andrews et al 2009, forthcoming).

Situated learning is learning that occurs within a particular context relating to workplace or vocational activity. Field trips and studies, clinical practice, etc are all examples of this. Rich media technologies added to mobile devices are already being used in remote areas to engage isolated learners and this potential will grow as the capacity for audio and videoconferencing, social networking and spontaneous informal connectivity increase.

Authentic learning is learning that engages the learner in "real world" activities and that students engage in genuine tasks as part of these activities (Kushla-Hulme and Traxler, 2007). Mobile devices are well suited to s support authentic learning through their ability to enable to enable 'content creation, data-capture, location awareness and collaborative working in real-world settings' (Kushla-Hulme and Traxler, 2007, p186.)

There is overlap between these approaches to learning which could well be further expanded and blurred through the application of mobile learning in developing contexts. In fact sociocultural considerations can be complementary to authentic and personalized learning approaches which situate and place control in the learners' hands. Furthermore situated learning could be particularly useful for rural and remote communities.

Linking Mobile Learning to Existing Teaching and Learning Theories

Naismith et al (2004) linked mobile learning to six broad theories of teaching and learning practice: behaviourist; constructivist; situated; collaborative; informal and lifelong learning; learning and teaching support (Naismith et al 2004, pp 2-3).

Table 1. Applications of learning theories to mobile learning activities

Theoretical approach	Activity
Behaviourist learning	quick feedback or reinforcement element, facilitated by mobile devices
Constructivist learning	immersive experiences such as those provided by mobile investigations or games
Collaborative learning	mobile devices provide a handy additional means of communication and a portable means of electronic information sharing
Informal and Lifelong learning	mobile devices accompany users in their everyday experiences and become a convenient source of information or means of communication that assists with learning, or records it on the go for future consultation
Learning and teaching support	availability of mobile technologies at all times for monitoring attendance or progress, checking schedules and dates, reviewing and managing, activities that teachers and learners engage in at numerous times during the day

Adapted from Traxler, (2009, p.4).

Traxler, (2009) provides an excellent summary of examples of what applications of these theories might look like in practice.

These theories also relate to the concepts of personalised, situated, authentic and informal learning outlined earlier.

Implications for Practice in Developing Countries

There is a significant opportunity for creativity and innovation in pedagogy with the rapid increase in usage of mobile technologies. However, Motlik (2008) gives us a stern warning about elearning implementation which we should heed as this discussion unfolds:

In Korea and China, Web-based learning has been plagued by poor instructional design, poor instructor training, poor matching of technology with learner skills, a lack of technology, and a lack of accessibility. Korea is one of the most wired nations in the world and Internet-based learning has been strongly supported in China; so the inability of these nations to develop efficient elearning methods should act as a deterrent for the developing nations considering a move to Web-based learning (p.5)

This view reinforces the notion that it is important to resist the temptation to fit old pedagogies to new technologies, despite the lag before innovations become generalized (Smyth, 2005). However, the opportunity to shed the old and be creative should be grasped for the benefit of improved student outcomes. If we acknowledge the place of planned learner-content interactions typical of current online practice and grasp the potential for planned and serendipitous learner-learner interactions now possible with mobile devices (Smyth, in press), we have the opportunity to enhance student cantered pedagogies beyond the scope limited by current mass education techniques. Furthermore there is a requirement to ensure socio-contextual issues are carefully investigated in any considerations. The application developed-world pedagogy for example should be questioned.

As connectivity improves across the globe, accessibility issues will diminish rapidly, providing greater opportunities for learning designers to assume less responsibility for planning person-to-person interaction (tutorials, discussion boards etc) since learners will have greater opportunity to initiate such communications independently of the teacher's control. This is increasingly obvious as young people especially create ways new ways of communicating. It will be important that learning is increasingly situated in spaces mobile or otherwise which are contextually driven. This will challenge long held notions of the nature of pedagogy and the roles of teacher and student and will require support from a range of stakeholders.

The role of the teacher will change towards increasing facilitation for independent learning and student-directed interaction. The role of the student as an independent director of own personal learning will well be beyond the experience of many learners in the developing world who will need guidance and support via transitional learning strategies. Interaction between both content and peers will support learning through teachers' decisions about the accessible resources in content repositories, stimulus to peer-to-peer interaction and opportunity for serendipity to encourage learning as we consider how to incorporate and preface newer mobile and video communications in learning design.

Forecasting Issues for Higher Education Institutions Adopting Mobile Technology and Rich Media

An Australian project which is focused on rich media technologies has been monitoring their implementation across universities there and has potential to provide some insight into the major issues arising as these technologies, including mobile devices, roll out in higher education globally.

The intent of the Leading Rich Media Project1 is to contribute to improved teaching and learning outcomes and increased institutional efficiency and effectiveness by investigating current policy and practice across the Australian Higher Education Sector. No comprehensive investigation of rich-media technologies which integrates their pedagogical and administrative uses, sustainability, viability and scalability across the sector has been published for Australia and there is a similar gap for many developing nations.

The project methodology is a mixed method one, comprising online surveying, case studies and comparative analysis of international trends and business plans from governmental and corporate users. At the point of initial data collection, use of mobile technologies was not widespread but the rapid expansion of these and social networking technologies indicates that the project now needs to address these specifically.

From our data, it appeared that rich media technologies were being used for learning and teaching, administration and research in almost equal proportions with most institutions indicating that they were considering pedagogical models in their decision making about rich media technologies. Responsibility for rich media technologies was spread across many directorates but in almost half of all institutions primary responsibility for maintenance, bookings policy development and facilities was held by the Information Technology Directorate. This is unsurprising given the digital nature of the technologies and their place in many institutions as core operating technologies. There was little evidence of strategic planning or policy making specific to the implementation of rich media technologies as separate from ICT infrastructure because rich media technologies were seen as enabling or support technologies and mentioned in teaching and learning or other strategic plans but not in ICT plans. Thus the initial analysis of data indicated a strategy and policy void. This provides an opportunity for implementers in developing nations to learn from our evidence and to be creative adopters, pressuring for policy and strategy development to underpin the roll out of technology.

The drivers for the adoption of rich media technologies flagged by respondents indicated that campus growth, reduction in travel budgets, staff and student demand were driving the adoption and proliferation of these technologies. This is supported by the analysis of data concerning the funding of rich media technologies. Respondents reported that initial purchase is predominantly from short term sources (government grant/project funds) with upgrade or replacement and maintenance from university and project funds. Staff were supported in their use/adoption of rich media technologies by professional development workshops and some seed funding but most respondents reported that ongoing financial support was the greatest barrier to adoption. For mobile technologies, we see student demand becoming the predominant driver with its inherent implications for upgrading of institutional infrastructure, depth of staff training and stimulus for development of creative pedagogies. This is predicted in our data which shows that the primary preconditions required to support the implementation of rich media technologies were network infrastructure, staff awareness and preparedness, provision of technical support and innovation in pedagogy. Since ongoing plans for funding rich media technologies were not well known amongst respondents this adds support to the conclusion that coherent institutional policies about implementation had not been formulated but were necessary.

Overall, the researchers concluded, from their initial data, that adoption of rich media technologies is still in its infancy and as yet its full potential to support the business of universities is unknown and therefore, unrealized. We suspect that this is not an uncommon problem globally and in particular where funding is project or grant based and therefore unable to sustain longer term maintenance, upgrade and broad-scale rollout issues. These too, will be significant issues for implementers in developing countries where funding relies on donor sources.

A second online survey was conducted with Chief Operating Officers from 13 universities who responded to the financial survey that focused on core linkages between their university's strategic business goals for campus wide teaching and learning and the degree of integration of the business planning processes adopted by them for the acquisition of rich media technologies and lifecycle support services to underpin or assist their university's delivery of teaching and learning. In particular, this online survey was seeking to uncover incidents of "Best Practice" in this vital element of fiscal policy given the increasing expenditure demand and the need to consider sustainability, viability and scalability when planning such acquisitions.

Although only three of the thirteen respondents acknowledged that they had a master plan for campus wide implementation of rich media technologies, there was a strong correlation between this group and the maturity and integration of the business planning processes that directly link to their university's strategic goals for teaching and learning. This group had formal processes for investment evaluation and funding the acquisition and life cycle support services and all respondents from this group stated that they their plans incorporate the concept of unifying communications and that their plans were underway. The potential here is that these three institutions will become lighthouses for other adopters to imitate, including those in developing countries.

There was a mixed response from the remaining respondents to the question of incorporating the concept of unifying communications which would include mobile devices. One confirmed that they had plans and were underway; one indicated that this had been identified as a business requirement; another had intentions to implement and the remaining seven respondents had no plans. This evidence confirms that the general data and conclusions drawn to date are equally applicable to emerging applications such as mlearning technologies.

Implications for Developing Countries

The risk for early adopters in tertiary institutions is that use of rich media technology particularly will become unsustainable into the long term without recurrent funding especially where infrastructure development, capacity and deployment need improvement. This represents a considerable concern for developing countries but also provides great opportunities for creative linkages with telecommunications providers, particularly in respect of dedicated network and satellite access for distance students. Of similar concern is the lag time in developing appropriate pedagogies which could cause ill-preparedness of teachers to impact negatively on student experiences and learning outcomes. Partly, rich media technologies provide greater opportunities for student centered pedagogies which may not be well aligned to cultural and institutional milieu where social interaction for learning is not a feature of existing pedagogies. The potential of rich media technologies to aid business processes through cost savings and aid teaching through scalability to increase individualized connectivity is acknowledged as providing value that can be planned for. Sustainability and viability of rich media technology appear to be reliant on explicit planning derived from institutional strategic plans and robust business processes which enable effective implementation into the longer term. These conclusions are consistent with the vast body of literature concerning the successful achievement of significant educational change so colleagues in developing countries would be advised to consider this wisdom in their deliberations.

Future Directions

The area of open source development for mobile phone usage is in its infancy. However this is an important area of development which could provide many opportunities for developing countries in terms of availability of software applications for mobile phone usage. Similarly, the development of appropriate content and pedagogies that are context relevant for educational purposes in developing countries provide opportunities for innovation and leadership from developing countries.

The ongoing development of phones with touch screens, voice recognition technology, improved processing ability and longer battery life also offers many opportunities for expanded educational applications in the near to mid future. Along with this the rapid expansion of access to mobile devices provided by entrepreneurs in developing nations means that the use of mobile devices for teaching and learning holds much more promise for real outcomes than other technologies and developments in this should be continued to be pursued and researched in relation to supporting teaching and learning in developing countries, hopefully with the aid of lessons learned by other early adopters.

REFERENCES

Aderinoye, R. A., Ojokheta, K. O., & Olojede, A. A. (2007). Integrating Mobile Learning into Nomadic Education Programmes in Nigeria: Issues and perspectives. *International Review of Research in Open and Distance Learning*, *8*(2), 1–17.

Ally, M. (2007). Guest Editorial: Mobile Learning. *International Review of Research in Open and Distance Learning*, *8*(2), 1–4.

Andrews, T., Smyth, R., Tynan, B., Vale, D., & Caladine, R. (2008). Rich Media Technologies and uncertain futures- developing sustainable, scalable models. In *Proceedings ASCILITE 2008 Hello! Where are you in the landscape of Educational Technology*, Melbourne November 30-December 3.

Baecker, R. (2003). *A principled design for scalable internet visual communications with rich media, interactivity, and structured archives.* Paper presented at the Proceedings of the 2003 conference of the Centre for Advanced Studies on Collaborative research Toronto, Ontario, Canada.

Caladine, R., Andrews, T., Tynan, B., Smyth, R., & Vale, D. (in press). New communications Options: a renaissance in IP use . In Veletsianos, G. (Ed.), *Using Emerging Technologies in Distance Education*. Athabasca University Press.

Donner, J. (2008). Research Approaches to Mobile Use in the Developing World: A Review of the Literature. *The Information Society*, *24*(3), 140–159. doi:10.1080/01972240802019970

Escobar, M., & Fernandez, A. L. (1994). *Paulo Freire on Higher Education: A Dialogue at the National University of Mexico*. New York: State University of New York Press.

Esselaar, S., & Stork, C. (2005). Mobile cellular telephone: Fixed-line substitution in sub-Saharan Africa. *South African Journal of Information and Communication*, *6*, 64–73.

Fraunholz, B., & Unnithan, C. (2004). Critical Success factors in mobile communications - a comparative roadmap for Germany and India. *International Journal of Mobile Communications*, *2*(1), 87–101. doi:10.1504/IJMC.2004.004489

Hodge, J. (2005). Tariff structures and access substitution of mobile cellular for fixed-line in South Africa. *Telecommunications Policy*, *29*(7), 493–505. doi:10.1016/j.telpol.2005.05.001

Ibarguen, G. (2003). Liberating the radio spectrum in Guatemala. *Telecommunications Policy*, *27*, 543–554. doi:10.1016/S0308-5961(03)00050-8

Katz, J., & Aakhus, M. (2002). *Perpetual contact: Mobile communication, private talk, public performance*. Cambridge: Cambridge University Press. doi:10.1017/CBO9780511489471

Kim, S. H., Mims, C., & Holmes, K. P. (2006). An introduction to current trends and benefits of mobile wireless technology use in higher education. *AACE Journal*, *14*(1), 77–100.

Kreutzer, T. (2009). *Generation mobile: online and digital media usage on mobile phones among low-income urban youth in South Africa*. Cape Town: University of Cape Town.

Kukulska-Hulme, A., & Traxler, J. (Eds.). (2005). *Mobile Learning: A Handbook for Educators and Trainers* (F. Lockwood, Ed.). London: Routledge.

Kushla-Hulme, A., & Traxler, J. (2007). Designing for mobile and wireless learning . In Beetham, H., & Sharpe, R. (Eds.), *Rethinking Pedagogy for a digital age: Designing and delivering e-learning*. London: Routledge.

Lehr, D. (2007). *Going Wireless for Development: How Mobile Devices are transforming development at the base of the pyramid*. Working paper. Acumen Fund. Retreived July 5, 2009 from Dialling for Development_Working paper_jG2w63Q2.pdf.

MacDonald, B. (2009). Appropriate Distance Education Media in the Philippines and Mongolia. *International Review of Research in Open and Distance Learning*, *10*(2).

Maddrell, J. (2008). *The effective of backchannel interactions on cognitive load*. Research paper. Retrieved July 2, 2009, from http://www.scribd.com/doc/3727549/Backchannel-Research-Paper-Jennifer-Maddrell.

Miller, D. (2007). Mobile Phones and Development: The future in new hands? iD21 insights. Retrieved May 20, 2009, from www.id21.org/insights/insights69/insights69.pdf

Motlik, S. (2008). Mobile Learning in Developing Nations. *International Review of Research in Open and Distance Learning*, *9*(2), 1–7.

Mulholland, P., Collins, T., & Zdrahal, Z. (2005). Bletchley Park Text: Using mobile and semantic web technologies to support the post-visit use of online museum resources. *Journal of Interactive Media in Education*. Retrieved June 28, 2009 from http://jime.open.ac.uk/2005/24.

Naismith, L., Lonsdale, P., Vavoula, G., & Sharples, M. (2004). *Literature Review in Mobile Technologies and Learning. Future lab Series, Report 11*. University of Birmingham.

Natavidad, Sc. D. (n.d.) Summative evaluation of The ELSA text2teach project: Final report. Retrieved July 2, 2009, from pdf.usaid.gov/pdf_docs/PDACK830.pdf.

Prensky, M. (2001). Digital Natives, Digital Immigrants. *On the Horizon, 9*(5). Retrieved May, 25 2009, from http://www.marcprensky.com/writing/default.asp

Sife, A. S., Lwoga, E. T., & Sanga, C. (2007). New technologies for teaching and learning: Challenges for higher learning institutions in developing countries. *International Journal of Education and Development using Information and Communication Technology, 3*(2), 57-67.

Smyth, G. (2006) Wireless Technologies Bridging the digital divide in Education. In *Proceedings, 4th World Conference on mLearning - Cape Town, South Africa*, 25-28 October 2005.

Smyth, R. (2005). Broadband videoconferencing as a tool for learner-centred distance learning in higher education. *British Journal of Educational Technology*, *36*(5), 805–820. doi:10.1111/j.1467-8535.2005.00499.x

Smyth, R. (in press). Enhancing learner-learner interaction using video communications in higher education: Implications from theorising about a new model. *British Journal of Educational Technology*.

Traxler, J. (2008). *Mobility, Modernity and Development*. Paper presented at first m4d Karlstadt, Sweden. Retrieved July 2, 2009, from http://wlv/academia.edu/JohnTraxler/Papaers/83101/Mobility--Modernity--Development

Traxler, J. (2009). Learning in a Mobile Age. *International Journal of Mobile and Blended Learning*, *1*(1).

ENDNOTE

[1] Funded by ALTC, the Australian Learning and Teaching Council.

Chapter 9
The Role of M-Government in Western China Development

Jesper Schlæger
University of Copenhagen, Denmark

ABSTRACT

This chapter explores the role of m-government in Western China development. Bureaucracy has been identified as a barrier for economic development in Western China. It is discussed how applications of m-government embedded in a larger reform agenda has reduced some of the growth stifling effects of bureaucracy. This chapter describes the content of ideas, institutions, and technologies of m-government to understand how fitting these levels has led to such an improvement in governance. Recommendations are made concerning continued emphasis on simple applications, rethinking government organisation, and increased discussion of the role of private service providers.

INTRODUCTION

This chapter explores the role of mobile government in Western China development. Mobile government, or m-government, is understood broadly as the use of mobile information and communication technologies (ICT) in the public administration to change government working procedures. The scarce research available on m-government in China emphasises the opportunities in service provision and organisational innovation created by mobile ICT (Li, 2005; Fan & Li, 2006; Song & Cornford, 2006), and political practice is reflecting this enthusiasm.

China is a one-party authoritarian state keeping control with development of critical infrastructure such as ICT. Government use of ICT is thus important, as government has historically been the first, and necessary, mover in the spread of new technologies. And the government has moved. Use of ICT as communication platform in governance, informatisation, has become a corner stone of economic and administrative reforms clearly reflected in central policy documents (Central Government, 2006). Informatisation is perceived as intrinsic to modernisation and development

DOI: 10.4018/978-1-61692-818-6.ch009

processes, and large investments have been made to build ICT infrastructure with the long term objective of maintaining momentum in China's unprecedented economic growth.

M-government applications entered the Chinese scene around 2003, when the government used SMS (Short Message Service) to communicate with people during the SARS (Severe Acute Respiratory Syndrome) outbreak. The short time-span considered, m-government has already seen some important applications, primarily in the more developed coastal provinces. Large cities such as Beijing, Shanghai, and Guangzhou have implemented urban management networks that imply fundamental changes in both ways of thinking and doing government (Song & Cornford, 2006). In rural areas things have also changed, even if on a more modest scale. SMS services such as weather forecasts and news about agricultural technology have helped farmers optimise their production and reduce waste of crops resulting from unforeseen weather changes. These m-government applications do not stand alone, but come in a multi channel package with other ICT mediated government services such as radio and TV broadcasts and telephone hotlines.

The economic growth in China has been unevenly distributed. Coastal China has seen impressive growth rates, while Western China has been left behind. Policies have been implemented to enhance development of Western China, and new ICT adds important dynamics to these policies. Researchers have been discussing the "e-transformation" of Western China referring to the use of new ICT to promote more flexible markets and enhance economic development potential of the region (Davison, et al., 2005). One of the barriers to development they found was that "bureaucracy can be stifling" (Davison, et al., 2005: 65). This chapter takes up the discussion of how bureaucracy has been made more service oriented and less stifling through the application of m-government solutions, thereby diminishing a barrier for development in Western China.

The structure of this chapter is as follows: First, a digital governance analytical framework is presented. Then, the analytical background of Western China is described. Following, the technologies, ideas, and institutions shaping the context for m-government are analysed. In each part it is discussed how fit between these levels can be attained. On this basis questions for further research are raised. Finally, conclusions on the role of m-government in Western China development are made.

DIGITAL GOVERNANCE MODEL

In this section, we discuss how to understand the role of mobile ICT in processes of institutional change. The theoretical challenge is to capture the importance of ICT without claiming that application of a particular technology will lead to a particular outcome. This can be achieved by constructing a theory that focuses on how the interplay between ideas, institutions, and technologies shapes the opportunity structures for powerful actors. Below, such a theoretical model will be proposed. We take the point of departure in technologies.

An apparent feature of m-government is the use of a particular set of technologies namely mobile ICT. Technology is "the use of scientific knowledge to specify ways of doing things in a reproducible manner" and the artefacts used in the process (Castells, 2000: 28). In this definition technology encompasses technical infrastructure such as e.g. mobile handsets, and transmitting devices, as well as the specialised knowledge about how mobile communication is achieved exemplified by technical knowledge held by engineers and technical personnel of government and in particular in telecommunication companies.

A theory of m-government should lead to an understanding of the mobile technologies. On the other hand, focusing solely on the technological component would likely lead to the technological determinism mentioned in the beginning of this section, claiming that a particular technology leads to particular social outcomes. As is well established, this is not the case (see e.g. Bijker, 1995). Instead, technologies are shaped by powerful actors to suit their interests or beliefs (Danziger et al., 1982).

When discussing m-government in Western China a powerful actor may be a public servant in a key position in the bureaucracy. Being a powerful actor does not allow complete freedom to act, though. In a bureaucracy, the freedom to act according to own interests and beliefs is constrained by the rules and practices within the organisation (March & Olsen, 1989). So, technologies constrain and enable action, and so do the rules and practices. We will call these rules and practices institutions and add them as a second element to build a theory.

More specifically, institutions are "...relatively enduring collection of rules and organized practices, embedded in structures of meaning and resources that are relatively invariant in the face of turnover of individuals and relatively resilient to the idiosyncratic preferences and expectations of individuals and changing external circumstances" (March and Olsen, 2006: 3). In other words, institutions are "sticky". Nonetheless, institutions can change even if the process of institutional change is often slow and incremental. It is the roots of this process of change that we are interested in finding, as it will allow identification of handles that can be pulled to change institutionalised misbehaviour in bureaucracy.

Does the theory constructed so far allow us to understand how rules and practices can change to allow development? Availability of new technologies can put a pressure on the usual way of doing things e.g. if they promise better efficiency. The technological level can thus take us some of the way towards understanding such changes. Nevertheless, new institutional theory (Peters, 2005) also points to another important enabler of institutional change that should not be overlooked, namely ideas.

New ideas, "taken-for-granted assumptions and articulated concepts and theories" (Campbell, 2004: 93), also under some conditions enable institutional change. For instance an idea of marketisation may enable bureaucracy to outsource government tasks to private service providers if it is accepted among the actors. So, to explain institutional changes our theoretical model must be able to account for the role of changes in ideas as well as institutions and technologies (Hoff, 2000).

Observations have lead us to believe that institutional changes fail to take place when there is a lack-of-fit between these three levels. For instance, failing to perceive the opportunities created by mobile ICT will impede innovative uses thereof. Similarly, even the most innovative ideas can falter on the grounds of lacking technological infrastructure.

Turning this observation around leads to a claim, that fitting ideas, institutions, and technologies will create the basis for institutional change. In the analysis that follows, we will show how the Chinese government in many cases has succeeded in fitting the levels. This has allowed government to create new ways of communication with citizens and organising working procedures in new ways. At the same time, aspects of m-government application have been less successful. The present analysis points to the particular lacks-of-fit existing between ideas, institutions, and technologies as a means to understanding why things are not changing in the way intended by government.

In the end opportunity structures created by fits and lacks-of-fit between the three levels are exploited by powerful actors. And as the following analysis will show, it is often in the last instance the interests and beliefs of these actors that lead the change process in a particular direction.

Together these theoretical elements provide a foundation for a nuanced analysis of both the institutional environment for m-government development, the interplay between the opportunities generated by new mobile ICT, actors' ideas about how to exploit them, and the sticky institutions the actors are at the same time embedded in and try to change.

ANALYSING MOBILE GOVERNMENT IN WESTERN CHINA

Western China is the least developed region of the country and merits attention concerning development issues, not least pertaining to the role of m-government in development processes. Western China, following the official geopolitical division, covers six provinces: Gansu, Guizhou, Qinghai, Shaanxi, Sichuan, Yunnan; five autonomous regions: Guangxi, Inner Mongolia, Ningxia, Tibet, Xinjiang; and one municipality: Chongqing. For purpose of presentation these administrative units will all be called provinces. Institutional differences that apply to the degree of local self-governance are assumed to be insignificant regarding m-government development, at least at the current level of development.

Neither culturally nor economically is Western China a homogeneous unit. There are large minorities of Muslim and Tibetan people, and large relatively well developed cities are interspersed with some of China's poorest rural areas. It is in these rural areas that the need for economic development is most acute. Development policy is often directed towards promoting industries in the slightly more developed villages and counties to absorb workers from rural areas. Through reform of land ownership the rural areas are attempted turned into more profitable and efficient large scale agricultural businesses. Redundant labour is then expected to seek employment in the cities.

It is in the midst of such large scale processes of urbanisation and modernisation that new ICT adds dynamics worth examining further. The present chapter attempts to analyse the development of m-government in the intersection between processes of administrative reform, informatisation, and economic development policies. This is attained by an interpretation of the role of m-government in the provincial informatisation plans. These documents are interpreted through the lens of policy documents and academic articles describing the creation of a "service oriented government". Service oriented government is the official key concept for administrative reform and as such the institutional background to which all other policies with an influence on administrative structures must relate and which they must support. As this chapter will show interpretation of this concept is fluid, and support has taken multiple forms.

In the following we will apply the digital governance framework presented above to analyse the role of m-government in Western China development. First, we describe actual availability of mobile ICT in Western China, and discuss the asymmetry of technological know-how existing between private service providers and the public administration. Then, we look at ideas of m-government to understand official accounts of what m-government is, why it should be used, and how it is going to be implemented in the context of Western China. Finally, we turn to the institutional side of the matter and describe the organisations built to develop m-government, and analyse the impacts of this way of organisation for crafting informatisation policy.

Technologies of M-Government

Under the heading of technologies we will following examine the availability of technological infrastructure in the shape of mobile ICT. This is important for understanding what can be achieved on the ground concerning m-government use. Next, we discuss the role of technological knowledge in m-government development and the fact that most of this knowledge rests with private service providers. Something, which may be a problem from the point of view of government control, but which may also be positive in the sense that innovation can be spurred by market forces.

Table 1. Mobile phones and internet users in Western China

Province	Mobile phones pr. 100 persons 2007	Mobile phones pr. 100 persons 2008	% Internet users 2008
Gansu	26.3	34.2	12.5
Guizhou	22.2	22.7	11.5
Qinghai	40.5	44.8	23.6
Shaanxi	43.2	51	21.1
Sichuan	29.4	35.1	13.6
Yunnan	30	36.2	12.2
Guangxi	29.3	34.1	15.4
Inner Mongolia	43.7	55.9	16
Ningxia	44.7	53	23.6
Tibet	26.2	30.6	16.4
Xinjiang	39.4	50.2	27.1
Chongqing	41.9	45.5	21.2

Sources: http://www.miit.gov.cn/n11293472/n11295057/n11298508/11979497.html
http://www.cnnic.cn/uploadfiles/pdf/2009/3/23/153540.pdf
http://www.gzca.gov.cn/news/20080429/20080429114221489261763_0.html

Availability of Mobile ICT

A necessary precondition for the use of m-government is the availability of mobile ICT, in particular mobile phones. Western China is sparsely populated and in some areas spread of ICT infrastructure is thus expensive if calculated on a per capita basis. This has to some extent been offset by government policies and investment in network infrastructure. In the following Table 1 the mobile phone density is described for the years 2007 and 2008 to give an image of the dispersion and growth rate of mobile phones. These numbers are also compared to internet users. This illustrates the much larger spread of mobile phones and consequently the enhanced importance that this medium can be ascribed in particular regarding government-to-citizen interaction.

The statistics presented above only accounts for absolute numbers of mobile phones. It does not discern city from rural areas. The author's observations from visits in the countryside in Shaanxi, Ningxia, Gansu, and Yunnan during 2008 is that mobile phones are readily available in even remote places. This is of course only indicative, as no statistical data collection was pursued. The numbers of mobile phones per 100 persons in 2008 also exhibit large differences from 23 in Guizhou to 56 in Inner Mongolia. Concerning the internet use rates, it should be noted, that they refer to internet use within the last half year. So, the rate does not reflect how many people have access to internet in their homes or use internet regularly.

From the figures it should be evident that the internet based services are still not usable as a main channel for government-citizen interaction whereas mobile phones have reached penetration rates making it attractive as a communication channel.

M-government is developed through market based dynamics, leaving the build-up to the three largest mobile service providers, China Mobile, China Unicom, and China Telecom. This contrasts with government acting as driver of internet development in China. Internet build-up has taken place as a centrally co-ordinated strategy built around twelve so-called "Golden Programs" (Holliday

& Yep, 2005) targeted on implementation of different types of internet infrastructure. Concerning m-government a question remains whether the decentralised and marketised infrastructure build-up will impede a friction-less functioning of future m-government applications, or whether creation of networks with different 3G communication standards such as CDMA2000, WCDMA, and the Chinese 3G standard TD-SCDMA will lead to compatibility problems and the need for different m-government platforms.

Availability of technologies described above is not always taken into account by policy makers. In official Chinese discussions about the potentials of m-government it is 3G solutions that dominate the agenda. Yet, such solutions are at the time of writing only available in the large cities in Western China, and only on a test basis. Speeds of network development continue to amaze, and can also readily be stimulated by government investment, but it will take years before 3G use has become widespread in rural areas of Western China. Furthermore, not only network availability but also the price of handsets to use the network is an important factor that influences use rates. Handsets are prohibitively expensive for most rural residents at the current stage. This creates a lack-of-fit between e.g. an idea that advanced 3G solutions are best, and the fact that the networks are not available and handsets too expensive to buy for the intended users in rural areas. Finally, the technological skills of rural residents are still limited, so complex applications have not had a large uptake so far. In the mean time, as will be argued below in relation to concrete examples from the provinces, it is beneficial to focus attention on how to optimise simple 2G information services based on GSM technology, in particular SMS based services.

Based on the author's observations in Ningxia in 2008 rural residents are hesitant to use even SMS based services requiring interaction. Remedies for this include education of rural habitants, already a part of government policy, and readjustment of research to include focus on simple 2G applications. Research on and user-friendly 3G applications can also be pursued bearing in mind that uptake cannot happen before networks have been established, and handsets as well as network fees decline to a level where a substantial part of the intended users can afford it.

There are on the other hand also examples of government being aware of the technological limitations. Some policies reflect a nuanced multiple channel approach taking into account the disparities in access to ICT. Reflections of such an approach to rural informatisation can be found in the Ministry of Agriculture's electronic unification program (san dian he yi) referring to TV, computer, and phone (Yunnan, 2009). Users can watch a menu on their TV containing points about e.g. job vacancies, weather forecast, agricultural news, agricultural product prices, etc. The information can then be accessed either by interactive control through the TV receiver or by dialling a phone number on the screen from a mobile or fixed phone. The same service can also be accessed from an internet connected computer. Finally, the information can be sent as SMS to a mobile phone. This kind of public service provision is new in China, because it is more personalised than previous modes of service delivery as the mobile phone is exactly mobile. Information endpoint thus follows the user, and services are interactive in the sense that the user can require certain information, providing a personalised information solution. An example of fitting the institutions to available technologies to achieve development.

Technological Know-How and Private Service Providers

As described above, in contrast to the spread of the internet which was very much government driven, the mobile networks are spread out on a commercial basis. This could mean a clear cut profit orientation of network build-up. Notably, this is not the case as the large telecoms are strongly influenced by the state. Network coverage thus reaches places that are not necessarily interesting from a profit point of view. The large telecommunication companies used to be public, so they are in a grey zone with informal obligations to provide special services for the public sector. Furthermore, as is also the case in the large former state-owned-enterprises, the top-executives in the telecom sector have to be approved by the Communist Party. This provides the state with a potential control lever. The institutions through which the government is able to control the sector and network development would be an interesting field to study further (Lu & Weber, 2007).

The definition of technology included the technological know-how. Not only the know-how and actual implementation of networks but also the the know-how of m-government to a large extent lies outside government. A low level of in-house expertise in government produces a potential problem of oversight, because many m-government related problems and solutions require an understanding of the technical aspects. A common problem in China in general, and Western China in particular, is the lack of qualified technical personnel to take care of development and practical implementation of informatisation programs (Qiang, 2007). Lacking the technical expertise effectively disables parts of the oversight capacity and subsequently also influences government capacity to macro co-ordinate m-government implementation.

Government is taking a more active role in strengthening its in-house capacity for m-government development. This is a result of government realising that framing of m-government has to some extent been captured by commercial interests. Considering international comparison private service providers set free to create public value seems as a dubious long term solution (Dunleavy et al., 2004). The remedies for this imbalance is basic education in ICT of local level public servants, which is also currently undertaken by the government, but it meets constraints in terms of lack of instructors.

Seen from another point of view, the commercialisation of mobile networks has benefits in enabling technological know-how to flow from Coastal China and inland through knowledge exchange within the companies. There is a profit motive for the service providers to provide competitive and innovative solutions to governments in Western China. To ensure that they maintain a competitive edge, telecommunication companies have an incitement to bring new technologies to Western China. The pace of this transfer of technological know-how may be faster than government in-house expertise would guarantee, as the companies need to be innovative to get the government deals and make a profit, whereas incitements for government in-house experts may be less direct. Here the question is, how to balance the fit between organising with or without in-house expertise with incitement structures to ensure continued technology innovation. As should also be clear, this is connected to which ideas lie behind, whether one believes in market incitements, a benevolent government, both, or neither.

Ideas of Service Oriented Mobile Government

Within government there is a strong belief in markets but also in a benevolent service oriented government administration. Below we will analyse the role of these ideas in shaping opportunity structures for m-government development. Before we do that, we need to clarify, what m-government development means: What is m-government seen from the point of view of the Chinese government?

Reading government informatisation plans from the Western provinces reveals that the concept of m-government covers a series of related empirical phenomena such as mobile government (yidong zhengwu) and wireless government (wuxian zhengwu) referring primarily to the services provided. This parallels the official use of the concept e-government (dianzi zhengwu) which refers primarily to service provision through government internet websites. More general, but related, concepts include wireless cities (wuxian chengshi), digital cities (shuzi chengshi), and digital districts (shuzi shequ) which all refer not only to government applications but also to societal and commercial uses of mobile ICT. Furthermore, these concepts focus on provision of a diversity of mobile TV, mobile internet and so on (Shaanxi, 2009). The last group of phenomena provides the infrastructural preconditions for m-government.

M-Government and Development

M-government alone is not by the government envisioned to create growth or remove barriers for development. Even in the provincial informatisation programs m-government is often not explicitly mentioned. Mobile (yidong) or wireless (wuxian) is mentioned accompanying discussions of basic infrastructure build-up referring to the networks and not to concrete m-government applications. M-government is primarily conceived as a means to improve public service provision. There is no direct link in the government policies between service and development aspects. Indirectly there are some links because more flexible procedures for business approval applications, SMS weather warnings, market information, and crop disease control all potentially lead to economic development by reducing waste.

For instance, there is a clear benefit in providing SMS information about which documents are needed for the processing of a particular application to avoid citizens going in vain to the administration. Moreover a notification when the final approval can be picked up from the administration. In rural areas the reduction of numbers of visits to the administration at different levels can save substantial amounts of time and money, as transportation is often inconvenient with little or no public transportation available. This kind of services have been adopted in more places in Western China as part of a larger package of e-government related administrative reforms including internet based office automation. The m-government solutions are thus interwoven with the computer systems used for e-government.

The above mentioned case exemplifies how a fit between an idea of service orientation, the available technologies in shape of internet and mobile phones, and actual institutional changes in the shape of new ICT mediated administrative practices have produced a solution conducive of development by reducing waste. The close connection between e-government and m-government is not without challenges, though. In particular, there is a risk of overlooking the simple applications described above.

M-government is namely often perceived in official documents as something put on top of e-government. It echoes definitions often used in the academic literature that m-government is "an extension or supplement of e-government" (Trimi & Sheng, 2008: 54). This idea puts a premium on developing 3G m-government solutions as m-government is supposed to be something better, more advanced than e-government. This idea presents a lack-of-fit with actually available technology in Western China. Solutions based on 3G applications are as a rule complex and present difficulties to use for less educated citizens in the region. On top of that, the costs of network access and mobile handsets are prohibitively high for average users in rural areas as described above.

Differences between ideas of m-government in the different case provinces apply to inclusionary aspects of informatisation and the role of private service providers. For instance, accessibility options are only mentioned in some of the informatisation programs, referring to e-government programs in general, and reducing the digital divide is explicitly mentioned as a purpose of informatisation (Shaanxi, 2009). Private service providers play a large role in some programs such as e.g. the one of Ningxia (Zhang, 2005). Such difference in ideas also lead to a need for different institutionalisations in form of policies and laws to support the direction and organisations to implement these formal institutions. Differences are also found in the choice of either informatisation office or informatisation bureau, and in the local choice of drafting informatisation plans or simply developing the programs on a piecemeal basis. We will return to a more detailed discussion of these points in the section on institutions.

The imprecise policy formulation means that implementation of programs are left to the operational level. That is not a problem if general directions are clear and there is close co-ordination between administrative levels. As mentioned, in Western China most implementation of m-government on the operational level is performed by private service providers that albeit to some extent responsible to government and susceptible to government control and sanctions are still not formally are part of the administrative apparatus. This effectively means that the responsibility for not only m-government implementation but also m-government development and innovation is outsourced. Mechanisms for macro co-ordination remain unclear. The role of in-house experts frequently participating in central policy co-ordination meetings could be a possible solution. On the other hand, it is not impossible to imagine, that part of these would be susceptible to incitements from private service providers interested in keeping their market share and doing business as usual. Clear, standardised regulations of the relations between government and private service providers are under all circumstances desirable to maintain transparency in the administration.

Service Oriented Government

Chinese public administration has been undergoing reform, adapting to the changing requirements from a society fundamentally marked by the last 30 years of economic reforms creating unprecedented economic growth. Many of the ideas for reform have been imported from abroad, but in most cases adapted to suit Chinese conditions. This is also the case of a concept which in the last few years has become the core guideline for administrative reform, namely: "service oriented government" (fuwuxing zhengfu).

The ideas behind the concept of "service oriented government" are identical to the ideas behind New Public Management, which is the well known ideological basis of "reinventing government" in the OECD countries (Hood, 1991). Consequently, there is a drive to make government lean, effective, and responsive to "customer citizens". In the Chinese context this implies openness in the administration, reduction of corruption, and reduction of abuse of administrative power. In central government policy an interpretation very close to the Western concept is visible, possibly because of many of the academics and public servants influencing the formulation of policy have received education in the OECD countries, or look to these for inspiration.

In some provinces, e.g. Ningxia, a slightly different focus can also be observed. The service orientation is not in the first hand aimed at reforming existing government tasks but just as much to provide new services to citizens. A selection criterion for these services is whether the program can finance itself. In other words, if there is a business-case for the program. This creates opportunities for public service innovations, and not only for the establishment but also for the continued working of the programs, as the program will have a revenue flow covering the running costs: "...it will not necessarily make a lot of profit, but the business model will at least provide for its own operation and development." (Zhang, 2005, author's translation). It is a rethinking of the role of private actors in public service provision. Private companies are invited to create services for the public administration in a fashion inspired by American cases where providers are allowed to charge a small fee for the service (Zhang, 2005).

Promising as it is, this approach to creating a service oriented government also entails pitfalls. Questions of how private service providers are regulated are so far left unanswered. This may be problematic from the point of information privacy, as private providers are not necessarily subject to administrative law, and regarding stable provision of services, considered that companies may disappear or find the service unprofitable, and hence discontinue provision. Also, as argued in the literature, creating public value is not something which can always be measured on a financial bottom line (Moore, 1995). The benefits of providing services to the public are often hard-to-quantify, yet require structural funding from government. This kind of services, e.g. swift processing of production permits, are also in danger of being overlooked if focus turns to services with a minimum break-even business case. This could be problematic, because exactly improvements in the core services of the government bureaucracy are potentially conducive for economic development as they make the business environment more flexible.

An important aspect not explicitly confronted by government in the informatisation programs is the question of user demand and uptake (Carroll, 2006). Generally, the citizen concept found in the policy documents refers to a relatively homogenous group, which they are not in reality. So, this provokes a series of questions of how to cater for different needs, which are the user demand for services, which services are most used, and so on.

The need for special solutions for vulnerable groups such as: "...old people, the disabled, people with low level of education, and people with low information technology skills..." (Shaanxi, 2009, author's translation) have been realised by the government. It is too early to say, how these ideas will turn into policy and practice, so assessment of this aspect will have to wait. Accessibility problems have been addressed theoretically from an early stage in the e-government program as well, but in actual practice websites tend to lack accessibility options (Shi, 2007). This indicates a need for close scrutiny of the implementation of accessibility also within m-government. There has to be a fit between the idea of accessibility and the software supporting it which there is not at present.

Institutions of M-Government

Taking point of departure in the ideas behind m-government in Western China the previous section has demonstrated how diverging ideas about m-government and different degrees of fit between ideas, institutions, and technologies present challenges on the use of m-government for development. Now, focus is turned to the institutional side of the equation, or how the ideas discussed above are translated into actual regularised practices. The organisations of informatisation offices and information bureaux are described as they are central for implementation of informatisation policy. Then, characteristics of the process surrounding the crafting of informatisation plans are analysed through focus on the interplay between institutions, ideas and technologies.

Institutionalising Ideas

Seen from the level of ideas, how is m-government development going to be achieved in terms of organisation? Which modes of government and governance does m-government enable? M-government is sometimes depicted as making the government employees mobile, or in other words affecting the modus operandi of government-to-employee as well as government-to-citizen relations, as the government starts roving. As working tasks of government are diverse, there cannot exist a one-size-fits-all solution and the predominance of governance modes will be different from service to service. Success stories of m-government application in e.g. Beijing's urban management network (cheng guan wang) (Song & Cornford, 2006), do not explicitly consider, how other kinds of public service provision through m-government are possible. As research on m-government is still scarce, there is an obvious lack of ideas about the interplay between mobile ICT and the types of organisation they enable.

The idea about a fundamentally changed mode of public service delivery has received attention among academics and policy makers in China. One thing is delivering government services to citizen's mobile devices; another is to use mobile technologies to enable public servants to work out of the office, in a kind of fluid organisation (Song & Cornford, 2006). Attempts to do this in shape of the urban management network in Beijing has generated attention. In Western China comparable systems are now also implemented on a trial basis in e.g. Chengdu City in Sichuan Province. The idea of these urban management networks is different from government in a bureaucratic mode because it focuses on creating government mobility, meaning that government actively searches for problems in urban management such as damaged roads in need of repair, traffic light failures, etc.

Although it has achieved attention among decision makers, this kind of governance is hard to imagine in all sectors of public administration. It will be contingent on the government function. Song & Cornford's (2006) claim of a paradigm shift in public service provision is thus premature, as it downplays the fact that government takes care of multiple functions. In some situations government mobility will be time consuming and not necessarily help in solving problems of e.g. applications for business permits. The idea of a changed mode of governance is exactly an idea in the sense of this chapter, and as such, we argue, it can enable change, when it fits with institutions and technologies. When powerful actors involved in policy making wish for such a change to happen.

Organising for M-Government

In terms of organisation changes have happened. New ministries and other organisational units have been created in response to the increasing importance of ICT in society. The newly formed Ministry of Industry and Information Technology is the central government organisation responsible for informatisation policy including development of m-government. M-government is thus institutionally considered a part of the general informatisation policy.

On provincial level two different types of organisations would be responsible for m-government development: The information bureau (xinxi ting) or the informatisation office (xinxihua bangongshi). There are differences among provinces concerning use of either one or the other. The choice of organisational set-up signals the importance of informatisation in provincial governance, with the provinces choosing to set up an information bureau providing informatisation policy additional administrative clout, as a bureau is a higher ranking organisation than an office. Nonetheless, choosing the informatisation office set-up can also reflect the idea that informatisation is not the job of the state but should be done primarily by private actors co-ordinated by the state (Zhang, 2005).

For the present discussion the important point is that particular organisations exist for informatisation policy implementation, but also that the responsibility of the organisations are broader than just m-government. The latter knowledge also promotes an understanding that m-government is seen not as a self-sufficient policy area but as a part of the overall informatisation policy and the government organisation reflects this.

An important task for these organisations is the co-ordination and standardisation of technological solutions. Forced co-ordination has been attempted from the central state level regarding e-government. This happened after it became clear that the locally produced, and incompatible islands of information (daoyu moshi) from the first wave of e-government implementation were hampering further development. M-government is potentially on the same path of fragmentation, as there is no strong and clear macro co-ordination. As such, it reflects the dilemmas faced in fitting institutions and technologies. Should the institutions be flexible allowing a larger room for technological innovation, or should there be institutionalised standards that ensure compatibility?

Apart from the above mentioned ministries, offices and bureaux, e-government and service orientation of government have resulted in other organisational restructuring of government affairs. Such restructuring is clearly visible from the upshot of government affairs service centres (zhengwu fuwu zhongxin) in many of the larger cities in Western China. Though, arguably an improvement in service provision, it is still an example of government in the bureaucratic mode where the citizen has to physically move to the relevant service centre. This stands in contrast to m-government in the sense of either direct access to government through mobile channels or mobilised government like the urban governance networks mentioned above. In a sense there is a lack-of-fit between the potentiality of mobile ICT and the organisational forms that do not take advantage of the new opportunities.

That being said, in most provinces there has been an upshot of local level "public service stations" (zhengwu fuwuzhan), local government one-stop-shops capable of handling most administrative affairs. If cases require higher level authorities' approval, this can be attained through the internet based e-government system. Through combination with an automated SMS service notifying citizens of case completion, a multi channel technological solution has been found to the institutional constraint that some documents are required by law to be presented in paper form. Such laws are not as flexible in the short term as technology, and therefore fitting the technology in this case has provided a fit, which improves governance.

This point underscores a characteristic of m-government application in Western China, namely the integration of m-government in a larger reform agenda. Even though proponents of m-government solutions and academics doing specific m-government research tend to be very optimistic about the promises of the new technology, the government is taking a more balanced approach. M-government is applied it in cases when profit or service improvement is considered realistically attainable, and where it supports existing legal structures rather than demands changes of these structures.

Corruption and M-Government

One of the most persistent, although informal, institutions in Chinese public administration is corruption. In small villages the "sky is high, and the emperor is far away" as an old Chinese saying goes, or in other words, the provincial and central authorities lack control mechanisms in relation to local government. Unfettered corruption and abuse of administrative discretionary power thus becomes possible. In countering such tendencies e-government, and as a subset thereof m-government, is seen by some provincial level governments as a battering ram to dismantle the lowest level of government by allowing direct contact between citizens and higher levels of government in both directions (Zhang, 2005). This is supposed to counter the fact that in many rural areas there is a considerable distrust towards the local public servants. Information from and communication with higher echelons can empower citizens' vis-à-vis these local public servants. But as an unintended side effect of empowerment of rural residents existing conflict can be aggravated. The rural residents may namely suddenly realise that some of their rights are being violated, yet they are unable to seek redress as they often lack channels of recourse.

The creation of "fairness, equality, and openness" (gongzheng, gongping, gongkai) in the public administration, which is an idea promoted by central government, seems to be what is sought after. E-government in general and m-government in particular in the rural areas may be a catalyst of the development of these three qualities of the public administration. Importantly, this is supposed to work through providing rural residents with access to laws and regulations through SMS, and opportunity to complain about administrative decisions. The full fledged implementation of such services through mobile ICT is still not a reality in Western China, even though elements of these ideas have been implemented.

In the government affairs service centres there are at present different technological measures, that could work to reduce corruption. For instance, all information entered into the computer system can be monitored from higher administrative levels within the province. Furthermore, cameras surveilling the work areas of the centres are plentiful and allow for control of the employees.

Even though m-government and e-government technological solutions thus fit with the idea of creating a service oriented government, and the formal institutions to punish corrupt practices are in place, the problem of corruption remains unsolved. Strong interests exist among the actors who hold the power to enforce the laws. So, solving issues of corruption requires a change of the informal institutions, which are closely related to individual interests of powerful actors. In a political system such as the Chinese without division of legislative, executive, and juridical powers such informal institutions are hard to change.

As corruption is costly in terms of public dissatisfaction the central government is making efforts at launching the idea that administration should take place "in sun-light" open to public scrutiny. Also, a number of strike hard campaigns have been made against corruption. In spite of these efforts, fundamental informal institutions have not changed.

Making Informatisation Policy

The government organisations mentioned above responsible for informatisation policy on a regular basis lay out new plans and strategies. Key policies to understand as the context for m-government are the informatisation programs issued both by the central state and by provincial governments. The current programs at provincial level at the time of writing are a specification of the general guidelines laid out in the "Eleventh Five Year Plan" and the "2006-2020 National Informatisation Strategy". The content of these policies (xinxihua zhuanxiang guihua) include a description of the results attained during the previous five year period, an analysis of the problems faced at the entry to the new five year period, and a list of general directions for informatisation work for the new period.

Creating a fit between such formal institutionalised plans and fast changing ICT's is a challenge. There is a trade-off between detailed implementation measures that can be evaluated and openness to new technological opportunities. Accommodating this pressure, government has formulated the programs in very general terms, casting their ballot in favour of policy innovation. The logic is well-known in Chinese experimental policy implementation. Test locations are allowed to go ahead with an innovative, but risky, project. If this experiment succeeds it can be rolled out to province or national level. On the downside, this approach makes only broad policy statements possible.

An example from the description of how to increase the general knowledge of informatisation reads: "Strengthen the publicity work concerning informatisation knowledge and cultivated and healthy online behaviour..." (Shaanxi, 2009). Broad concepts that can be interpreted in multiple ways. It is followed by suggestions to use the schools and adult education to promote this knowledge, but it does not get any more concrete. There is a large room for local differences in implementation including which technologies to use. Implementation success will also be hard to measure, as the goals of this concrete initiative are to: "...reduce the digital divide, increase the degree of participation from all groups of society, and strengthen the general ability to use information technology" (Shaanxi, 2009). Such statements provide a sense of direction, but are not usable as evaluation indicators.

This is an example of how the ideas of creating a service oriented government promoting development are institutionalised into relatively broad policy plans reflecting the fluidity of the technological environment in constant change, leaving room for local innovations. Subsequently, how this room for innovation creates a void in evaluation measures, which is a problem in an administrative system where corruption is common. Fitting ideas, institutions, and technologies in one situation thus creates potential lacks-of-fit in others. These in turn put pressures on development of technologies for improving control of lower levels of government by their superiors.

FUTURE RESEARCH DIRECTIONS

Resembling e-government research at the beginning of this decade, empirically grounded m-government research is scarce. This tendency is exacerbated by the fact that Western scholars are prohibited from producing their own surveys, and limited concerning access to the public administration. Even within local level government there is at some places little knowledge about the actual spread of communications infrastructure, little or no evaluation of m-government programs, etc.

A neglected research area of importance for development in Western China is comparative research of simple applications such as e.g. SMS-services concerning information provision. In particular, research regarding agricultural information and government openness would be valuable to suggest concrete policy innovations. Comparisons could be made on both township, city, and provincial level. This kind of research is already to some extent pursued by government organisations. At least study trips is a well-known phenomenon within the administration. In this case it would be valuable to collect and compare results and importantly make them publicly available and easily accessible online.

Concerning 3G applications more focus should be given to empirical studies that links the promising opportunities provided by the technology closely to the actual conditions on the ground in rural Western China. The benefit of such research would be to suggest operational solutions to development problems, such as e.g. what rural residents need in relation to their daily life and their farming, and if they are willing to pay for certain services.

The institutional level is also important to address with more concerted research efforts that can point to under-specifications in the legal or policy documents creating unintended room for local level selective implementation. Thereby, central and provincial government organisations would be better able to maintain the upper hand in macro co-ordination of m-government development, and will be better able to tailor m-government development to fit into existing reform programs.

On the level of ideas, conceptual changes applying to "service oriented government" found in the Ningxia model and their translation into political practices would be very interesting to examine. The changes could indicate further strengthening of the provinces vis-à-vis central state and local government generating a sort of "perverse federalism" (Mertha, 2005) where provincial power centres grow so strong that they are basically ungovernable from the central state.

On a more general note, the digital governance model could fruitfully be applied to empirical case studies of particular informatisation programs to generate a more detailed understanding of the technologically mediated mechanisms of institutional change, their influence on development of rural China, and the possibility of applying the programs in other settings.

CONCLUSION

Economists have warned against thinking of bridging of digital divides as the cure for economic underdevelopment. Instead, they argue, we should focus on the underlying causes of underdevelopment in areas such as education, health, environment, and market building and market protecting institutions. Although missed by some technology enthusiasts among researchers and policy makers, this point is nonetheless clearly reflected in m-government development policies in Western China. Yet, the focus in official documents and policies analysed in this chapter lack reference to some of the potentials of m-government to promote better governance in Western China, and as a consequence fail to under-gird other initiatives taken to promote economic development.

Mobile ICT provides a platform for information between government and citizens, and in rural areas of Western China availability of mobile phones is high compared to the internet. The opportunities provided by mobile ICT have been embedded in ideas of informatisation and are perceived as a means to create a service oriented government and promote development in rural areas. These ideas are reflected in the informatisation programs that provide general guidelines for informatisation development, but leaves much space for local variations in implementation. It has resulted in different applications, for the rural areas mainly as agricultural information services and weather forecasts.

The m-government solutions are seen as only one of several complementary channels also including TV, radio, and voice telephony. Together these channels has provided a basis for improved public services for rural residents and thereby reduced one of the barriers for economic development in the region. As such m-government is an important factor to think into development programs. It may serve to prevent losses among vulnerable groups such as small scale farmers by agricultural information services delivered through SMS and prevent unnecessary travel between administrative units. Solutions to improve m-government for development in Western China suggested by the approach in this chapter are down-to-earth and low-tech. As a benefit they are less expensive and technically easier to handle than more advanced m-government applications.

Challenges for use of m-government that have become clear through the argument include: Moving focus to simple applications, handling central government policy ambiguity, and reforming government organisation. As a precondition for government control with m-government development strengthening government in-house expertise or having a transparent system of control of private service providers was seen as requisite. Some more fundamental problems that need to be addressed include rethinking levels of government, and reducing corruption. Yet, the complexity of these problems does not allow for any conclusions. It was discussed how these challenges can be met using the logic of fitting ideas, institutions, and technologies to each other.

How to fit the levels of ideas, institutions, and technologies is ultimately a political question with no simple solution. The present chapter hopes to provide a basis for a more nuanced discussion that can avoid technological determinism, and to raise discussion about how e.g. the provincial governments are apparently using new technologies as a tool to dismantle lower levels of government which may be, but is not necessarily, a good thing for citizens.

REFERENCES

Bijker, W. E. (1995). *Of bicycles, bakelites, and bulbs toward a theory of sociotechnical change*. Massachusetts: MIT Press.

Campbell, J. L. (2004). *Institutional Change and Globalization*. Princeton: Princeton University Press.

Carroll, J. (2006). 'What's in it for me?': Taking m-government to the people. Paper presented at the 19th Bled eConference, Bled, Slovakia.

Castells, M. (2000). *The Rise of the Network Society* (2nd ed.). Malden: Blackwell Publishing.

Central Government (The Central Government of the People's Republic of China). (2006). The Eleventh Five-year Plan. Retrieved June 12, 2009, from http://www.gov.cn/ztzl/2006-03/16/content_228841.htm

Danziger, J. N., Dutton, W. H., Kling, R., & Kraemer, K. L. (1982). *Computers and politics: High technology in American local governments*. New York: Columbia University Press.

Davison, R. M., Vogel, D. R., & Harris, R. W. (2005). The e-transformation of Western China. *Communications of the ACM*, *48*(4), 62–66. doi:10.1145/1053291.1053320

Dunleavy, P., Margetts, H., Bastow, S., & Tinkler, J. (2004). Government IT performance and the power of the IT industry: A cross-national analysis. Paper presented at the Annual meeting of American Political Science Association, Chicago.

Fan, Y., & Li, D. (2006). Mobile government development research (yidong dianzi zhengwu fazhan yanjiu). E-government, 2006(9), 48-52.

Hoff, J. (2000). Technology and social change: The path between technological determinism, social constructivism and new institutionalism . In Hoff, J., Horrocks, I., & Tops, P. (Eds.), *Democratic governance and new technology: Technologically mediated innovations in political practice in Western Europe*. London: Routledge.

Holliday, I., & Yep, R. (2005). E-government in China. *Public Administration and Development*, *25*(3), 239–249. doi:10.1002/pad.361

Hood, C. (1991). A Public Management for All Seasons? *Public Administration*, *69*(1), 3–19. doi:10.1111/j.1467-9299.1991.tb00779.x

Li, M. (2005). Mobile government, extending government services (Yidong zhengwu, kuozhan zhengfu fuwu). China Computer User (Zhongguo jisuanji yonghu), 2005(31), 45-46.

Lu, J., & Weber, I. (2007). State, power and mobile communication: A case study of China. *New Media & Society*, *9*(6), 925–944. doi:10.1177/1461444807082640

March, J., & Olsen, J. P. (2006). Elaborating the "New institutionalism" . In Rhodes, R. A. W., Binder, S. A., & Rockman, B. A. (Eds.), *The Oxford Handbook of Political Institutions*. Oxford: Oxford University Press.

March, J. G., & Olsen, J. P. (1989). *Rediscovering Institutions: The Organizational Foundation of Politics*. New York: Free Press.

Mertha, A. C. (2005). China's "Soft" Centralization: Shifting Tiao/Kuai Authority Relations. *The China Quarterly*, *184*(1), 791–810. doi:10.1017/S0305741005000500

Moore, M. H. (1995). *Creating Public Value: Strategic Management in Government*. Cambridge: Harvard University Press.

Peters, B. G. (2005). *Institutional Theory in Political Science: The 'New Institutionalism* (2nd ed.). London: Continuum.

Qiang, C. Z.-W. (2007). *China's Information Revolution: Managing the Economic and Social Transformation*. Washington, D.C.: The World Bank. doi:10.1596/978-0-8213-6720-9

Shaanxi (Shaanxi Provincial Government). (2009). Shaanxi Province people's economy and societal development informatization special item program (Shaanxi sheng "shiyi wu" guomin jingji he shehui fazhan xinxihua zhuanxiang guihua). Retrieved June 12, 2009, from http://www.cia.org.cn/information/syw_dqzx_15.htm

Shi, Y. (2007). The accessibility of Chinese local government Web sites: An exploratory study. *Government Information Quarterly*, *24*, 377–403. doi:10.1016/j.giq.2006.05.004

Song, G., & Cornford, T. (2006). Mobile Government: Towards a Service Paradigm. In Proceedings of the 2nd International Conference on e-Government (pp. 208-218). Pittsburgh: University of Pittsburgh.

Trimi, S., & Sheng, H. (2008). Emerging trends in m-government. *Communications of the ACM*, *51*(5), 53–58. doi:10.1145/1342327.1342338

Yunnan (Yunnan Provincial Government). (2009). Yunnan Province people's economy and societal development informatization special item program (Yunnan sheng guomin jingji he shehui fazhan xinxihua zhuanxiang guihua). Retrieved June 12, 2009, from Http://Www.Cia.Org.Cn/Information/Syw_Dqzx_14.Htm

Zhang, L. (2005). Integrate the social resources to accelerate the pace of e-government construction (Zhenghe shehui ziyuan jia kuai dianzi zhengwu jianshe bufa). Retrieved June 12, 2009, from http://www.nx.gov.cn/structure/zfxx/zfld/fsjhxx_5124_1.htm

Chapter 10
Exploring Civil Servant Resistance to M-Government: A Story of Transition and Opportunities in Turkey

Ronan de Kervenoael
Sabanci University, Turkey & Aston University, UK

Mark Palmer
Aston University, UK

N. Meltem Cakici
Gediz University, Turkey

ABSTRACT

The concept of mobility, related to technology in particular, has evolved dramatically over the last two decades including: (i) hardware ranging from walkmans to Ipods, laptops to netbooks, PDAs to 3G mobile phone; (ii) software supporting multiple audio and video formats driven by ubiquitous mobile wireless access, WiMax, automations such as radio frequency ID tracking and location aware services. Against the background of increasing budget deficit, along with the imperative for efficiency gains, leveraging ICT and mobility promises for work related tasks, in a public administration context, in emerging markets, point to multiple possible paths. M-government transition involve both technological changes and adoption to deliver government services differently (e.g. 24/7, error free, anywhere to the same standards) but also the design of digital strategies including possibly competing m-government models, the re-shaping of cultural practices, the creation of m-policies and legislations, the structuring of m-services architecture, and progress regarding m-governance. While many emerging countries are already offering e-government services and are gearing-up for further m-government activities, little is actually known about the resistance that is encountered, as a reflection of civil servants' current standing, before any further macro-strategies are deployed. Drawing on the resistance and mobility literature, this chapter investigates how civil servants' behaviors, in an emerging country technological environment, through their everyday practice, react and resist the influence of m-government transition. The findings points to four main type of resistance namely: i) functional resistance; ii) ideological resistance; iii)

DOI: 10.4018/978-1-61692-818-6.ch010

market driven resistance and iv) geographical resistance. Policy implication are discussed in the specific context of emerging markets.

INTRODUCTION

Resistance across history has remained a powerful symbol of human territoriality. In turn, territoriality is often described as the primary expression of social power (Sack, 1986). In the context of this chapter, resistance is associated with the possibility of territorial changes, power shifts and responsibility evolutions occurring. Territoriality traditionally involves some form of classification or assigning tasks to a category affecting ongoing relationships between territorial units. Territoriality displaces attention within relationships as a mean to exteriorize power and creates de-facto boundaries among territorial divisions. Examples include expression such "ours" and not "yours" and "it is the local regulation", "you may not do this there" (Sack, 1986). In particular, at individual civil servant level, the scope of knowledge is often graded according to different territorial spaces leading to a scale regarding the endowment of responsibilities. The span of control of any individual is then devised to an extensive level of details.

One of the most recent and significant shifts in territoriality to occur has been the advance in Information Communication Technology (ICT thereafter) and the creation of digital and virtual spaces (Castells, 2001). At the same time, the concept of mobility has evolved dramatically over the last two decades and transformed many aspects of business management and day-to-day activities (Ling, 2008). Accordingly, recent studies have begun to address the role of resistance in the particular context of mobile technology (Venkatesh et al., 2003; Lapointe & Rivard, 2003). While many studies have used a positivist, deterministic framework, one further recurrent theme, in social sciences, has been the difference between mobility being driven by personal motivation and mobility driven by societal and ideological factors (Avgerou et al., 2004; Goggin, 2008; Ling, 2008). Significantly, several studies have analyzed ICT and mobility in the context of m-government primarily from an end user perspective (Kushchu, 2007). However, leveraging ICT and mobility promises within public administration in emerging market point also to multiple other factors such as civil servant resistances, organizational culture clashes and capacity absorption issues (Sheng & Trimi, 2008; Oxford Internet Institute, 2006; Vincent & Harris, 2008). M-services, in public administration, are linked to actions and functions involving services such as instant up to date information release, error free data collection, mobile transaction (e,g. taxes, benefits), crime prevention (e.g. use of video/photographic elements of mobile phones) and the way digital technologies are used within the socio-political context to allow more transparency, the inclusion of minorities, and the civil society at large (Welsh & Wong, 2001; Hallin & Lundevall, 2007; Yu and Kushchu, 2004; Kushchu et al., 2007).

M-government is defined as "a strategy and its implementation involving the utilization of all kinds of wireless and mobile technology, services, application and devices for improving benefits to the parties involved in e-government including citizens, businesses and all government units" (Kushchu & Kuscu, 2003, p.3). The main heralded benefits include: cost reduction, efficiency mainly in data collection, processing and regarding redundancies and errors, transformation/modernization of public sector organizations, added convenience and flexibility (e.g. urgency factor), better services to the citizens and civil servants, ability to reach a larger number of people through mobile devices than would be possible using wired internet only. Mobility is understood in the general sense of being mobile. Examples involve not returning to a fix point between appointments, working in the field or from home rather than a fixed location.

Others aspects of mobility encompass the notion of inter-jurisdictional mobility and sphere of power, equal competence and service provision, standards, protocols and interoperability (EC., 2003), privacy and security issues (Kumar et al., 2008; Garson, 2006b). Concurrently, depending on the context mobility is also virtual allowing workers in different location to communicate extending civil servant reach of interaction. Time in turn in an 'always on' environment becomes relative with dynamic reconfiguration and rescheduling (Roggenkanp, 2004). In particular, three main steps for m-government and level of complexity are observed including: (a) mobile access, (b) mobile content and (c) mobile applications. In other words, a genuine mobile application allow for complete transactional services and interactivity.

In this context this chapter analyzes: how civil servants' behavior, in an emerging country technological environment, through their everyday practices, react and resist the influence of m-government transition.

Based on this, resistance is defined following Pile and Keith (1997 pp 8-14) as "mass mobilization in defense of a common interests, where resistance is basically determined by the action: the strike, the march, the formation of community organizations [...] potentially the list of act is endless –everything from foot dragging to walking, from sit-ins to outings, from chaining oneself up in treetops [...] the definition of resistance have become bound up with the ways that people are understood to have capacities to change things".

This chapter draws on a number of intellectual traditions, including organizational development, strategic management, innovation studies, development studies, and evolutionary theory of the firm (Marcelle 2004)[1]. Putting aside the methods and models of management science,[2] we use a practice perspective (Jarzabkowski et al. 2007) and come closer to everyday life of civil servants. Strategy as practice utilizes three interrelated concepts (i) practitioners investigating what people do individually or in group, (ii) practices which represent the large array of tools and mechanisms used to implement day to day strategy and (iii) praxis representing the stream of activities (Whittington, 2007; Palmer & O'Kane, 2007; Jarzabroski et al, 2007).

While many emerging countries are already offering e-government services and are gearing-up for further m-government activities, little is actually known about the forms of resistances that are encountered, as a reflection of civil servants' current standing, before any further macro-strategies are deployed. In addition, despite the conceptual and empirical development relating to resistance little is known regarding resistance for emerging new services in the specific context of public service. Specifically, while traditional m-government studies have created a picture of the factors influencing successful digital government outcomes, we identify three partial gaps in the understanding:

First, published work has almost exclusively focused on citizens (end-users) rather than civil servants compelled to deliver the new services and activities (Kushchu, 2007). In doing so, we investigate innovation diffusion while the diffusion process is underway limiting possible pro-innovation bias (Rogers, 1995). Second, while country level strategies exist, there has been a minimal analysis of the impact across the competing daily activities treated in the literature as non-events. Hence, we investigate the broader context in which change occurs. Third, we argue that, in our context, there has been a lack of consideration regarding resistance spatial sources of resistance in particular not from an individual perspective but from a system bias (Sack, 1986).

As resistance emerges a critical issue in the forming stage of m-government projects, we aim at offering insights to politicians, regulators and government strategists, in the specific case of emerging markets on: (i) what type of resistances to m-government are currently being experiences by practitioners and (ii) how do these different types of resistances influences and shape the likelihood of m-government adoption?

This chapter proceeds in six steps. Section two provides a brief description of Turkey's m-government activities. Section three reviews the literature on resistance mobility, and forming stage as setting concepts to the theoretical analysis. Section four underlines our methodology. Section five presents the analysis of the findings followed by discussion. Lastly, section six offers some conclusions, limitations and directions for future research.

M-GOVERNMENT BACKGROUND IN TURKEY

Over the last decade, the Internet has increasingly become the channel for delivering government services to citizens. E-Government is the provision of traditional government services to citizens through electronic tools such as telephone, computers and other digital devices. Based on this, digital databases have become center to the strategy and support back end operations able to provide instant up to date services. Recently, the developments in mobile technologies with the introduction of Internet enabled mobile phones (3G), and others PDA's, have created extra methods to access and provide government services to citizens in an even more effective and cheaper way. These developments are opening a new era in public service delivery practices.

M-government strategies and implementations of those strategies increasingly involve the utilization of all kinds of mobile wireless technologies, services, applications and devices for improving benefits to the parties involved with governmental agencies including, for example, citizens, businesses and all others governmental units. Significantly, mobile government is foreseen to create extra value added when compared to former ways of public service delivery in areas such as: (i) costs (Melitski et al., 2005; Holden et al., 2003), (ii) efficiency of work practices, transformation/modernization of public sector organizations (Larsen & Rainie, 2002; McIvor et al., 2002), (iii) democracy and inclusion strategies (Foley, 2004; Klein, 1999), and (iv) convenience and efficiency through an increased range and type of application delivered anywhere at anytime (Kushchu & Kuscu, 2003).

In line with the importance of technologies, it is estimated that governments' ICT spending for the 2000-2010 period stands at over $3 trillion. Nevertheless, it is anticipated that between 60 to 85 percent of e-government projects fail (Gubbins, 2004). For instance, in emerging markets, limited access by the majority to digital infrastructure remains high. Interestingly, Turkey is ranked 76th out of 182 countries regarding E-government Readiness by the UN in 2008. As expected, advance in ICT choices are restricted by capability absorption issues related to: (i) ICT skills, (ii) relatively low-income levels, and (iii) investment climate (Evci et al., 2004; Torenli, 2006). Indeed, Turkey can be considered as a good example for such tendencies marred by regular financial crisis and IMF loans, civil unrest at its border, a protracted process of integration conditions to enable entry to the European Union and a country with great regional disparities (Dedeoglu, 2004; Ozcan & Kocak, 2003).

Nonetheless, in November 2008, the E-Government Gateway initiative, which represent the effort to create a single portal enabling provision of all e-government applications, was successfully launched (Kumas, 2007; Sungu et al., 2008). *"Once the e-gateway or the portal is created, the service will be extended to include other communication devices like cell phones and pocket PCs, and users will be able to access the system with smart cards or imprinted digital certificates for a secure transactions"* (TURKSAT, 2008). Table 1 provides a snap shot summary related to all current government e-activities.

Regarding mobile application in particular, the leading initiatives in the field, are mainly

Table 1. E-Government Gateway Action Plan

	Activity	Responsible Agency	Date TURKSAT Plans for the Service
1	Exam Results Inquiry	Ministry of Education	Dec 2008
2	Complaint Results Inquiry	Ministry of Industry and Trade	Dec 2008
3	Complaint Number Inquiry	Ministry of Industry and Trade	Dec 2008
4	Advertisement Complaint Submission	Ministry of Industry and Trade	Dec 2008
5	Consumer Judge Commission Inquiry	Ministry of Industry and Trade	Dec 2008
6	Consumer Association Inquiry	Ministry of Industry and Trade	Dec 2008
7	• Warranty document • After sales services • Competence document • Authorization document for door-to-door sales • Turkish user manual inquiry	Ministry of Industry and Trade	Jan 2009
8	e-Military Drafting: Updating residency address information	Ministry of Defence	Jan 2009
9	Vehicle Inquiry	Security General Directorate	Jan 2009
10	Driver Penalty Points Inquiry	Security General Directorate	Jan 2009
11	e-Tax Registration Certificate	Revenue Administration Department	Jan 2009
12	Citizen Information System: Access to Summaries of Cases seen in Courts	Ministry of Justice	Feb 2009
13	e-Military Drafting: Online delivery of information for classification	Ministry of Defence	Feb 2009
14	e-Military Drafting: Choosing drafting period	Ministry of Defence	Feb 2009
15	e-Military Drafting: Military mobilization manoeuvres	Ministry of Defence	Feb 2009
16	Access to Civil Servant Payrolls	Ministry of Finance	Feb 2009
17	Consumer Complaint Submission	Ministry of Industry and Trade	Feb 2009
18	e-Military Mobilization: Processing Assignments	Ministry of Defence	Mar 2009
19	e-Military Mobilization: Roll Calls	Ministry of Defence	Mar 2009
20	e-Military Mobilization: Processing Logistics Assignments	Ministry of Defence	Mar 2009
21	Information Desk for Traffic Accidents	Security General Directorate	Mar 2009
22	Driver License Permanent Cancellation Inquiry	Security General Directorate	Mar 2009
23	e-Passport Application and Control System	Security General Directorate	1st Quarter of 2009

source: (e-DTR, 2008)

infrastructure enabling tools, including the Mobile Electronic System Integration, Traffic Information System, Mobile information Project (Ministry of education), G2G applications (mobile intranet for ministry personnel) and the National Judicial Network (Court system integration) (Cilingir & Kushchu, 2004; Kuran, 2005; Ghyasi & Kushchu, 2004). These leading initiatives are orchestrated at national level. In contrast, at local level considerable efforts have been made in creating and developing initiatives that provide day to day applications in areas such as SMS for public an-

Table 2. Examples of m-government activities in Turkey

G2C	Government Agencies	Description
UYAP SMS alerting	Ministry of Justice	Informs citizens about their ongoing trial dates etc.
İBB Cep trafik	İstanbul Metropolitan Municipality	Enables to control traffic flow with online camera systems
İstanbul Mobile City Guide	İstanbul Metropolitan Municipality	Provides various information about İstanbul
Wap Beşiktaş	Beşiktaş Municipality	Is a wap module which includes applications for particular municipality services.
SMS alerting	Turkish Statistical Institute	Informs citizens about new released statistics like monthly consumer price index etc.
TCDD Mobil	Turkish Republic State Railways	Provides TCDD services digitally
Mobile Customs House	Undersecretariat of Customs	Enables users to access recorded information on their current custom declarations
İSKİ Wap	İstanbul Municipality Waterworks	Provides information on water bills
MOBESE	İstanbul Police Department	Provides various services from car following to car number sensors
Tax Pay Module	Şişli Municipality&Turkcell	Enables citizens to make their tax payments

Source: Compiled by authors.

nouncements, giving information about services status, local taxes and utilities, road traffic information, benefits for low income and elderly and to give a voice to citizens (Ghyasi & Kushchu, 2004). Table 2 provides a series of current m-application by local authorities and Table 3 shows contrasted examples of private activities in Turkey. Despite the great potential and positive expectation by many people, m-government is still in its early phase of development.

LITERATURE REVIEW

The importance of effectively dealing with any new m-government strategy from a civil servant perspective is likely to be vital for successful future engagement. In this section, three key concepts related to m-government strategy implementation are evaluated: (i) resistance, (ii) mobility and (iii) strategy forming stage. In investigating the emerging various forms of resistance to m-government,

Table 3. Examples of company activities in Turkey

Application	Company	Industry	Description
Turkcell-im	Turkcell	GSM	A mobile application for young people, enables users to download music, share pictures and videos
Patlican	Avea	GSM	
Vodafone Live!	Vodafone	GSM	
İşcep	T.C. İş Bankası	Banking	Mobile bank applications providing mobile transactions
Garanti	Garanti Bankası	Banking	
Fortis Mobile	Fortis	Banking	
Cepte TEB	TEB	Banking	
Mobile signature	Turkcell/Avea	GSM	Enabling users to complete and verify their transactions in digital environment
Bilyoner	Bilyoner.com	Entertainment	A mobile bet application

Source: Compiled by authors.

in emerging market context, we make a contribution to the literature on two main fronts: (i) we contribute to theory development by identifying in practice the different forms of resistance which civil servants are experiencing when facing strategic changes; (ii) we also take a first step towards understanding why m-government is perceived as such a radical change and what are the factors stakeholders can use to minimize drawbacks as early as possible.

Resistance

From a deterministic perspective, Sheth (1981) developed a typology of innovation resistance, which included the concept of 'Habit Resistance'. This stream of work was often grounded in Mittelsteadt et al.'s (1976) symbolic adoption or rejection model whereby users may have symbolically accepted an innovation but not sufficiently enough to try, which in turn do not lead to adoption. Reflecting this, Gatignon and Robertson (1991) suggested that non-adoption of an innovation may be explained by either rejection or postponement depending on the context. Context is understood as "the set of environmental states and settings that either determines an application's behavior or in which an application event occurs and is interesting to the user" (Chen & Kotz, 2000). In other words, as defined by Lee and Kun (2005) "any personal and environmental information that may influence the person when he/she is using Mobile Internet". Furthermore, Szmigin and Foxall (1998) building on the general principles described above have categorized innovation resistance into three varieties: rejection, postponement and opposition.

Concurrently, from within other social sciences areas, resistance has been analyzed in various settings including: political resistance (Watts, 1997), power resistance (Moore, 1997), identity resistance (Banister & Hogg, 2004), salesperson management (Lee & Cadogan, 2009), consumption resistance (Clarke et al, 2006), space of resistance, regulatory resistance and cultural resistance (Kozinets & Handelman, 2004).

In particular, resistance was further defined by Collinson [1994:49] in strategic alliance settings as seeking to "challenge, disrupt or invert prevailing assumptions, discourses and power relations. It can take multiple material and symbolic forms, and its strength, influence and intensity are likely to be variable and to shift over time..." As quoted from Palmer et al (2010), p9 and drawing on the work of Ackroyd and Thompson (1999), and Fleming and Spicer (2007), four main expressions of resistance are emerging including: refusal, voice, escape and creation:

Resistance as *refusal* does not follow instruction and overtly blocks the effects of power by undermining the flow of domination rather than attempting to change it. Resistance as *voice* is to let one be heard by those in control in order to change particular aspects of power relations in favour of those being affected by them. Resistance as *escape* reflects those mechanisms used to disengage mentally from the workplace such as cynicism, skepticism and dis-identification. Resistance as *creation* refers to the way that alternative identities and discursive systems of representation emerge within the broader flows of domination (Fleming & Spicer, 2007: paraphrased from pages 29-43, emphasis added).

Each expression is now discussed in the context of mobile technology and public administration. Technology refusal is a common site especially among lower ranked civil servants. Different reasons are put forwards for such a situation including lack of initial awareness, limited investment in communication, hardware and software maintenance, upgrading, standards, training and at individual if used privately running costs (Kuschu, 2007). Technological devices are also perceived as reserved for certain category of individuals especially when they are newly introduced creating boundaries and subsequently refusal. Refusal also reflects the constant imposition of certain technologies and systems that greatly disrupt day-to-day practices. Thereafter, voice is often portrayed as a lack of choice in the decision making process regarding the type of tools, timing, control and intensity for examples. Escape is reflected in the current use of technologies that are often offering, in the early stages, a more robust and reliable service understood by the majority. Consequently, little efforts are undertaken to attempt and solve early glitches and issues both technology and processes related. Lastly, resistance as creation represent the parallel individual systems that are being deployed both at local level to solve local needs, wants and issues and at individual level whereby a civil servant is going to utilize its own personal technology for work purposes. Following these, it is clear the government changes are not only originating from internal explicit strategies but also as a result of individual innovative resistances.

Interestingly, one constant argument in the literature remains the divergence connecting resistance motivated by personal motivations (internally driven) and motivated by societal and ideological factors (externally driven). Studies tend to analyze resistance and barriers as a negative bipolarization with a winner and loser, many actors are considered as non-participant undermining the implementation aspects and practices at the local level in day to day activities (Knights & McCabe, 2000). Beside, barriers and resistances are often portrayed as fixed and detached from the dynamic environment of ongoing activities (Palmer et al, 2010). Significantly, many resistance acts are often indirect, underlying the importance of other factors in the decision making process (Prasad & Prasad, 2000).

Economic geographers have also contributed to further aspects of resistance related to space and geographies. Here, actors are positioned differently leading to "unequal and multiples power relationships, that more or less powerful people are active in the constitution of unfolding relationships of authority, meaning and identity, that these activities are contingent, ambiguous and awkwardly situated, but that resistance seeks to occupy, deploy and create alternative spatialities from those defined through oppression and exploitation" (Pile & Keith, 1997 p3). In this context, resistance is re-defined as "looking at a spatiality of resistance understood in terms of those defined by structural power relations, second by thinking about ways in which spaces of resistance are distinct from spaces of domination; finally,[…] by suggesting that resistance may involve spatialities that lie beyond 'power'" (p5). Importantly, actors appearing in similar positions to re-interpret the terrain on which resistance takes place are reaching different outcomes. In other words, spaces of resistance can be described as dis-located from the traditional sources of power. In particular, spaces can be visible but are often constituted of hidden and ulterior motives showing the arbitrary of practices.

Under the umbrella of modernization, new places are occupied as traditional location decrease in importance. Actors are coming up with new often un-intended and possibly contradictory conclusions about the reasons for resisting. A corner stone of spatial resistance is embedded within the daily politics of lived spaces. Importantly, resistance is then re-defined by Haynes and Prakash (1991 p3) as "those behaviors and cultural practices by subordinates groups that context hegemonic social formation, that threaten to unravel the strategies of domination. Consciousness needs not be essential to its construction". Resistance as contestation is situated "betwixt and between the structural certainties of any fixed binary, such as a perspective explore the continuous play of history, culture and power" (Hall, 1990, p 225). Consequently, a different relation to power is required that is transitory, flexible and ambivalent tracking previously lock-in conflicts (Bhabha, 1990; Law, 1997). At this point, the significance of apparently apolitical standpoints cannot be overstressed. In other words, following Canel, (1992) the traditional state centered locations of politics and new emerging civil society voices should not be seen as separated.

Lastly, a further stream of research has centered attention on anti-choice behaviors reflecting voluntary anti-consumption as another facet of resistance. These studies suggest that sources of resistance are multiple ranging from (a) simplifying decision mechanism (Shaw & Newholm, 2002; Hogg & Michell, 1996); (b) prioritization of better options (Craig-Lees & Hill, 2002, Piacentini & Banister, 2009); (c) resistance as a reflection of hidden intention including ethical concerns and environmental principles (Shaw & Newholm, 2002; Iyer & Muncy, 2009; Gabriel, 1999; Knights & McCabe, 2000; Prasad & Prasad, 2000). These types of resistances are also described as moral avoidance (Lee et al., 2009 a,b); and (d) resistance as a coping mechanism (Lazarus & Folkman, 1984).

A central assumption to resistance related specifically to ICT lies in the fact that m-ICT brings its own ritual, myths, values and basic conjectures (Schein, 2004). Civil servants' local day to day context and histories with previous government ICT projects have often shaped opinions and attitudes towards future ICT adoption or rejection. Indeed, the impact of mobile technologies on individual life ought not to be misjudged (Ling, 2008). Mobile technology are usually (a) personally carried by each individual, (b) used for both business and personal task (c) never disconnected from the network even at night, (d) ubiquitous for all type of users including children and (e) expected to allow instant connection (Solomon & Buchanan, 1991; Wilska, 2003).

Decisions regarding technologies are usually a tradeoff between increased efficiency and convenience on the one side and decreased personal control over work practice and multitasking expectations on the other side (Castells, 2001; Giddens, 1991; 1992). Contradictory elements of mobile technologies such as access anywhere anytime have to be integrated within civil servants' working rituals and histories.

Moreover, the vast majority of civil servants are expected to be pragmatic and conservative. The concept of consideration then becomes relevant as a basis for understanding the contested interactions between stakeholders. Consideration is understood as Ettlinger (2003) as "*Why do people [...] often make decision that result in suboptimal productivity, competitiveness or effectiveness?*". Many new technologies and practices will then ultimately remain only under consideration but will never be engaged with. A threshold period is *de-facto* imposed upon user whereas the future intention to use is considered. This portrays a more fluid approach to m-ICT and government services engagement. It is important to examine what multiples actors *do* in view of the interactions and interactivity that m-government will require. To this end, other functional factors also require particular attention, but remain largely beyond this study, such as network infrastructure and security and legal requirements regarding data collection, storage and retrieval practices. These also include regulations to prevent usage in certain circumstance (e.g. mobile phone and driving or usage during flights) and enforcement. Attention is now turned to the multifaceted concept of mobility and technology.

Mobility

Sociologists have related the phenomenon of mobility to three main types of capital that place a person in a certain social category. These are economic capital, social capital, and cultural capital. Mobility is often defined in terms of change within individual social status over time. In other words, mobility command control over economic resources such as the ability to access greater sources of money and accumulate assets (Almeida & Kogut, 1999; Argote & Ingram, 2000). In turn, mobility enhances social capital by allowing individual to reach upon wider group membership, relationships, and networks of influence. Furthermore, mobility is shaping cultural capital towards a post modern society based on global knowledge. In addition, mobility needs to be considered within a dynamic continuum whereby individuals regularly reassess their specific mobility constrains within the given context. Absolute vs relative mobility is providing a true reflection *in-situs* of users practices. In particular, it is important to remember that other macro-societal factors such as gender, race, ethic, and culture create *de-facto* boundaries for mobility.

A second area of research led by economists encompasses areas such as labor and capital mobility. Labor mobility is defined as the ability of workers or worker groups to benefit various economic opportunities. Mobility is depicted as key for efficient allocation of resources, capital and overall productivity (Maliranta & Rouvinen, 2004). In particular, new sub-categories are emerging in the specific area of mobile pricing models (Courcoubetis & Weber, 2003)

A third approach re-defins the meaning of mobility *in-situ*. These ranges from studies analyzing lifestyle of mobile phone adapter (Leung & Wei, 2000), satisfaction with mobile devices (Woo & Fock, 1999), learning (Honold, 1999), gender relation (Hjorth, 2009), structuring of public space (Kopomaa, 2000), motivation and patterns of usage (Leung & Wei, 2000), and negative consequence (Moisio, 2003). These highlight the importance of m-technologies in shaping new social capital creation.

Lastly, regarding m-government, mobility in public administration has been centered on the debate of citizen inclusion and exclusion in public life (Foley, 2004; Vincent & Harris, 2008). Key debates on the ability and capability of government to create a technology driven knowledge society, where citizens and public administration are actively involved and truly interacting together are directly linked to mobility (Moon, 2002). Sub-components of these studies have been echoed by many international organizations and NGOs broadly calling for greater governance (Heeks, 2001; Dawes, 2009), voice for the civil society (Klein, 1999), transparency (Larsen & Rainie, 2002; McIvor et al., 2002), and democratic arguments (Musso et al., 2000). Significantly, in recent years many governments have changed the access of services towards an inclusive multi-channel solution centered on citizens needs. In particular, services such as instant up-to-date information release, error free data collection, mobile transaction, crime prevention (e.g. use of video/photographic elements of mobile phones), taxes payment and the way digital technologies are used within the socio-political context have rapidly emerged (Hallin & Lundevall, 2007; Yu & Kushchu, 2004; Kushchu, 2007; Sheng & Trimi, 2008). Yet, as with any changes mobile technologies are particularly vulnerable at the forming stages of any strategy whereby many questions remain to be answered. Attention is now turned to defining forming stage in the case of public administration.

Forming Stages

In the case of m-ICT and government services suitability, including explicit higher value added for individual civil servants' working processes, clear long term applications, and personal incentives prospect, ought to be paramount in creating a dynamic comparative advantage over current practices and processes. On one hand, the institutional capacity to shift and demonstrate higher value added services and processes (e.g. standard operating procedures) need to be explained and engaged with by the potential users. On the other hand, mutually supportive roles of different civil service corps leading to a clear cumulative sequence of change are required to provide the necessary investment (e.g. time and effort) in civil servants' human capital development. In addition, the maturity or skills to allocate resources productively (i.e. capability absorption) to make progress may not be present and may require further investment.

Following on the work of Tuckman (1965) regarding team development, four classical stages namely forming, storming, norming and performing are particularly relevant to good engagement of agents or team (s) to any new project. This study concentrate on the first stage also called 'preliminary/preparation needs' as crucially important to legitimize engagement in future collaboration. Areas such as project remit, requirement expectations, leadership roles, time frame, benefits, and external and internal supports for example are assessed. Reflecting this, micro-strategies that are deployed to encourage civil servants engagement with new technologies at an early stage ought to increase dramatically the likelihood of overall outcome success.

Front loading as a micro-strategy is defined as processes that seek to improve pre-service activity performance by shifting the identification and the solving of problems to earlier phases in the process. With the front loading problem approach, all users of a system are concerned with the early identification of obstacles. This task usually involves understanding the non-homogenous approaches and experiences of ICT with particular attention paid to past failures. The aims of front loading are to reduce the number of substandard experiences and eliminate bottlenecks in the process, to increase loyalty and trust in the system, utilize opportunities offered by new technology to reduce cycle time and increase convenience (Thomke & Fujimoto, 2000). Significantly, pre-services activities are traditionally not fully acknowledged by stakeholders. Indeed, public administration theorists have not given sufficient attention to imperfect and incorrect processes and how this shape transition and future opportunities in the case of emerging markets.

Front loading activities typically consist of a number of different steps. The first step involves learning to know civil servants as enablers of government services. Importantly, there are different types of civil servants' categories and cultures in any public administration and within these, individuals are also different. For instance, the way and the manner in which ICT is currently available and used can be considered as a direct reflection of how civil servants interpret where the technology frontier lies and how others public administration departments are dealing with the same resources. At this stage, civil servants try to minimize unpleasant experiences and increased workload and replicate the things that have affirmed or provided them with further control over their work. In turn, hierarchical setting and historical relationships among services and individuals have an impact on learning opportunities and communication style directly shaping the transition and acquiring of novel communication competences (Gudykunst, 1993). The second step involves being surprised by the familiar. In particular, the frenetic nature of government activities within civil servants workflow may leave little time for structured critical reflection on how deal with constant ICT upgrade. For example, without an indication to the contrary, civil servants may assume that diverging from the norm and challenging accepted practices may be perceived as a sign of misunderstanding. The third step usually involves solving problems collaboratively. Civil servants have many pressing needs both regarding the type of task they are required to deal with and the way they are expected to do their jobs. Yet, coming together with the whole civil service community is not a usual occurrence. In addition, bottom-up determination of priorities issues/problems to be solved is not traditionally part of the current practices. Consequently, front loading activities should not be considered as fixed but rather unfolding as a proactive or reactive recognition of the practices displayed within the dispersed nature of the stakeholders. These in turn reflect the struggle towards the dominant political rhetoric and ICT dynamic settings. Examples of front loading criticality, in other domains, can be found in areas such as behavior towards smoking and alcohol (Prochaska et al., 1992), encouragement towards greener disposition of garbage (Pieters, 1993); systematic wear of seat belt (Wittenbraker et al., 1983), and use of drugs (Drugfree.org, 2008), dispossession of products (Roster, 2001; Jacoby et al., 1977).

METHODOLOGY

Our paper presents qualitative evidences from 30 interviews with civil servants adopting the notable people as employed by Kempster, 2007 working in the IT section of local municipalities in Turkey. These data are part of a wider data set collected in 2008-9. This period corresponds with the initiation of the first comprehensive national e-government Gateway portal (Kumas 2007; Tutkun 2007). The research questions were investigated using a series of phone interviews. Each participant was contacted more than once for each of the four waves of questions. Our data collection strategy was designed around a set of sequential semi-structured interview guides all recorded, translated and transcribed. This allowed busy civil servants to engage with the research without imposing too much on their time and pays attention to the pro-innovation bias. The four waves were conducted over a year and a half period. Respondents were not feedback the results of previous questioning. The primary objective was to shed more light on the multifaceted impact of early resistance underpinning m-government initiatives in Turkey including: (i) Wave one investigated the current state of digital government including e-government and m-government emerging activities. (ii) Wave two explored organizational processes, within day to day activities of civil servants regarding possible resistance to ICT. (iii) Wave three unpacked the multifaceted concept of mobility. (iv) And, wave four investigated the concept of capability building link with personal and organizational resistance regarding m-government in particular (see survey instrument in appendix 1). The interviews were designed to explore participant personal experiences and actions as well as those they had observed in their day to day work. Each wave of interview lasted between 20 and 60 minutes.

Following Yin's (1989) pattern matching our analysis was guided by theoretical issues identified in the resistance literature. Our design was also based on Strauss (1987) and Strauss and Corbin (1998) linking for each stage of the process some theoretical appreciation of the area to the interview guide, in contrast with Glaser (1992) whereby theory must emerge naturally from the data. As described in Lee and Cadogan (2009 p.359) the "analysis proceeded concurrently with data collection and theorizing, with each having an impact on the procession of the others". In other words, analysis of early interviews was allowing later interviews to benefit from preliminary insight, as well as to help to decide on the moment of "theoretical saturation" (Strauss & Corbin, 1998). Furthermore, according to Henwood and Pidgeon (2003) the notion of theoretical agnosticism is also applied by considering overall congruence in 'fit' and 'workability'. Following Yin's (1989) to address internal validity issues the ideas of 'pattern matching' is used. While analyzing the data as a whole, peer evaluation method of coding and conceptual discussion was applied where emerging salient themes were debated and justified between the authors (Miles & Huberman, 1994; Merriam, 1998). The logic behind our design was to allow highly exploratory themes to emerge based on grounded theory followed by a more 'confirmatory phase' based on constructs from the resistance literature (Yin, 1994).

For each respondent a separate within case analysis was conducted. This was followed by cross cases comparisons to offer: (i) consistency on the issues, themes and (ii) a holistic appreciation of the data (Miles & Huberman, 1994). Four main salient themes emerged including: i) functional resistance, ii) ideological resistance, iii) market driven resistance and iv) spatial / geographical resistance. The first three themes are subsequently structured from a personal and organizational perspective while the last theme explores internal vs external spaces of power.

FINDINGS AND DISCUSSION

Four broad dimensions leading to resistance to m-municipality practices and activities emerged as the most enduring aspects namely: i) functional resistance which relates to lack of current technical equipment and infrastructure that induces resistance to new technologies engagement, ii) ideological resistance which refers to the conflicts faced by the civil servants related to the current cultural values and perception of change, iii) market driven resistance explained as external factors that have an impact on the working conditions of the municipalities and (iv) spatial /geographical resistance concerned with the different layers of activities and stakeholders in multiple location and positions. Each of them is now discussed. W=wave number, R= respondent number.

Functional Resistance

The first dimension explaining civil servants resistance to m-government is grounded in functional issues. Functional resistance was described as representing a long list of technical incompatibilities among the large variety of technological equipment and infrastructures, both within and between municipalities, leading to many giving up trying new ICT. System blame factors were recognized before individual resistance to innovation. Functional resistance was depicted as having two main drivers and foundations namely: (i) personal functional resistance but to a larger extent (ii) organizational functional resistance. Personal functional resistance was portrayed as driven by individuals' norms, beliefs, attitude and educational background. Organizational functional resistance was expressed as technology level embededness within any given location related to the services offered. A lack of fit between change strategies and financial resources for example was underlined.

Significantly, it was explained that both type of resistance had not been seen and experienced in isolation, only for technological issues, but were omnipresent. Here, the collective memory of the civil service corps, past experiences related to change; past negotiation of powers were having a large recurring impact. The diversity of services processes and individuals within the civil service corps was perceived as the greatest source of conflict and a key driver for generating resistance. In particular, it was emphasized that every situation related to technology had different interpretations and meanings depending on the local service history and type of activities. Processes were described as fragmented and lacking flexibility. Furthermore, the interrelated connections with private technology providers and the, often negative, experiences and anecdotes attached, were very strongly anchored in our respondent minds. Significantly, functional change was overall perceived as positive for example improving work condition, allowing better, error free output as well as allowing creativity to resolve common issues. In contrast, creating, organizing and sharing technology (often imposed and chosen from the top) and software materials were perceived as ongoing concern especially regarding the allocation of responsibilities. Technology resistance was illustrating other ongoing concerns about practices change and tools used within public administration especially in regard to individual responsibilities, control and power relations.

"There are still many problems with basic instrument such as printers and connections to PC"W2R4; "We are just starting to master some e-government services, how do you expect us to change and learn so much about technological tools in such a short time?W2R6; "I still don' know how to use many of the current technical software programs"W4R2; "We don't have the internal required technical infrastructure

for offering m-municipality services to our citizens"W2R1; "who is going to take responsibility and correct errors?"W4R8; "We don't work with multiple service providers. There should be more alternatives"W2R21; "at the end of the day who is going to be responsible for m-technology errors?W4R12".

Ideological Resistance

The second important dimension in terms of civil servants' resistance to m-municipality services is ideological resistance. This type of resistance reflects the conflicts faced by civil servants as a consequence of the current value and reward system in place that has created certain type of working procedures, conditions and rules. Conflict is defined following Pondy (1966) p 246 as "a concept of many meanings. It is quite legitimate to think of conflict as interpersonal hostility; but it is equally legitimate to think of it as a disagreement or perception of disagreement between two persons on some choice or preferences; or as the inability to resolve such disagreement; or merely as incompatibilities among several formally defined jobs". This form of resistance also includes both personal ideological resistance and organizational ideological resistance over tasks, interpersonal relations or function of the innovation. One the one hand, personal ideological resistance was made explicit in the questioning of the procedures employed by municipalities described as coercing or forcing civil servant to adopt the new technologies. Efficiency gains were not described as the main issue rather a divergent dualism between individuals' opinions over the future of public administration and the role of technologies. Indeed, the processes and intensity at which the change was taking place was described as key feature fostering resistance behaviors and putting in jeopardy future potential engagement and commitment. Many civil servants had doubts about their career paths and criticized the management style in their municipalities whereby their skills were quickly becoming obsolete without the possibilities to learn new ones. This was also perceived as a political game by the government to allow new preferred entrant by in effect get rid of older members. On the other hand, organizational ideological resistance was openly found regarding promotion and pay related to performance. Discrepancies linked to unwillingness to re-assess individual job description are undermining further progress. Furthermore, the safeguarding of current processes as long as possible through exemptions, derogations and non-application of the new processes was also identified to foster tensions. In turn, public administration was portrayed as a complex organization with a multi-party structure. Apart from these characteristics, aggregation of activities under one unit or on process is often contrary to the relative decentralization process and autonomy recognized as key to deal efficiently with the local political conjectures.

"I think there should be a better way to conduct the business here"W3R2; "we have been doing good work for a long time, do we really need to change everything?"W3R21; "I think they are thinking for the few not the many, few new position created many of us not knowing how to do things anymore"W4R8; "They do not know how to control civil servants in current condition when they are face to face, how are they going to do it remotely?"W1R7; "an entire culture has to be modified this will take time"; "there is always a good reason not to apply the new process"W1R6.

Market Driven Resistance

Market driven forms of resistance were also articulated by our respondents. Again they included traditional personal and organizational dimensions. Market driven resistance were defined as the set of external market forces influencing internal processes and how these external entity' powers were perceived by the civil servants individually and as a whole as shaping their municipality management of ICT. In these perspectives, technological changes are echoing a wider debate on privatization of services and regionalization of activities. As more and more services are outsourced, technology is allowing to clearly identify what each civil servant responsibilities are and where the responsibility lies when a system is not working as planned. The private sector role was currently described as ambiguous. The autonomy of the public sector and the interdependencies required for the technology to work were not perceived as congruent to organizational development and change. Overall, the market (including the IMF in the case of Turkey) is requiring a reduction of the number of personnel with a clearer allocation of tasks. On a constructive note, much duplication of services has been made explicit with the emergence of databases allowing real time tracking of activities. Many stakeholders are now asking difficult questions about individual differences in productivity and allocation of resource. This newly found transparency is yet to be aligned with political practices and equitable distribution of system resources among the various government layers.

"why such and such m-service should be delivered by only these people?"W1R13, "is there an overall better m-way to provide our current activities? beyond the mobile technological tools, power is the key issue"W1R9; "Mobile technologies are a hot topic but how much is sustainable from how much is hot air?"W1R1; "I will be really curious to really know the return on investment compared to face to face services? After all we are talking public money here"W1 R25.

Geographic Resistance

A deeper look at the above concepts demonstrates that a relation to local spaces and geographies is emerging from our findings. Spatial resistance to technological mobility, in public administration, refers to not only the extensive movement of civil servants in completing their daily tasks but also consider the resistance to mobility of objects, symbols and space itself. In particular, resistance to carry mobile handsets often out of work time; the symbolic stress of oneself being videoed and hearing one's own voice and the dissolving of 'here' and 'there' locations with virtual communities for example are triggering new patterns of resistance. M-government resistance is found to be greatly influenced by two types of local spaces: (i) the internal spaces (e.g. offices, buildings), the geographically located, multi-level, unbalanced distribution of tasks and powers (e.g. supra-national, national, regional, local) and the changing ongoing practices (e.g. access to databases, international phone line) are re-defining core spaces and power control instruments among the diversity of government and public-private partnerships units; (ii) the external impact of mechanisms that (i) were used and are becoming redundant re-defining cultural expectations (e.g. red tape, need for signatures and stamps, notary papers and other physical evidences such as utilities bills, property deeds), (ii) are dominating the outsourcing of technologies leading to opportunistic behaviors and information asymmetries. M-government was described as quickly becoming the focal point in evaluating and authenticating macro ICT strategies against current micro-level of performance with an unbalanced appropriation of power often by the center under the cover of economic efficiency and transparency. In doing so, ICT was positioned at the crossroad of multiples relations of powers including simultaneously offering the possibility of both liberation and oppression, together with a range of other experiences. In short, the personal modes of identification which form around ICT issues such as skills, compatibility, learning, intensity of usage and distributed collaboration were creating a polarized fault line between civil servants (user and non-user) and between each type of service offered by local public administration.

From an internal perspective, civil servant prospects were being re-negotiated, re-shaped and sometime physically re-allocated. M-government is opening a space for alternatives subjectivities capable of differentially engaging the relations of the civil servants and public administration. As described in the literature, a new conception of power is emerging more transitory, flexible and ambivalent leading to different impacts. This re-conception was often equated with becoming more modern where dominant codes can be transgressed. Still, different statuses are given to the range of technology available, often based on technical fashion, while downgrading the importance of micro-practices requirements and knowledge depository held by individuals. The following up of cases requiring the local knowledge of historical idiosyncrasies was being diluted. Moreover, it was felt by our respondents that un-intended consequences streaming from national m-ICT strategies will have to be dealt at local level without any formal support together with cross jurisdictional effects. For example, the level of interoperability and integration among the different services such as tax collection, benefits and land registry were still too compartmented, legacy of older strategies. The gap between 'ICT processed' allowing to cross-check data and the direct intervention on suspicious cases and the current reality creating resistance of their own. In other words, the details of the resource management have yet to convince local civil servants of the worthwhile of the long term investment. Consequently, they are so far skeptical that they should provide dynamic creative responses to the new circumstances.

M-government initiatives were described as in a state of 'play'. Play is defined as "a process of performative experiment: the ongoing underlying process of off-balancing, loosening, bending, twisting, reconfiguring, and transforming the permeating, eruptive/disruptive energy and mood below, behind and to the side of focused attention" (Schechner, 1993 p 43). M-government initiatives were depicted as still relatively un-regulated, hence, free to develop following multiple paths. Initiatives were not perceived as cumulative but fragmented providing hints of potential future services and searching for sustainable activities. Resistance was evident between the emerging reality and civil servants personal vision. While located within the local environment m-initiatives were perceived as difficult to command and control with little mean of integration, resource sharing, joint goals and benefits. This last factor was exacerbated by the perceived lack of service homogeneity offline and the heterogeneity of the public service. This argument was illustrating a potential source of exaggeration related to the potential negative aspects of mobility over traditional civil servant service delivery practices. The state of flux of the technology with regular upgrading, cancellation of services and emergence of new activities was leading to instabilities, and unrest lowering long-term trust and loyalty to the channel.

"We are working at a big municipality and we are supposed to lead smaller municipalities? They do not like to be told how to run their business"W4R29; "Small municipalities do not have the same resources nor the same needs"W4R22; "Do we like for the e-government need to all have the same level of provision for m-services at the same time? W4R27".

In consequence, the external impact was also described as changing and challenging. Indeed, m-government is expected to breach dominant codes. New approaches must be negotiated by traditional gate keepers. According to this view, the new change agents may ultimately be unsuccessful in understanding the emerging space and eliminated or excluded from the system. For example, retired civil servants have traditionally kept their internal social network and developed a parallel set of services allowing quicker navigation of the traditional system for a fee. Their knowledge of the regulations and internal spaces of power may now become redundant. M-government is expected to provide greater transparency and clearer playing field for all bypassing formal and in-formal power structures. These events are described as key turning points whereby m-government through the reconfiguration of practices is creating a de-facto new genre of public administration.

These new kinds of politics are contrary to traditional government continuity with top-down rather than bottom-up initiatives. This reinforces the notion that power is getting more local. Put differently, the definition of power and success becomes one of identifying how highly contingent practices and way of life undertaken by various civil servants are to a varying degree perpetuating injustices while attempting to fix them. This was further illustrated in the lack of processes for integrating public and private m-services providers which was described as diminishing the credibility and legitimacy of m-government initiatives while encouraging traditional politically motivated circuit of power towards fund allocation and choice of private partners. The result was also described as poor technology in term of cost-benefit and possible efficiency.

"We have developed some m-application but the central government does not take into account our progress"W4R3; "Authorities do not ask our opinions and there is no integration among different public services"W4R8.; "There is no real rule on how to choose a private partner this is especially important for m-application as many appear everyday"W4R22; "We need the signature of our chief for certain things. With m-government, we can not do this. Legal issues need first to be sorted then financial arrangement in function of the local needs"W4R29.

CONCLUSION AND FUTURE RESEARCH

This chapter has attempted to supply an answer on the array of resistance facing m-government in emerging market situations. In particular, we have tried to show that resistance is present, multifaceted, and at times contradictory. The findings provide insight into the reason why m-government initiatives may not be as effective in persuading civil servant to engage with the new tasks. Almost all of the discourse currently displayed by the central government addresses issues regarding final citizens benefits rather than civil servant opportunities and advantages in their day to day work. Indeed, in Turkey a lack of radical action tradition generated a range of discrete social relations between actors. In other words, the current m-government initiative may fail to capture the ongoing shifts between the central government and local public administration. This is echoed in other parts of politics for example with a clear geographical segmentation in 2009 of the electoral landscape between independent parties in the east of Turkey, conservative in the center and left on the coast.

Foremost, resistance is providing a novel way to signal and understand the antagonistic quality of modern democracy at work. As the above quotes indicate, resistance can generate different outcomes rather than absolute negative impacts. The alleged culture of complacency and inertia of top-down planning is quickly changing with many stakeholders failing to grasp the shifting relations between the state and local authorities. The findings demonstrate that the spatialized nature of m-government is dramatic and that many paths to m-readiness exist in a context where other supporting infrastructures do exist. The capacity of some actors to reconcile the gaps between locations, affecting different performances and strategies reveal the tenuous and instability of any macro-micro relationships in modern governments.

M-government activities illustrate concretely the problematic face by front loading activities aiming to encourage likelihood of adoption.

Contrary to traditional wisdom attention needs to be directed not only at end users but at convincing civil servants. This new direction in policy requires substantial changes in the way those involved as ICT providers and central government strategist approach m-government. Secondly, macro and micro practices need to be re-examined and re-conciliated. M-government initiatives naturally attract enthusiastic speculation regarding what it can do or will do, or what consequence it may have on individual civil servant jobs. This chapter goes beyond the dominant political discourse on governance, transparency and voice for the civil society. While end users will ultimately determine what combination of services is adopted, it is becoming essential to ensure that civil servants are fully integrated and part of the processes co-creation. However, it remain the opinion of the authors that m-government will provide a useful service but should not be portrayed as an exclusive tool but as a complementary access tool. It is also essential to anchor better public and private efforts to numerous other services, hence, creating a critical mass of activities making it worthwhile for stakeholders to invest resources and engage with innovation. Along these lines, further administrative reforms are require to align the potential of m-government with current practices including: m-signature, m-legal document (link with the multimedia possibility of 3G phone such as camera for insurance claims for example), m-political commitment relying on an ideological shift, m-approval system, m-accountability, m-financial regulations and a reduction of the overall digital divide within the population by continuous improvement of infrastructures. Future research ought to address the need for further in-depth understanding of the complicated structure of civil servant resistances within the growing mobile trends in multiple emerging countries. Further studies ought also to address the divergence between the intention to use and engage with m-government activities and the actual usage and services delivered at municipal level.

ACKNOWLEDGMENT

We are also very thankful to TUBITAK for funding this project (grant no 107K272). Part of this study was presented at the GBATA, 2009 conference.

REFERENCES

Ackroyd, S., & Thompson, P. (1999). *Organisational misbehaviour*. Thousand Oaks, CA: Sage.

Ajzen, I. (1985). From intentions to actions: A theory of planned behavior . In Kuhl, J., & Beckmann, J. (Eds.), *Action Control: From Cognition to Behavior* (pp. 11–39). Heidelberg: Springer.

Ajzen, I. (1991). The theory of planned behavior. *Organizational Behavior and Human Decision Processes*, *50*, 179–211. doi:10.1016/0749-5978(91)90020-T

Ajzen, I., & Fishbein, M. (1980). *Understanding attitudes and predicting social behaviour*. Englewood Cliffs, NJ: Prentice-Hall.

Almeida, P., & Kogut, B. (1999). Localization of knowledge and the mobility of engineers in regional networks. *Management Science*, *45*(7), 905–917. doi:10.1287/mnsc.45.7.905

Argote, L., & Ingram, P. (2000). Knowledge transfer: A basis for competitive advantage in firms. *Organizational Behavior and Human Decision Processes*, *82*(1), 150–169. doi:10.1006/obhd.2000.2893

Avgerou, C., Ciborra, C., & Land, F. (Eds.). (2004). *The social study of information and communication technology*. Oxford: Oxford University Press.

Banister, E. N., & Hogg, M. K. (2004). Negative symbolic consumption and consumers' dive for self-esteem: The case of the fashion industry. *European Journal of Marketing*, *38*(7), 850–868. doi:10.1108/03090560410539285

Bhabha, H. (1985). Signs taken for wonders: questions of ambivalence and authority under a tree outside Delhi, May 1917. *Critical Inquiry*, *12*, 144–165. doi:10.1086/448325

Canel, E. (1992). New social movement theory and resource mobilization: the need for integration . In Carroll, W. (Ed.), *Organizing Dissent: Social Movements in Theory and Practice* (pp. 22–51). Toronto: Garamond.

Castells, M. (2001). *The internet galaxy: Reflections on the Internet business and society*. Oxford: Oxford University Press.

Chen, G., & Kotz, D. (2000). *A Survey of context-aware mobile computing research* (TR2000-381). Hanover, NH, USA, Dartmouth College: 16.

Cilingir, D., & Kushchu, I. (2004). E-government and m-government: Concurrent leaps by Turkey. In D. Remenyi (Ed.) *Proceedings of European Conference on E-Government (ECEG 2004)* (pp. 813-821). Trinity College, Dublin, June 17-18. Department of Taoiseach, Dublin, Ireland. Reading, UK: Academic Conferences International.

Clarke, I., Hallsworth, A., Jackson, P., de Kervenoael, R., Del Aguila, R. P., & Kirkup, M. (2006). Retail restructuring and consumer choice 1. Long-term local changes in consumer behavior: Portsmouth, 1980-2002. *Environment & Planning A*, *38*(1), 25–46. doi:10.1068/a37207

Collinson, D. (1994). Strategies of resistance . In Jermier, J., Knoghts, D., & Nord, W. (Eds.), *Resistance and Power in Organizations*. Thousand Oaks, CA: Sage.

Courcoubetis, C., & Weber, R. R. (2003). *Pricing communication networks: Economics, technology and modeling*. Wiley. doi:10.1002/0470867175

Craig-Lees, M., & Hill, C. (2002). Understanding voluntary simplifiers. *Psychology and Marketing*, *19*(2), 187–210. doi:10.1002/mar.10009

Dacin, M. T., Oliver, C., & Roy, J.-P. (2007). The legitimacy of strategic alliances: An in-stitutional perspective. *Strategic Management Journal*, *28*, 169–187. doi:10.1002/smj.577

Dawes, S. S. (2009). Governance in the digital age: A research and action framework for an uncertain future. *Government Information Quarterly*, *26*, 257–264. doi:10.1016/j.giq.2008.12.003

Dedeoglu, A. O. (2004). The symbolic use of mobile telephone among Turkish consumers. *Journal of Euromarketing*, *13*(2/3), 143–163. doi:10.1300/J037v13n02_08

DiMaggio, P. J., & Powell, W. W. (1983). The iron cage revisited: Institutional isomorphism and collective rationality in organizational fields. *American Sociological Review*, *48*, 147–160. doi:10.2307/2095101

Ettlinger, N. (2003). Cultural economic geography and a relational and microspace approach to trusts, rationalities, networks, and change in collaborative workplaces. *Journal of Economic Geography*, *3*, 145–171. doi:10.1093/jeg/3.2.145

Evci, C., Ciliz, K., Anarim, E., & Sankur, B. (2004). Wireless networks in Turkey: A jewel in the crowd. *Alcatel Telecommunications Review 3rd Quarter*.

Fleming, P., & Spicer, A. (2007). *Contesting the corporation*. Cambridge, UK: Cambridge University Press. doi:10.1017/CBO9780511628047

Foley, P. (2004). Does the internet help to overcome social exclusion? *Electronic . Journal of E-Government*, *2*(2), 139–146.

Gabriel, Y. (1999). Beyond happy families: A critical reevaluation of the control resistance-identity triangle. *Human Relations*, *52*(2), 179–203. doi:10.1177/00187267990520020l

Gatignon, H., & Robertson, T. S. (1991). Innovative decision processes . In Robertson, T. S., & Kassarjian, H. H. (Eds.), *Handbook of Consumer Behavior* (pp. 316–348). Englewood Cliffs, NJ: Prentice-Hall.

Ghyasi, A. F., & Kushchu, I. (2004). Uses of mobile government in developing countries, unpublished, *Mobile Government Lab.*

Giddens, A. (1991). *Modernity and self-identity*. Cambridge: Polity Press.

Giddens, A. (1992). *The transformation of intimacy*. Cambridge: Polity Press.

Glaser, B. (1992). *Basics of grounded theory analysis: Emergence vs. forcing*. Mill Valley, CA: Sociology Press.

Goggin, G. (2008). *Mobile phone cultures*. London: Routledge.

Granovetter, M. S. (1985). Economic action and social structure: The problem of embeddedness. *American Journal of Sociology*, *91*, 481–510. doi:10.1086/228311

Gubbins, M. (2004). *Global IT spending by sector*. Computing.

Gudykunst, W. B. (1993). Toward a Theory of Effective Interpersonal and Intergroup Communication. In Wiseman & Koester (Eds.), *Intercultural Communication Competence*. Thousand Oaks, CA: Sage.

Hall, S. (1990). Culture identity and diaspora . In Rutherford, J. (Ed.), *Identity: Community, Culture, Difference* (pp. 222–237). London: Lawrence and Wishart.

Hallin, A., & Lundevall, K. (2007). mCity: User focused development of mobile services within the city of Stockholm. In I. Kushchu (Ed.), *Mobile Government: An Emerging Direction in E-Government.* New York: IGI Publishing.

Haynes, D., & Prakash, G. (1991). Introduction: The entanglement of power and resistance . In Haynes, D., & Prakash, G. (Eds.), *Contesting power: Resistance and everyday social relations in South Asia.*

Heeks, R. (2001). Understanding e-governance for development. I-government working paper series, 11. Retrieved December 200 from http://idpm.man.ac.uk/wp/igov/igov_wp11.htm

Heeks, R., & Stanforth, C. (2007). Understanding e-government project trajectories from an actor-network perspective. *European Journal of Information Systems*, *16*(2), 165–177. doi:10.1057/palgrave.ejis.3000676

Henwood, K., & Pidgeon, N. (2003). Grounded theory in psychological research in qualitative research . In Camic, P. M., Rhodes, J. E., & Yardley, L. (Eds.), *Psychology: Expanding Perspectives in Methodology and Design* (pp. 131–155). Washington, DC: American Psychological Association. doi:10.1037/10595-008

Hjorth, L. (2009). "It's complicated": A case study of women and mobile intimacy. gbc2009 proceedings/papers/P204

Hogg, M. K., & Michell, P. C. N. (1996). Identity, self and consumption: A conceptual framework. *Journal of Marketing Management*, *12*(7), 629–644. doi:10.1080/0267257X.1996.9964441

Holak, S. L., & Lehman, D. R. (1990). Intentions and the dimensions of innovation: An exploratory model. *Journal of Product Innovation Management*, *7*(1), 59–73. doi:10.1016/0737-6782(90)90032-A

Holden, S. H., Norris, D. F., & Fletcher, P. D. (2003). Electronic government at the local level: Progress to date and future issues. *Public Performance & Management Review*, *26*(4), 325–344. doi:10.1177/1530957603026004002

Honold, P. (1999). Learning how to use a cellular phone: Comparison between German and Chinese users. *Technical Communication*, *46*(2), 196–205.

Iyer, R., & Muncy, J. A. (2009). Purpose and object of anti consumption. *Journal of Business Research*, *62*(2), 160–168. doi:10.1016/j.jbusres.2008.01.023

Jacoby, J., Berning, C. K., & Dietvorst, T. F. (1977). What about disposition? *Journal of Marketing*, 22–28. doi:10.2307/1250630

Jarzabkowski, P., Balogun, J., & Seidl, D. (2007). Startegizing: the challenges of practice perspective. *Human Relations*, *60*(1), 5–27. doi:10.1177/0018726707075703

Klein, H. (1999). Tocqueville in cyberspace: Using the internet for citizen associations. *The Information Society*, *15*(4), 213–220. doi:10.1080/019722499128376

Knights, D., & McCabe, D. (2000). Ain't misbehavin? Opportunities for resistance under new forms of "quality" management. *Sociology*, *34*(3), 421–436.

Kopomaa, T. (2000). *City in your pocket: Birth of the mobile information society*. Helsinki: Gaudeamus.

Kozinets, R. V., & Handelman, J. M. (2004). Adversaries of consumption: Consumer movements, activism, and ideology. *The Journal of Consumer Research*, *31*(3), 691–704. doi:10.1086/425104

Kumas, E. (2007). *E-devlet kapısı ve risk değerlendirme metodolojisi. 12.* Ankara: Turkiye'de Internet Konferansi.

Kuran, N. H. (2005). *Türkiye icin e-devlet modeli*. Istanbul, Turkey: Bilgi Universitesi Yayınları.

Kushchu, I. (2007). *Mobile government: An emerging direction in e-government*. Hershey: IGI Publishing.

Kushchu, I., & Kuscu, H. (2003). From e-government to m-government: Facing the inevitable. In *Proceedings of European Conference on E-Government (ECEG 2003), Trinity College, Dublin, July 3-4* (pp. 253-260). Reading, UK: Academic Conferences International.

Lapointe, L., & Rivard, S. (2003). A multilevel model of resistance to information technology implementation. Retrieved from http://cat.inist.fr/?aModele=afficheN&cpsidt=15877245

Larsen, E., & Rainie, L. (2002). The rise of the e-citizen: How people use e-government agencies' web sites. Washington, DC: Pew Internet & American Life Project, 3.

Law, L. (1997). A matter of choice: discourses on prostitution in the Philippines . In Manderson, L., & Jolly, M. (Eds.), *Sites of Desire/Economies of Pleasure: Sexualities in Asia and the Pacific* (pp. 233–261). Chicago: University of Chicago Press.

Lazarus, R. S., & Folkman, S. (1984). *Stress, appraisal, and coping*. New York: Springer.

Lee, M. S. W., Fernadez, K. V., & Hyman, M. R. (2009a). Anti-consumption: An overview and research agenda. *Journal of Business Research*, *62*, 145–147. doi:10.1016/j.jbusres.2008.01.021

Lee, M. S. W., Motion, J., & Conroy, D. (2009b). Anti-consumption and brand avoidance. *Journal of Business Research*, *62*, 169–180. doi:10.1016/j.jbusres.2008.01.024

Lee, N., & Cadogan, J. W. (2009). Sales force social exchange in problem resolution situations. *Industrial Marketing Management*, *32*(3), 355–372. doi:10.1016/j.indmarman.2008.02.002

Lee, T., & Jun, J. (2005). Contextual perceived usefulness? Toward an understanding of mobile commerce acceptance. In *Proceedings of the International Conference on Mobile Business* (ICMB'05)

Leung, L., & Wei, R. (2000). More than just talk on the move: Uses and gratifications of the cellular phone. *Journalism & Mass Communication Quarterly*, *77*(2), 308–320.

Ling, R. (2008). *New tech, new ties: How mobile communication is reshaping social cohesion*. Cambridge, MA: MIT Press.

Maliranta, M., & Rouvinen, P. (2004). *Informational mobility and productivity - Finnish evidence. Discussion Papers 919*. The Research Institute of the Finnish Economy.

Marcelle, G. M. (2004). *Technological learning*. UK: Edward Elgar.

McIvor, R., McHugh, M., & Cadden, C. (2002). Internet technologies: Supporting transparency in the public sector. *International Journal of Public Sector Management*, *15*(3), 170–187. doi:10.1108/09513550210423352

Melitski, J., Holzer, M., Kim, S., Kim, C., & Rho, S. (2005). Digital government worldwide: An e-government assessment of municipal websites throughout the world. *International Journal of E-government Research*, *1*(1), 1–19.

Merriam, S. B. (1998). *Qualitative research and case study applications in education*. San Francisco: Jossey-Bass.

Miles, M. B., & Huberman, A. M. (1994). *Qualitative data analysis: An expanded sourcebook*. Thousand Oaks, CA: Sage Publications.

Mittelstaedt, R. A., Grossbart, S. L., Curtis, W. W., & Devere, S. P. (1976). Optimal stimulation level and the adoption process. *The Journal of Consumer Research*, *3*, 84–94. doi:10.1086/208655

Moisio, R. J. (2003). Negative consequences of mobile phone consumption: Everyday irritations, anxieties and ambiguities in the experiences of Finnish mobile phone consumers. *Advances in Consumer Research. Association for Consumer Research (U. S.)*, *30*, 340–345.

Moore, D. S. (1997). Remapping resistance: 'Ground for struggle' and the politics of place . In Pile, S., & Keith, M. (Eds.), *Geographies of Resistance* (pp. 87–106). London: Routledge.

Musso, J., Weare, C., & Hale, M. (2000). Designing web technologies for local governance reform: Good management or good democracy. *Political Communication*, *17*(1), 1–19. doi:10.1080/105846000198486

Oxford Internet Institute (2006). Breaking barriers to e-government online.

Ozcan, Y. Z., & Kocak, A. (2003). Research note: A need or a status symbol? Use of cellular telephones in Turkey. *European Journal of Communication*, *18*(2), 241–254. doi:10.1177/0267323103018002004

Pagani, M. (2004). Determinants of adoption of third generation mobile multimedia services. *Journal of Interactive Marketing*, *18*(3), 46. doi:10.1002/dir.20011

Palmer, M., & O'Kane, P. (2007). Strategy as practice: Interactive governance spaces and the corporate strategies of retail TNCs. *Journal of Economic Geography*, *7*, 515–535. doi:10.1093/jeg/lbm015

Palmer, M., Owens, M., & De Kervenoael, R. (in press). Paths of the least resistance: Understanding how motives form in international retail joint venturing. *The Service Industries Journal, 30*(8).

Piacentini, M., & Banister, E. M. (2009). Managing anti-consumption in excessive drinking culture. *Journal of Business Research*, *62*, 279–288. doi:10.1016/j.jbusres.2008.01.035

Pieters, R. (1993). Consumers and their garbage a framework and some experiences from the Netherlands with garbage separation programs. *European Advance in Consumer Research*, *1*, 541–546.

Pile, S., & Keith, M. (1997). *Geography of resitance*. London: Routledge.

Pondy, L. R. (1966). A systems Theory of Organizational Conflicts. *Academy of Management Journal*, *9*(3), 296–320. doi:10.2307/255122

Prasad, P., & Prasad, A. (2000). Stretching the iron cage: The constitution and implications of routine workplace resistance. *Organization Science*, *11*(4), 387–403. doi:10.1287/orsc.11.4.387.14597

Prochaska, J. O., DiClemente, C. C., & Norcross, J. C. (1992). In search of how people change: Applications to addictive behaviors. *The American Psychologist*, *47*(9), 1102–1114. doi:10.1037/0003-066X.47.9.1102

Roster, C. A. (2001). Letting go: The process and meaning of dispossession in the lives of consumers. In Meyers-Levy, J., & Gilly, M. (Eds.), *Advances in Consumer Research* (pp. 425–430). Provo, UT: Association for Consumer Research.

Sack, R. D. (1986). *Human territoriality: Its theory and history*. New York: Cambridge University Press.

Schechner, R. (1993). *The future of ritual: Writings on culture and performance*. London: Routledge. doi:10.4324/9780203359150

Schein, E. H. (2004). *Organizational culture and leadership*. San Francisco: Jossey-Bass.

Shaw, D., & Newholm, T. (2002). Voluntary simplicity and the ethics of consumption. *Psychology and Marketing*, *19*(2), 167–185. doi:10.1002/mar.10008

Sheng, H., & Trimi, S. (2008). M-government: Technologies, applications and challenges. *Electronic Government: An International Journal*, *5*(1), 1–18. doi:10.1504/EG.2008.016124

Sheth, J. N. (1981). Psychology of innovation resistance: The less developed concept (LDC) in diffusion research. *Research in Marketing*, *4*, 273–282.

Solomon, M. R., & Buchanan, B. (1991). A role-theoretic approach to product symbolism: Mapping a consumption constellation. *Journal of Business Research*, *22*(March), 95–110. doi:10.1016/0148-2963(91)90044-X

Strauss, A. (1987). *Qualitative research for social scientists*. Cambridge: Cambridge University Press. doi:10.1017/CBO9780511557842

Strauss, A., & Corbin, J. (1998). *Basics of qualitative research techniques and procedures for developing grounded theory*. Thousand Oaks, CA: Sage.

Sungu, E., Sungu, H., & Bayrakci, M. (2008). E-inclusion: Providing services towards an information society for all. *8th International Educational Technology Conference.*

Szmigin, I., & Foxall, G. (1998). Three forms of innovation resistance: The case of retail payment methods. *Technovation*, *18*(6/7), 459–468. doi:10.1016/S0166-4972(98)00030-3

Thomke, S., & Fujimoto, T. (2000). The effect of front-loading problem-solving on product development performance. *Journal of Product Innovation Management*, *17*, 128–142. doi:10.1016/S0737-6782(99)00031-4

Torenli, N. (2006). The 'other' faces of digital exclusion: ICT gender divides in the broader community. *European Journal of Communication*, *21*(4), 435–455. doi:10.1177/0267323106070010

Tuckman, B. W. (1965). Developmental sequences in small groups. *Psychological Bulletin*, *63*, 384–399. doi:10.1037/h0022100

TURKSAT. (2008). http://www.turksat.com.tr/english/index.php/e-Government-/e-Government-Gateway-Project/What-is-e-Government-Gateway-Project.html

Tutkun, C. (2007). *Devletin Kisa Yolu. In 12.* Ankara: Turkiye'de Internet Konferansi.

Venkatesh, V., Morris, M. G., Davis, G. B., & Davis, F. D. (2003). User acceptance of information technology: Toward a unified view. *Management Information Systems Quarterly*, *27*(3), 425–478.

Vincent, J., & Harris, L. (2008). Effective use of mobile communication in e-government: how do we reach the tipping point. *Information Communication and Society*, *11*(3), 395–413. doi:10.1080/13691180802025632

Watts, M. (1997). Black gold, white heat: State violence, local resistance and the national question in Nigeria . In Pile, S., & Keith, M. (Eds.), *Geographies of Resistance* (pp. 33–67). London: Routledge.

Whittenbraker, J., Gibbs, B. L., & Kahle, L. R. (1983). Seat belt attitudes, habits and behaviors: An adaptive amendment to the Fishbein model. *Journal of Applied Social Psychology*, *13*, 406–421. doi:10.1111/j.1559-1816.1983.tb01748.x

Whittington, R. (2007). Strategy practice and strategy process: family differences and the sociological eye. *Organization Studies*, *28*(10), 1575–1586. doi:10.1177/0170840607081557

Wilska, T. A. (2003). Mobile phone use as part of young people's consumption styles. *Journal of Consumer Policy*, *26*(4), 441–463. doi:10.1023/A:1026331016172

Woo, K. S., & Fock, H. K. Y. (1999). Customer satisfaction in the Hong Kong mobile phone industry. *The Service Industries Journal*, *19*(3), 162–174. doi:10.1080/02642069900000035

Yin, R. (1989). *Case study research: Design and methods*. Thousand Oaks, CA: Sage Publishing.

Yin, R. K. (1994). *Case study research: Design and methods*. Thousand Oaks, CA: Sage.

Yu, B., & Kushchu, I. (2004). The value of mobility for e-government. In *Proceedings of European Conference on E-Government (ECEG 2004)* (pp. 887-899). Trinity College, Dublin, June 17-18. Department of the Taoiseach, Dublin, Ireland. Reading, UK: Academic Conferences International.

ENDNOTES

1. Marcelle, 2004 page 26 Table 2.1: Key texts on organisational learning and technological capability building, provides a detailed examination and literature review of these theories providing insight in the case of emerging market on firms how firms operating in an age when generic technologies and paradigm changing technologies are pervasive, should go about technological learning and successful product/process innovation.
2. Rogers (1995) created the adoption of innovation framework which identified desirable factors for the success of a new product or innovation including relative advantage, compatibility, complexity, communicability observability and triability. Further models such as the Theory of Reasoned Action (TRA) and Theory of Planned Behaviour (TPB) (Ajzen, 1985, 1991; Ajzen & Fishbein, 1980) provide a theoretical framework for understanding better the human behavior factors. Other key work include Davis et al (1989) to guide IS/IT acceptance research and (Holak & Lehmann, 1990; Pagani, 2004; Venkatesh et al., 2003).

APPENDIX

SUMMARY OF SURVEY INSTRUMENT

Wave one investigated the current state of digital government including e-government and m-government emerging activities.

- Q0 How long have you been working in public sector? Educational level, type of work and general information on municipality characteristics regarding ICT
- Q1 Can you summarize your municipality main ICT activities over the last 5 years
- Q2 Set of question on ICT expertise location
- Q3 Set of question on Trust in other knowledge regarding ICT
- Q4 Set of Question on Coordination of ICT knowledge*

    *Q2-4 adapted from Kanawattanachai, P. (2002), "Formation and development of socially-shared cognition and its impact on performance of virtual teams over time", doctoral thesis, Department of Information Systems. Case Western Reserve University.

Wave two explored organizational processes, within day to day activities of civil servants regarding possible resistance to ICT.

- Q0 What impact do the following barriers have on the municipality's usage of ICT in general?
    - Errors/defects in supplied equipment hardware and software
    - Lack of flexibility for ICT suppliers
    - Lack of integration between applications
    - Lack of ICT qualified staff in the municipality
    - Difficulties to recruit or retain ICT qualified staff
    - Reluctance among staff to use ICT
    - Municipality lack of updated ICT strategy
    - Lack of commitment by management
    - ICT cost

Wave three unpacked the multifaceted concept of mobility.

- Q0 What does mobility mean to you in your line of work?
- Q1 Should there more than one way to gain access to electronic services and databases provided by the government? Why? And How?
- Q2 How do you evaluate if there is a demand for this sort of mobile services?
- Q3 How do you perceive the use of mobile phones for your daily tasks?
- Q4 What is the most important benefit of being mobile for a civil servant work?
- Q5 What may be the key problems related to using mobile phones for public administration work, and from an individual user perspective?

Wave four investigated the concept of capability building link with personal and organizational resistance regarding m-government in particular.

- Q0 What type of initiatives have been undertaken to introduce m-government to civil servants in your municipality
- Q1 At an individual municipality level how are you preparing for the future development of m-government services? Why? Why not?
- Q3 Can you describe the current ICT capabilities from a technical perspective then from an individual civil servant user skill perspective? Can you provide anecdote
- Q4 How do you think the political pressures are shaping m-government adoption? Why? Why not? Benefits/problems?
- Q5 Do you perceived any power-control issues arising from the new possibilities and opportunities offered by m-government

Section 3
M-Technology:
Economic and Social Effects

Chapter 11
Mobile Telephony and Economic Growth in Developing Economies

Heli Virta
Lappeenranta University of Technology, Finland

Kaisu Puumalainen
Lappeenranta University of Technology, Finland

Anni Tuppura
Lappeenranta University of Technology, Finland

ABSTRACT

This chapter investigates the influence of mobile phone penetration on economic growth in developing economies. It is widely agreed that telecommunications infrastructure has a positive effect on GDP growth. The empirical evidence concerns mainly fixed line telephony and is mostly conducted with samples from developed countries. Mobile telephony, on the other hand, may be particularly important in those low-income countries, where landlines are not accessible to all. As there also is some evidence that mobile technology may encourage innovative entrepreneurial activity, it is likely that mobile telephony has a great positive impact on welfare especially in developing economies. To examine the role of mobile telephone penetration in economic growth, we use difference and system GMM estimators with a sample of low-income and lower middle-income countries and find mobile telephony to boost economic growth. This result suggests that extensive mobile cellular network coverage facilitates economic development in developing countries.

INTRODUCTION

In this chapter, we assess how mobile phone penetration impacts on economic growth in developing economies. Mobile diffusion started later in low-income than high-income countries, but after that the growth of the penetration rate has been rapid. The diffusion is accelerated by the fact that it is cheaper and faster to build mobile infrastructure than fixed lines. Moreover, as penetration has already reached nearly 100 percent in high-income countries, future growth of

DOI: 10.4018/978-1-61692-818-6.ch011

mobile telephony will concentrate in developing economies (Donner, 2008).

It is widely agreed that telecommunications infrastructure has a positive effect on GDP growth because it lowers search and transaction costs, improves the quality of information and accelerates the diffusion of knowledge. Thus, mobile telephony should play a similar role in economic development. As the impact of mobile diffusion on access to communication services might be at its largest if fixed lines are lacking (ITU, 2003), mobile telephony may be particularly important in those low-income countries, where landlines are not available to all. There is already some evidence that the adoption of this innovation has increased the efficiency of small business activity, and generated new kinds of businesses (Frempong, 2009; Hahn and Kibora, 2008; Donner, 2007). A problematic aspect from the point of view of this study is, however, that the official GDP figures are unlikely to capture the impact of mobile telephony entirely.

Particularly the earlier literature on the relationship between telecommunications and economic growth uses the number of mainlines as a proxy for telecommunications infrastructure, while some studies use other measures such as the annual share of real GDP devoted to telecommunications industry (Röller and Waverman, 2001; Madden and Savage, 2000). Due to data availability, the bulk of the research concentrates on OECD countries. Nowadays, data on mobile telephony are also available even for developing economies, but the topic of mobile phone penetration and economic growth in less developed countries has received only limited attention. Moreover, many studies use cross-sectional data even if the use of panel data would make it possible to take country-specific fixed effects into account. Another shortcoming in the existing literature is that some of the studies treat telecommunications infrastructure or mobile telephony as an exogenous regressor or falsely conclude that endogeneity does not diminish the reliability of the results in question.

In this chapter, we aim to fill some of these gaps by focusing on low-income and lower-middle incomecountries. Furthermore, we use panel data and methodology that accounts for endogeneity. The chapter starts with a literature review followed by a methodology section also including estimation results. The chapter ends with directions for future research and conclusions.

LITERATURE REVIEW

Telecommunications in large has been claimed to have a positive effect on economic development since it drastically decreases transaction costs and accelerates the diffusion of knowledge. A positive relationship between investment on telecommunications, or teledensity, and economic growth has been tested with various empirical settings (e.g., Norton, 1992; Dholakia and Harlam 1994; Dutta, 2001; Röller and Waverman, 2001; Cieslik and Kaniewsk, 2004; Datta and Agarwal, 2004). Norton (1992) employs two different datasets with partly overlapping time periods and country samples (124 countries in the larger sample) and finds a positive and unidirectional relationship between growth and telecommunications (measured as stock of telephones per hundred persons). Based on his results, Norton (p. 192) concludes that telecommunications infrastructure "must be viewed as at least as important as conventional economic forces such as stable money growth, low inflation, and an open economy". Dutta (2001) uses data from 15 developing and 15 industrialized countries also finding a positive relationship between telecommunications (total number of telephones and number of telephones per hundred persons) and GDP. To check the direction of causality, Dutta applies Granger causality test and concludes that although there is some evidence for causation from economic activity to telecommunications, this relationship is substantially weaker than for opposite direction. Further he concludes that there is no difference between the industrialized and developing countries in this sense. Several other researches report evidence for two-way causal relationship between telecom investment and economic growth. Among these are Cronin et al. (1991) and Madden and Savage (1998). Cronin et al. use US data between 1958 and 1988 and find that an increase in gross national product increases telecom investment but also that such investment has a positive influence on overall growth. Madden and Savage, in turn, employ data from 27 Central and Eastern European transitional economies in 1991-1994. Also Datta and Agarwal (2004), using data from 22 OECD countries between 1980 and 1992, uncover a robust positive impact of telecom on economic growth and evidence for two-way causality. However, in light of the regression coefficients the effect of growth on telecom is smaller than telecom on growth.

While several studies find evidence in support of a positive relationship between telecommunications and economic growth, it may be that a certain amount of investment has different consequences in different environments. In a recent paper employing Chinese data from 1989-1998, Shiu and Lam (2008) find that the effects and direction of causation vary between different geographic regions. They suggest that the direction of causation (or the ability of telecom investment to boost economic growth) depends on "complementary factors" such as business environment, transportation network and level of higher education. Furthermore, there is some evidence that due to network effects, the cumulative amount of users may influence the strength of the relationship between telecom investment and growth. Based on their empirical study (21 OECD countries, 1970-1990), Röller and Waverman (2001) suggest that a critical mass phenomenon may be present in telecommunication investment and that the economies under the critical threshold may have slower growth rates than the economies above. Somewhat contradictory, however, Datta and Agarwal (2004) discover that the GDP gain of telecom investment is largest in countries where the telecom infrastructure is initially smallest.

In sum, the empirical evidence suggests that telecommunications has a positive influence on welfare, but that this influence is somewhat circumstantial. It is noteworthy, however, that most of this evidence is based on samples from developed or industrialized countries. Furthermore, because mobile phone is a fairly young innovation the bulk of the empirical evidence is carried out using fixed line telecom investments as a measure of telecommunications (with some, more recent, exceptions, e.g. Shiu and Lam, 2008). As discussed above, the mobile phone technology is not only substituting technology for fixed line phones, but holds potential for additional uses, and may even encourage new types of businesses.

METHODOLOGY

Data Sources and Measurement

We estimate our model with panel data from those low-income and lower middle-income countries for which data are available between years 1999-2007. Year 1999 was chosen for the first year of the period because mobile telephony started to become more common in developing economies only in the late-1990s. Moreover, the estimation method used below is only suitable for panels with a large number of units relative to the number of periods.

The data are collected from Global Market Information Database (GMID) (Euromonitor, 2009), World Bank (2008), International Telecommunication Union (ITU) (2008), Barro and Lee (2001) and International Country Risk Guide (ICRG) (2006). Figure 1 lists the variables used as well as their sources. The list of countries included in the analysis is presented in Appendix. The descriptive statistics of Figure 2 indicate large differences in the adoption of mobile technology

Figure 1. Variable descriptions and sources

Series	Variable
gdp_ppp	GDP PPP
gdp	Total GDP, constant 2000 $
pop	Population: national estimates at January 1st ('000)
govc	Government expenditure, constant 2000 $
exports	Exports, constant 2000 $
imports	Imports, constant 2000 $
mtcov	Coverage of mobile cellular network, % of population
i_icrg_c	Corruption
gfcf_share	Gross fixed capital formation, % of GDP
i_tyr	Average schooling years in the total population (over 25)
i_tyr15	Average schooling years in the total population (over 15)

among the developing countries. A more extensive survey is presented in Donner (2008).

To measure human capital, we use average schooling years in the total population aged 25 and over from Barro and Lee (2001). As the original data set covers years 1960-2000 at five-year intervals, missing values have to be intrapolated and extrapolated to get a yearly series between 1999 and 2006. This can be done without a large impact on the estimation results because the level of education changes slowly. In addition, because the citizens entering the labor market in developing economies are typically younger than in developed economies, we test for the robustness of the results by using the average schooling years in the total population aged 15 and over as the proxy for human capital.

Because earlier literature has indicated that trade openness might be beneficial for economic development (see e.g. Sachs and Warner, 1995 and Datta and Agarwal, 2004, in the context of telecommunications), we include trade openness in some of the regressions. We measure the variable by the ratio of imports plus exports to GDP.

Furthermore, e.g. Mauro (1995) and Gyimah-Brempong (2002) claim that corruption has a negative impact on GDP growth. The issue has gained plenty of attention in the recent years and might be connected to the diffusion of mobile technologies in developing countries because building infrastructure often offers rent-seeking opportunities. Therefore, we also control for the level of corruption. To do that, we use the corruption index from International Country Risk Guide (ICRG) (2006). The index is based on subjective evaluation of the level of corruption within the political system of a country and has been popular in cross-country studies because it is the only index that provides a consistent time series back to the early 1980s. The index ranges

Figure 2. Summary statistics

Variable	Obs	Mean	Std. Dev	Min	Max
Log of GDP (PPP) per capita	766	0.60	1.19	-7.68	3.22
Growth of GDP (PPP) per capita	764	0.06	0.05	-0.31	0.33
Gross fixed capital formation, % of GDP	680	21.76	8.22	3.00	62.00
Average schooling years in the total population over 25	432	4.13	1.85	0.76	8.69
Government expenditure, % of GDP	451	24.89	14.60	0.02	125.14
Coverage of mobile cellular network, % of population	400	61.99	29.90	0.00	100.00
Trade openness	709	0.76	0.48	0.11	3.69
Corruption	340	2.07	0.80	0.00	4.00

between 0 and 6, with 6 assigned to the countries with the lowest level of corruption.

Model Specification and Method of Estimation

To find out how mobile telephony impacts on economic growth, we regress GDP per capita growth on a set of variables in the spirit of Barro (1991). The most important theoretical foundations for our model are Solow (1956) and Mankiw et al. (1992). Our specification also relies on Datta and Agarwal (2004), who use panel data to examine how telecommunications affect economic growth. In a general form, the relationship we are estimating is

$GRTH_{it} = f(GRTH_{it-1}, GDP_{it-1}, INV / Y_{it}, H_{it}, G / Y_{it}, MTCOV_{it})$ (1)where i indexes countries and t time. $GRTH$ is the rate of growth of GDP per capita measured in purchasing power parity (PPP). The lagged growth rate $GRTH_{i,t-1}$ is included as an explanatory variable in order to account for the potential correlation between previous and subsequent growth rates (see e.g. Datta and Agarwal, 2004). We also control for GDP per capita (PPP) in the previous year, $GDP_{i,t-1}$, to account for the convergence of income levels predicted by neoclassical growth models. Because these models predict that countries with low GDP per capita levels should grow faster than the countries where GDP per capita is higher, the expected sign of $GDP_{i,t-1}$ is negative. In accordance with Mankiw et al. (1992), who find that an augmented Solow model with physical and human capital accumulation fits cross-country data well, we include investment and human capital: INV/Y is the share of gross fixed capital formation of GDP, and H represents human capital. Both variables should play a positive role in economic development. The ratio of government expenditures to GDP, G/Y, is expected to reflect the possibly distortionary effects of taxation or government-expenditure programs (Barro, 1991). To capture the impact of mobile telephony on economic growth, we use data on the coverage of mobile cellular network, $MTCOV$. As discussed above, this variable can be expected to impact on GDP per capita growth positively.

If mobile telephony facilitates economic growth, the same holds for the reverse relationship: economic development is likely to accelerate mobile diffusion. Therefore, we control for the endogeneity by using the dynamic panel general method of moments (GMM) estimator by Arellano and Bond (1991), also called difference GMM. It is an instrumental variables estimator designed for panel data. As the name implies, difference GMM uses lagged predetermined and endogenous variables as instruments for those predetermined and endogenous variables in first differences. Another option is Blundell and Bond system GMM estimator (Blundell and Bond, 1998), which adds the original equations in levels to the system of first-differenced equations. The validity of the additional instruments hinges on the assumption that changes in the instrumenting variables are uncorrelated with the fixed effects. Roodman (2009) discusses difference and system GMM more extensively.

The equation we are estimating is

$$\Delta y_{it} = \alpha + \beta \Delta y_{i,t-1} + \sum_{j=1}^{k} \gamma_j x_{jit} + \eta_i + \varepsilon_{it},$$
$$E[\eta_i] = E[\varepsilon_{it}] = E[\eta_i \varepsilon_{it}] = 0 \qquad (2)$$

where Δy_{it} denotes the growth rate of the GDP per capita (PPP) in country i at time t, is the set of control variables including GDP per capita in the previous year as well as the coverage of mobile cellular network, is an unobserved country-specific time-invariant effect, and is a disturbance term.

Differencing eliminates the country-specific effects, resulting in

$$\Delta y_{it} - \Delta y_{i,t-1} = \beta\left(\Delta y_{i,t-1} - \Delta y_{i,t-2}\right) + \sum_{j=1}^{k} \gamma_j \left(x_{jit} - x_{ji,t-1}\right) + \varepsilon_{it} - \varepsilon_{i,t-1}. \quad (3)$$

Although predetermined variables become endogenous in first differences, deeper lags of the regressors remain available as potential instruments. Difference GMM uses this attribute and instruments the predetermined and endogenous variables in first differences with suitable lags of their levels. System GMM (Blundell and Bond, 1998) adds the original equations in levels into the system. The predetermined and endogenous variables in levels are then instrumented with their lagged first differences. The main assumption is that $E\left[\eta_i \Delta\varepsilon_{it}\right] = 0$, implying that the country-specific effects should be uncorrelated with disturbance-term changes. The difference-in-Sargan test can be used to test the validity of a subset of instruments. (Roodman, 2009)

The number of instruments is quadratic in T, which is problematic because too many instruments tend to overfit endogenous variables. The number of lags of the instrumenting variable is thus restricted to lags 1 and 2 in the transformed equation and lag 0 in the levels equation for predetermined variables. For endogenous variables, lags 2 and 3 are used in the transformed equation and lag 1 in the levels equation. The set of instruments is collapsed in a manner described in Roodman (2009) to further reduce the number of instruments.

The one-step GMM estimator is efficient only under homoskedasticity and uncorrelated error terms, while the two-step estimator is asymptotically more efficient. However, the two-step estimates of the standard errors tend to be biased downward (Arellano and Bond, 1991; Blundell and Bond, 1998). Therefore, a finite-sample correction for the asymptotic variance of the two-step GMM estimator of Windmeijer (2005) is used below in two-step estimations.

We use the Stata module xtabond2 by Roodman (2005) both to correct for the downward bias in the two-step estimates of the standard errors and to reduce the number of instruments. Small sample statistics are reported throughout.

Estimation Results

Figure 3 depicts the one-step estimation results based on difference (uneven columns) and system (even columns) GMM. Regressions 1 and 2 correspond to equation (1), regressions 3 and 4 add trade openness and regressions 5 and 6 the level of corruption.

The difference GMM estimations support the convergence hypothesis: higher GDP per capita in the previous year seems to lower the subsequent growth rate. Somewhat surprisingly, investment and human capital are statistically insignificant, although the coefficients of both are of the expected sign. The same holds for government expenditure. Trade openness and corruption do not seem to affect economic development. Instead, the coverage of mobile cellular network seems to have a positive impact on economic development. This positive effect can also be seen from the system GMM results, which do not, however, support the convergence hypothesis. Instead, they indicate that a higher share of gross fixed capital formation of GDP boosts GDP per capita growth. Other variables remain statistically insignificant.

According to the Hansen J statistic test for over-identifying restrictions depicted at the bottom of Figure 3, the set of instruments is valid, and the Arellano-Bond test for AR(2) in first differences implies that the error terms in levels are not serially correlated, as required. Moreover, the difference-in-Sargan test indicates that the additional instruments of system GMM estimation are valid.

The results hold with the average schooling years in the total population aged 15 and over as the proxy for human capital.

Figure 3.One-step estimation results for GDP growth

	(1) difference	(2) system	(3) difference	(4) system	(5) difference	(6) system
Lagged GDP per capita growth	-0.181	-0.025	-0.213	-0.027	-0.233	-0.106
	(-1.04)	(-0.12)	(-1.29)	(-0.13)	(-0.96)	(-0.40)
Lagged GDP per capita	**-0.511****	-0.028	**-0.555****	-0.024	-0.584	-0.022
	(-2.38)	(-0.85)	**(-2.77)**	(-0.79)	(-1.21)	(-1.63)
Gross fixed capital formation, % of GDP	0.001	**0.004****	0.001	**0.003***	0.001	**0.005***
	(0.97)	**(2.25)**	(1.13)	**(1.83)**	(0.54)	**(1.79)**
Average schooling years in total over 25 pop.	0.016	0.004	0.012	0.004	0.013	0.006
	(0.53)	(0.35)	(0.40)	(0.40)	(0.32)	(1.05)
Government expenditure, % of GDP	-0.003	-0.000	-0.001	-0.000	-0.001	-0.001
	(-0.52)	(-0.24)	(-0.29)	(-0.33)	(-0.17)	(-0.57)
Coverage of mobile cellular network, % of pop.	**0.001***	**0.001***	**0.001****	**0.001*****	**0.001***	**0.001****
	(1.94)	**(1.84)**	**(2.40)**	**(2.73)**	**(1.71)**	**(2.37)**
Openness			0.019	-0.004	0.019	-0.007
			(0.68)	(-0.38)	(0.92)	(-0.77)
Corruption					0.000	-0.015
					(0.02)	(-1.24)
Constant		-0.095		-0.076		-0.066
		(-1.56)		(-1.27)		(-1.34)
Year 2001	-0.158**	-0.003	-0.164**	-0.010	-0.175	0.001
	(-2.45)	(-0.15)	(-2.33)	(-0.54)	(-1.08)	(0.03)
Year 2002	-0.145**	-0.010	-0.147**	-0.015	-0.156	-0.010
	(-2.62)	(-0.60)	(-2.38)	(-0.89)	(-1.16)	(-0.47)
Year 2003	-0.123**	0.003	-0.125**	-0.002	-0.133	-0.001
	(-2.58)	(0.19)	(-2.21)	(-0.11)	(-1.15)	(-0.04)
Year 2004	-0.085**	0.009	-0.088**	0.006	-0.095	0.007
	(-2.26)	(0.71)	(-2.12)	(0.48)	(-1.00)	(0.54)
Year 2005	-0.058*	0.007	-0.062*	0.005	-0.065	0.007
	(-1.93)	(0.84)	(-1.94)	(0.54)	(-0.93)	(0.76)
Year 2006	-0.022	0.017**	-0.024	0.015**	-0.027	0.014**
	(-1.34)	(2.73)	(-1.39)	(2.69)	(-0.70)	(2.43)
t statistics in parentheses (* significant at 10%; ** significant at 5%; *** significant at 1%)						
Observations	94	138	92	136	87	127
Number of groups	27	35	27	35	25	32
Number of instruments	17	23	19	26	21	29
Arellano-Bond test						
for AR(2) in first differences (P-value)	0.973	0.258	0.762	0.228	0.721	0.252
Hansen test of overidentifying restrictions (P-value)	0.182	0.807	0.541	0.680	0.384	0.481
Difference-in-Sargan test (P-value)		0.542		0.374		0.209

Notes:
Null hypotheses of the tests are: Arellano-Bond test for autocorrelation: null = no autocorrelation;
Hansen test of overidentifying restrictions: null = the set of instruments is valid;
difference-in-Sargan test: null = the level moment conditions imposed
in system GMM are valid.

Datta and Agarwal (2004) incorporate the square of telecommunications infrastructure to examine the returns to scale of telecom investment. In the same spirit, we included the coverage of mobile cellular network squared in our estimations, but found the variable insignificant in all cases. Because the number of instruments is such a crucial issue and because including the variable squared might result in multicollinearity, we leave the squared term out of the regression results presented here.

The results of asymptotically more efficient two-step estimations are presented in Figure 4. They are mainly in line with Figure 3, implying that countries with initially lower GDP per capita levels experience faster growth. Also a high investment rate and an extensive mobile cellular network coverage seem to facilitate economic development. Thus, the results presented here indicate that mobile telephony has a statistically significant positive role in economic growth of low-income and lower middle-income countries.

Figure 4. Two-step estimation results for GDP growth

	(1) difference	(2) system	(3) difference	(4) system	(5) difference	(6) system
Lagged GDP per capita growth	-0.081	-0.036	-0.189	-0.041	-0.305	-0.023
	(-0.39)	(-0.15)	(-0.75)	(-0.17)	(-0.54)	(-0.08)
Lagged GDP per capita	**-0.529***	-0.046	-0.433	**-0.044***	-0.278	**-0.031***
	(-1.77)	(-1.69)	(-1.60)	**(-1.95)**	(-0.31)	**(-2.01)**
Gross fixed capital formation, % of GDP	0.001	**0.004****	0.002	**0.004****	0.003	0.004
	(0.73)	**(2.56)**	(1.45)	**(2.60)**	(0.82)	(1.21)
Average schooling years in total over 25 pop.	0.003	0.007	-0.008	0.006	-0.017	0.005
	(0.03)	(0.47)	(-0.15)	(0.53)	(-0.13)	(0.48)
Government expenditure, % of GDP	0.002	0.001	0.002	0.001	0.000	-0.000
	(0.20)	(0.34)	(0.40)	(0.46)	(0.03)	(-0.38)
Coverage of mobile cellular network, % of pop.	0.001	0.001	**0.001****	**0.001*****	0.000	**0.001****
	(1.62)	(1.06)	**(2.07)**	**(3.15)**	(0.65)	**(2.70)**
Openness			0.048	-0.006	0.025	-0.009
			(0.96)	(-0.54)	(1.05)	(-1.04)
Corruption					-0.008	-0.014
					(-0.45)	(-0.99)
Constant		-0.104*		-0.113**		-0.058
		(-2.02)		(-2.33)		(-0.86)
Year 2001	-0.160	-0.006	-0.126	-0.007	-0.101	-0.001
	(-1.16)	(-0.29)	(-1.08)	(-0.53)	(-0.28)	(-0.10)
Year 2002	-0.145	-0.013	-0.112	-0.013	-0.095	-0.008
	(-1.22)	(-0.59)	(-1.09)	(-0.89)	(-0.32)	(-0.68)
Year 2003	-0.115	0.003	-0.085	0.004	-0.074	0.003
	(-1.23)	(0.18)	(-0.99)	(0.29)	(-0.31)	(0.24)
Year 2004	-0.083	0.008	-0.063	0.007	-0.046	0.010
	(-1.06)	(0.50)	(-0.96)	(0.63)	(-0.23)	(0.91)
Year 2005	-0.062	0.007	-0.048	0.006	-0.033	0.011
	(-1.00)	(0.68)	(-0.92)	(0.74)	(-0.21)	(0.97)
Year 2006	-0.025	0.016**	-0.016	0.016***	-0.010	0.015**
	(-0.80)	(2.51)	(-0.57)	(3.07)	(-0.11)	(2.37)
t statistics in parentheses (* significant at 10%; ** significant at 5%; *** significant at 1%)						
Observations	94	138	92	136	87	127
Number of groups	27	35	27	35	25	32
Number of instruments	17	23	19	26	21	29
Arellano-Bond test						
for AR(2) in first differences (P-value)	0.590	0.473	0.887	0.506	0.904	0.521
Hansen test of overidentifying restrictions (P-value)	0.182	0.807	0.541	0.680	0.384	0.481
Difference-in-Sargan test (P-value)		0.542		0.374		0.209

Notes:
Null hypotheses of the tests are: Arellano-Bond test for autocorrelation: null = no autocorrelation;
Hansen test of overidentifying restrictions: null = the set of instruments is valid;
difference-in-Sargan test: null = the level moment conditions imposed
in system GMM are valid.

FUTURE RESEARCH DIRECTIONS

While the relationship of telecommunications and economic growth has been studied from many angles, the developing country point of view has been somewhat neglected. Moreover, some recent papers imply that mobile technology may generate totally new businesses and ways of doing business particularly in developing economies. Therefore it is necessary to further investigate the role of mobile telephony in economic development. Case studies may be especially useful in analyzing this multifaceted phenomenon.

Empirical studies have indicated that various environmental factors, such as policies and infrastructure, impact on the relationship between telecommunications and economic growth. Thus, another interesting direction for future research would be to study which circumstantial factors are necessary for mobile telephony to boost economic growth in low-income countries.

CONCLUSION

This chapter focused on the consequences of mobile telephone diffusion in developing countries. The literature review implied that the subject is of non-negligible importance. First, earlier literature has found telecommunications to boost economic growth. However, the limited coverage of landlines may hinder economic development in low-income countries, particularly in rural areas. The introduction of mobile telephony is thus even more important. In addition to providing a means of communication, this innovative technology may result in new business opportunities.

Despite the potential significance of the topic, the number of studies examining the connection of mobile telephony and economic development in low-income countries is still relatively low. Moreover, some of the studies neglect to account for the endogeneity of mobile telephony even if there is plenty of evidence that mobile penetration increases with income level (as measured by GDP per capita). This chapter tried to fill this void.

The empirical results based on difference and system GMM estimations implied that the positive effect of mobile telephony on growth is strong enough to impact on the official GDP figures. In accordance with the earlier literature, we also found evidence for the convergence hypothesis and the significant role of physical capital accumulation in economic development.

The main contribution of this chapter is that it provides evidence to that mobile telephony facilitates economic growth. In other words, it seems that promoting mobile diffusion in low-income countries truly is beneficial for their economies. It seems therefore important to strive to remove barriers for mobile telephony infrastructure and its different applications that boost small businesses and ease the everyday life. Liberalization of mobile operator competition, and facilitating the basic infrastructure (roads, energy supply) to enable the setting up and maintenance of the base stations are examples of actions that would facilitate the mobile diffusion. However, it is also important to gather more information on the phenomenon with robust econometric techniques and up-to-date data in order to shed more light on this phenomenon.

REFERENCES

Arellano, M., & Bond, S. (1991). Some Tests of Specification for Panel Data: Monte Carlo Evidence and an Application to Employment Equations. *The Review of Economic Studies*, *58*(2), 277–297. doi:10.2307/2297968

Barro, R. J. (1991). Economic Growth in a Cross Section of Countries. *The Quarterly Journal of Economics*, *106*(2), 407–443. doi:10.2307/2937943

Barro, R. J., & Lee, J. (2001). International Data on Educational Attainment: Updates and Implications. *Oxford Economic Papers*, *53*(3), 541–563. doi:10.1093/oep/53.3.541

Blundell, R. W., & Bond, S. (1998). Initial Conditions and Moment Restrictions in Dynamic Panel Data Models. *Journal of Econometrics*, *87*(1), 115–143. doi:10.1016/S0304-4076(98)00009-8

Bond, S. R. (2002). Dynamic Panel Data Models: A Guide to Micro Data Methods and Practice. *Portuguese Economic Journal*, *1*(2), 141–162. doi:10.1007/s10258-002-0009-9

Cieslik, A., & Kaniewsk, M. (2004). Telecommunications infrastructure and regional economic development: The case of Poland. *Regional Studies*, *38*(6), 713–725. doi:10.1080/003434042000240996

Cronin, F. J., Parker, E. B., Colleran, E. K., & Gold, M. A. (1991). Telecommunications infrastructure and economic growth: An analysis of causality. *Telecommunications Policy*, *15*, 529–535. doi:10.1016/0308-5961(91)90007-X

Dalgaard, C.-J., Hansen, H., & Tarp, F. (2004). On the Empirics of Foreign Aid and Growth. *The Economic Journal*, *114*(496), 191–216. doi:10.1111/j.1468-0297.2004.00219.x

Datta, A., & Agarwal, S. (2004). Telecommunications and economic growth: a panel data approach. *Applied Economics*, *36*(15), 1649–1654. doi:10.1080/0003684042000218552

Dholakia, R. R., & Harlam, B. (1994). Telecommunications and economic development: Econometric analysis of the US experience. *Telecommunications Policy*, *18*(6), 470–477. doi:10.1016/0308-5961(94)90015-9

Donner, J. (2007). The use of mobile phones by microentrepreneurs in Kigali, Rwanda: Changes to social and business networks. *Information Technologies and International Development*, *3*(2), 3–19. doi:10.1162/itid.2007.3.2.3

Donner, J. (2008). Research Approaches to Mobile Use in the Developing World: A Review of the Literature. *The Information Society*, *24*(3), 140–159. doi:10.1080/01972240802019970

Dutta, A. (2001). Telecommunications and economic activity: An analysis of Granger causality. *Journal of Management Information Systems*, *17*(4), 71–95.

Euromonitor (2009). *Global Market Information Database*. Euromonitor International.

Frembong, G. (2009). Mobile telephone opportunities: the case of micro- and small enterprises in Ghana. *Info - The journal of policy, regulation and strategy for telecommunications, 11*(2), 79-94.

Gyimah-Brempong, K. (2002). Corruption, economic growth, and income inequality in Africa. *Economics of Governance*, *3*(3), 183–209. doi:10.1007/s101010200045

Hahn, H. P., & Kibora, L. (2008). The domestication of the mobile phone: oral society and new ICT in Burkina Faso. *The Journal of Modern African Studies*, *46*(1), 87–109. doi:10.1017/S0022278X07003084

International Country Risk Guide. (2006). *The ICRG Researchers Dataset*. The PRS Group, Inc.

International Telecommunications Union. (2003). *Mobile Overtakes Fixed: Implications for Policy and Regulation*. Retrieved April 15, 2009, from http://www.itu.int/osg/spu/ni/mobileovertakes/Resources/Mobileovertakes_Paper.pdf

(2008). *International Telecommunications Union* (11th ed.). World Telecommunication/ICT Indicators Database.

Madden, G., & Savage, S. J. (1998). CEE telecommunications investment and economic growth. *Information Economics and Policy*, *10*, 173–195. doi:10.1016/S0167-6245(97)00020-6

Madden, G. & Savage, S.J. (2000). Telecommunications and economic growth. *International Journal of Social Economics, 21*(7/8/9/10), 893-906.

Mankiw, N. G., Romer, D., & Weil, D. N. (1992). A Contribution to the Empirics of Economic Growth. *The Quarterly Journal of Economics*, *107*(2), 407–437. doi:10.2307/2118477

Mauro, P. (1995). Corruption and Growth. *The Quarterly Journal of Economics*, *110*(3), 681–712. doi:10.2307/2946696

Norton, S. W. (1992). Transaction costs, telecommunications, and the microeconomics of macroeconomic growth. *Economic Development and Cultural Change*, *41*(1), 175–196. doi:10.1086/452002

Röller, L.-H., & Waverman, L. (2001). Telecommunications Infrastructure and Economic Development: A Simultaneous Approach. *The American Economic Review*, *91*(4), 909–923. doi:10.1257/aer.91.4.909

Roodman, D. (2005). *XTABOND2: Stata module to extend xtabond dynamic panel data estimator*. Center for Global Development, Washington. Retrieved from http://econpapers.repec.org/software/bocbocode/s435901.htm

Roodman, D. (2009). How to do xtabond2: An introduction to difference and system GMM in Stata. *Stata Journal*, *9*(1), 86–136.

Sachs, J. D., & Warner, A. M. (1995). Economic Reform and the Process of Global Integration. *Brookings Papers on Economic Activity*, *1*, 1–118. doi:10.2307/2534573

Shiu, A., & Lam, P.-L. (2008). Causal relationship between telecommunications and economic growth in china and its regions. *Regional Studies*, *42*(5), 705–718. doi:10.1080/00343400701543314

Solow, R. M. (1956). A Contribution to the Theory of Economic Growth. *The Quarterly Journal of Economics*, *70*(1), 65–94. doi:10.2307/1884513

Windmeijer, F. (2005). A Finite Sample Correction for the Variance of Linear Efficient Two-Step GMM Estimators. *Journal of Econometrics*, *126*(1), 25–51. doi:10.1016/j.jeconom.2004.02.005

World Bank. (2008). *World Development Indicators Online*. Retrieved in 2008 from http://devdata.worldbank.org/dataonline/

APPENDIX

Countries used in estimation:

- Algeria
- Bangladesh
- Bolivia
- Cameroon
- China
- Democratic
- Republic of the Congo
- Dominican Republic
- Egypt
- El Salvador
- Gambia
- Ghana
- Guatemala
- India
- Indonesia
- Iran
- Jordan
- Kenya
- Lesotho
- Mali
- Nepal
- Nicaragua
- Pakistan
- Papua New Guinea
- Philippines
- Republic of the Congo
- Senegal
- Sierra Leone
- Sri Lanka
- Swaziland
- Syria
- Thailand
- Togo
- Tunisia
- Uganda
- Zambia

Chapter 12
Understanding Mobile Phone Usage While Driving:
Mini-Bus and Taxi Drivers' Experiences In Istanbul

Ronan de Kervenoael
Sabanci University, Turkey & Aston University, UK

Canan Devletkusu
Dogus University, Turkey

ABSTRACT

In emerging markets, the amount of mobile communication and the number of occasions mobile phones are used are increasing. More and more settings appropriate or not for mobile phone usage are being exposed. Although prohibited by many governments, there is evidence that use of new mobile devices while driving are somehow becoming current everyday practice, hence legitimatizing usage for many users. Dominant dangerous behavior in the absence of enforced legal framework is being deployed and has become routine for many m-users. This chapter adopts a qualitative case study approach (20 cases) to examine the public transport drivers' motives, logic and legitimacy processes. The question which these issues raise in the light of advancing m-technologies is: How do, in the context of emerging market, undesired emerging routines enactment get to be reflected upon and voluntarily disregarded to maximize the benefits of m-technologies while minimizing their drawbacks? Findings point out at multiple motives for usage including external social pressure through the ubiquitous 24/7 usage of m-technology, lack of alternative communication protocol, real time need for action and from an internal perspectives boredoms, lack of danger awareness, blurring of the boundaries between personal and business life and lack of job fulfillment are uncovered as key factors. As secondary dynamic factors such as education, drivers work' histories, impunity, lack of strong consumer opposition appear central in shaping the development of the routines.

DOI: 10.4018/978-1-61692-818-6.ch012

INTRODUCTION

Dominant dangerous behaviour in the absence of an enforced legal framework is often described as becoming routine practice for many mobile phone users (Hussain, Al-Shakarchi, Mahmoudi, & Marshall, 2006). Along with advances in mobile technologies (e.g. Phone, GPS, DVD in cars, PDAs), as many emerging countries, since its liberalization period in the mid 1980s, Turkey has experienced a rapid increase in motorization without having adequate road traffic safety mechanisms in place to control the growing number of emerging unwanted behaviours. Characteristics which are common to a number of developing countries include a relatively high proportion of accident involving pedestrian and children and many fatal accidents involving trucks, buses and other public service vehicles (Downing, Baguley, & Hills, 1991). Contributory individual factors observed in Pakistan, Egypt, Jamaica, Thailand, Kenya, Chile, India and Zimbabwe often involve a combination of road user errors, ranging from crossing continuous no-overtaking lines, not stopping at stop signs, lack of knowledge of stopping and following distances, overloading of goods and passengers, passenger transported in open pick-ups, lack of safety education and overall less disciplined users. Chronic institutional system failure point out to a range of issues including mix road usage (e.g. shared with non-motorized vehicles), deficiencies in the road environment, abnormal number of junctions per kilometer, lack of traffic calming designs, ill equipped police, lax attitude of the authorities, unresponsive judicial system to lack of licensing and training to ensure standards especially for professional drivers (Downing, 1985; Jacobs, 1976; Kosasih, Robinson, & Snell, 1987; Sayer & Downing, 1981). In addition, political issues involving populist approaches, amnesties and conflicting signals between the various government agencies are also often present (Baharoglu & Leitmann, 1998).

This study examines the formal and informal institutional frameworks for road safety enforcement regarding mobile phone usage by professional drivers in emerging markets through an in depth case study of Istanbul min-bus and taxi experiences. The objective of the research is to improve the understanding of individual and institutional constraints that contribute to inadequate behaviours and mobile phone usage while driving. The understanding of incentives and behaviours that are motivated by existing institutional rules and personal habits, both formal and informal, currently result in failure to adequately meet expected behaviour and require new approaches.

Numerous models of driver behaviour are present in the literature focusing mainly on driver decision-making (Janssen & Tenkink, 1988). Models ranging from prospect theory, utility theory and homeostasis theory have been applied centered on the perception of risk against benefits (Calisir & Lehto, 2002). Behaviours are found to be judgement or rule based depending on the level of task performance (Lehto, 1991). Social sciences have employed other models including theory of planned behaviour (Ajzen, 1991), and theory of reasoned action (Fishbein & Ajzen, 1975). Behaviours' intentions are determined by attitudes, subjectives norms and perceived behavioural control (Rothengatter & Mansted, 1997). An extended TPB model further included the constructs of moral norms, anticipated regret and habit (Aberg, 2001). Anticipated affective consequence of breaking internalized moral rules (regret) and habits were recognized as particularly important in the case of driving behaviour. From a social psychological perspective, health belief models have also been used to understand how attitude, beliefs and norms influence drivers (Simsekoglu & Lajunen, 2008; Stroebe, 2000).

In this context, self-regulation and legitimacy as counter balancing forces for deviant behaviour appear to be part of the complex multifaceted process. Legitimacy is defined by Dowling and Pfeffer (1975) as "players that seek to reach a consensus on social values closely associated with specific behaviours and norms within the system that they are part of. Accordingly, if two systems are congruent, we speak of ...legitimacy". Legitimacy is also described as conforming to recognized principles or accepted rules and standards. However, legitimacy can also be interpreted as a reaction and resistance to dominant discourse and behaviour. Counter hegemonic forces are regularly described as representing a key characteristic of democracy at work. Hegemony is characterized as a 'unity existing in a concrete social formation' (Laclau & Mouffe, 1985 p7). Hegemony is realized when 'a particular social force assumes the representation of a totality that is radically incommensurable with it' (Laclau & Mouffe, 2001). That is, hegemony is ascertained when a dominant behaviour strategy is taken for granted and is routinely displayed within the whole population.

Interestingly, a number of studies have shown that consumers are not rational and are finding it difficult to self-regulate their behaviours and legitimize their actions. In retailing for example, cases ranging over compensatory shopping behaviours, compulsive buying and addictive behaviours to illegal activities such as shop lifting demonstrate that consumers do not always act in their and others' 'best' interests (Elliott, 1994; Hirschman, 1992; Hoyer & MacInnis, 2001; Singh & Pandya, 1991; Woodruffe, 1997). Other prominent works where self-regulation difficulties are encountered include attitude and behaviour towards smoking, alcohol and use of drugs (Prochaska, DiClemente & Norcross, 1992), encouragement towards greener disposition of garbage (Pieters, 1993), systematic wearing of seat belts and bicycle helmet (Lajunen & Rasanen, 2004; Wittenbraker, Gibbs & Kahle, 1983), dispossession of products (Roster, 2001; Jacoby et al, 1977) and fishing quotas (Jentoft, 2000; Nielsen, 2003).

In parallel with these trends, mobile Information Communication Technologies and the number of occasions and contexts in which mobile devices are used in everyday life have created further novel challenges. Indeed, mobile technologies have attracted interdisciplinary interest from across management (Bruner & Kumar, 2005) and an extensive body of work discusses consumer usage of mobile technologies concentrating typically on three main areas including domestication (Fortunati, 1998; Palen, Salzman & Youngs, 2001; Weilenmann, 2001); mobile technology and user practices (Fortunati, Katz & Riccini, 2003; Katz, 2003) and at a societal level adoption with engagement in the wider political sphere (Brown, Green & Harper, 2002; Rheingold, 2002). Further related aspects include the psychological impact of mobile consumption by children and adolescents (APS, 2004; Doring & Gundolf, 2005; Kubey, Lavin & Barrows, 2001; Leopoldina & Manganelli, 2002), status symbols (Dedeoğlu, 2004; Özcan & Koçak, 2003), the social and pragmatic aspects of MMS/SMS (Chapman & Schofield, 1998; Lin & Tong, 2008; Sharma & Sturges, 2007; Rettie, 2007), the networked society (COST269, 2003; Ling, 2004), gender relations (Shade, 2008), fashion items (Hulme & Peters, 2003; Katz & Sugiyama, 2006); entertainment (Ha, Yoon & Choi, 2007; Harmer, 2003) and M-technology consumption (Pagani, 2004; Vrechopoulos, Constantiou, Sideris & Doukidis, 2003).

These bodies of work are complemented by studies concentrating on the mainly negative impact of mobile technologies covering issues such as health (Malayapa et al., 1998; WHO, 2000), addiction (Bianchi & Phillips, 2005), the impact on concentration while driving or multitasking (Brusque & Alauzet, 2008; Esbjornsson, Juhlin, & Weilenmann, 2007; Jessop, 2008; Walker, Williams, & Jamrozik, 2006; Walsh, White, Hyde, & Watson, 2008), increased reaction time (Alm & Nilsson, 1995), and mis-appreciation of cognitive and/or physical demand (Goodman, Tijerina, Bents, & Wierwille, 1999; McEvoy et al., 2005; Redelmeier & Tibshirani, 1997).

As a result, more and more settings are being re-negotiated and re-evaluted for their appropriateness to mobile phone usage. Although prohibited by many governments (McEvoy et al., 2005), there is evidence that the use of new mobile devices while driving is somehow becoming current everyday routine, hence legitimatizing usage for many users. Significantly, a strong regulatory framework exists in Turkey including the Road Traffic Act, law code 2918, Section 73, that prohibit the use of handheld mobile and vehicle phones whilst driving. Regulation of this law made it mandatory to use hands-free equipment. Heavy fines are part of the regulation with the possible loss of driving license for multiple offenders.

Amid these research agendas, we have identified three partial gaps in the literature. Firstly, because using a mobile phone while driving (without hand free equipment) is illegal and contrary to commonsense little research has been conducted into these novel and undesirable emerging routines. Secondly, the impact for professional drivers of such legitimization and routine creation over the medium term has been left unanswered, especially in the case of emerging markets where resources for law enforcement are minimal. Thirdly, the personal self-regulation mechanisms linked to users' legitimacy *in-situs* need to be unpacked to explore potential solutions.

This qualitative study unpacks the legitimacy factors advocated by public transport professionals for using their mobile phone when driving, through a series of 20 self-reported mobile users while driving interviews. We aim to offer insights to regulators and civil society groups on what types of, and arguments for, legitimacy of using mobile phone devices are currently being deployed by professional drivers and why these different routines are only partially mediated and counteracted by self-regulation.

This chapter consists of five sections. It begins with an overview of the literature on legitimacy and of self-regulation in the context of mobile technologies. In section 3, we present the methodology. In section 4 the findings are presented along with discussion and analysis. Lastly, section 5 provides a conclusion and points up future research.

LITERATURE REVIEW

Conceptualizing Legitimacy

The multifaceted concept of legitimacy is usually grounded in different theoretical areas such as transaction cost theory (Barney & Hesterly, 1996; Ireland, Hitt, & Vaidyanath, 2002); resource based view theory (Chung, Singh, & Lee, 2000; Eisenhardt & Schoonhoven, 1996); organizational learning theory (Kale, Singh, & Perlmutter, 2000; Khanna, Gulati, & Nohria, 1998); social network theory (Ahuja, 2000; Gulati, 1999; Kenis & Knoke, 2002) and ICT capacity building theory (Burkink, 2002). Legitimacy for our purpose is defined as "a generalized perception or assumption that the actions of an entity are desirable, proper, or appropriate within some socially constructed system of norms, values, beliefs, and definitions" (Suchman, 1995, p. 574).

Early Greek philosophers were the first to argue on the power of legitimacy grounded as a way to justify the authority of often the strongest, in turn, defining the rules to obey in public life (Zelditch, 2001). Following this perspective, Clegg (1975) argued that legitimacy of formal rule-based authority cannot be accepted or taken for granted. Socially constructed norms and values are also portrayed as important. Clegg (1989) further argued that every domain of one's life was influenced by the dominant discourse understood as the set of tacit understanding on how people were supposed to act and relate to each other. Indeed, it is recognized that the stronger the dominant powers of control, the less powerful individuals or marginal groups were at shaping their values, norms and beliefs.

Building on this long tradition of legitimacy, many authors have explored the concept unpacking various key dimensions. Tyler (1990) noted that legitimacy was the reservoir of loyalty which leaders were given to govern effectively. In addition, Suchman (1995) attached legitimacy to the cognitions and evaluations of an entity to conform to the appropriate and socially acceptable. In other words, legitimacy is very similar to "institutionalism". Furthermore Courpasson (2000) claimed that sub-agencies of authority which have been formally recognized represent and determine the culture of what is socially acceptable at the local level (e.g. arms-length relationships in public-private partnerships). Thus, those sub-entity's values, norms and beliefs shape the masses values, norms and beliefs attributed to them as legitimate by the centre. Significantly, many central authorities are legitimacy-producing while at the same time being legitimacy-seeking entities for decentralized agencies and stakeholders as a whole.

Another justification for legitimacy to be established includes the attraction of sufficient resources necessary for survival, such as political approval and patronage (Hearit, 1995). In particular, a lot of effort is devoted to maintaining legitimacy (Ashford & Gibbs, 1990; Deegan, Rankin, & Voght, 2000), extending legitimacy (Ashford & Gibbs, 1990), defending legitimacy (Hearit, 1995), explaining the legitimacy of power (Hollander, 1993) and preserving the distribution of justice (Deutsch, 1985).

Another strand of research is related to the concept of illegitimacy highlighting further processes that prevent stigmatization, ex-communication and other exclusions from main-stream society (Goffman, 1963; Elsbach, 2003). It is important for an entity to be legitimate because this justifies the privileges and trust accorded by the public. Significantly, management dedicates a large part of its time to prevent illegitimate news from reaching key stakeholders (Elsbach & Sutton, 1992), since being perceived as conducting illegitimate business has adverse effects on individual's and organization's alike (Crocker, Major, & Steele, 1998; Fombrun, 1996). Specifically, it is important to note that stigmatization depends on the values and norms of a society that are often independent of the legal system which is enforced. For instance, injustices towards minorities are frequently tolerated and legitimized by the public.

In particular, social identity theory has demonstrated that besides permeability of group boundaries and stability of group status, legitimacy of the status structure is also important (Tajfel, 1974). Stability of the status-quo is dependent on low-status members finding the system to be legitimate or illegitimate. Importantly, research has shown that even those in disadvantaged groups are often attempting to protect the status-quo within a system, because change outcomes are too uncertain (Jost, Banaji, & Nosek, 2004). People tend to preserve their cognitive and ideological state often opting for inertia, in turn perceived as 'more' legitimate (Lerner, 1980). From another point of view, Becker (1968) argues, people pursue personal gains obtained by bypassing legitimate regulations when the likelihood of detection and being caught is low. Supporting that, equity theory proposes inequitable treatment to employees, for example, often leads to sub-group 'justified' deviant behaviour, partially explaining the justification for employee's withheld effort, absenteeism (Kemper, 1966), and stealing habits (Greenberg, 1990). Interestingly, inequity theory predicts that employees would act out their perceived relative injustice as deviant behaviours (Adams, 1965; Goodman & Friedman, 1971).

Furthermore, Ashforth and Anand (2003) have proposed a three stage model for explaining normalization of corrupt behaviour by groups including: (i) institutionalization, making the corrupt behaviour embedded and routinized in the organization; (ii) rationalization, as a self-serving mechanism to justify the act; (iii) socialization, whereby new incoming group members are reframing and re-defining corruptive acts as permissible. By using these rationalizing techniques, individuals retain their self-image avoiding negative appraisals, guilt and shame, and justify their deviances (Sykes & Matza 1957). Those various tactics help to explain how entities routinely exercise deviant behaviours without experiencing conflicts. With such normalization operating, the deviant actors seem surprised to be accused of guilt (Coleman & Ramos, 1998; Hunt & Manning, 1991).

Lastly, from a strategic alliances perspective, five type of legitimacy are analysed (Dacin, Oliver, & Roy, 2007). Firstly, *Market legitimacy* - that is when a company expands or maintains its place in a market. Secondly, *relational legitimacy* defined as the perceived worthiness of one particular firm to act as provider to attract alliance partners. Thirdly, *social legitimacy* described as the approval from others to access and use resources. Fourthly, *investment legitimacy* obtained by increasing one's investment in mobile communication devices, in our case, and getting better at multitasking with the device while driving. This renders drivers more likely to be approved and supported by external parties such as potential new supervisors and patrons, thus making their future goals more feasible to accomplish. Lastly, *alliance legitimacy* seeks endorsement of other firms in the practice presenting a consistent array of practices to consumers. Alliance legitimacy in the transport sector affects the legitimacy of the illegitimate behaviour in the first place (Dacin, Oliver, & Roy, 2007).

Self-Regulation

As previously mentioned, self-regulation can be considered as a balancing force while analysing the multifaceted concept of legitimacy. Self-regulation has mainly been analysed from three perspectives - social and cognitive theory (Baumeister & Heatherton, 1996), consumer behaviour research (Hoch & Loewenstein 1991) and motivation (Bandura, 1997). Self-regulation is defined as the process by which people initiate, adjust, terminate or alter actions to promote goals or plans (Baumeister, Heatherton, & Tice, 1993; Carver & Scheier 1998). Self-regulation failures create a number of problems for individuals and the society as a whole. Self-regulatory processes are embedded in people's behaviours to pursue positive goals and being unable to regulate one's impulses, thoughts, emotions and actions may undermine future lifestyle.

People's choices and actions are largely affected by future potential outcomes from their current behaviour. According to Carver and Scheier (1998), self-regulation involves several steps; (i) having standards and reference points; (ii) comparing one's current state to a desired state; (iii) motivation and willingness to override certain responses to bring about a desired change if current behaviours fall short of expected standards. Following this, Bandura's (1997) self-efficacy theory states that people motivate themselves according to their expectancies from future behaviours. Significantly, anticipated consequences conveyed from specific behaviour provide information on intentions to act or not to act that shape possible self-regulation mechanisms (Fishbein & Ajzen 1975). Outcome anticipation is the foundation of self-regulation. The feedback-loop model developed by Carver and Scheier (1998) explains how the self-regulation process occurs to the extent of anticipating the potential desired or undesired outcome.

From a cognitive approach, self-regulation requires setting "deliberative mindsets" while reflecting about the potential outcome of certain behaviour. Pre-decision deliberation creates a cognitive effort that influences long-term outcomes and helps to weigh positive and negative consequences (Gollwitzer & Bayer 1999; Heckhausen & Gollwitzer 1987). Therefore, individuals are less likely to show estimation bias.

METHODOLOGY

The logic followed in our design was exploratory to allow salient themes to emerge (Yin, 1994). Our analysis follows the spirit of grounded theory (Glaser & Strauss, 1967) complemented by a more confirmatory phase unpacking in particular the concepts of legitimacy and self-regulation (Strauss and Corbin, 1998). We examine the legitimacies underpinning the routine usage of mobile phone while driving with a multiple case study method. This method has been tested as appropriate for capturing the ambiguous, conflicting, unpredictable and subjective context of individual decision making (Merriam, 1998). The sampling method was purposive and consisted of a random sampling of 10 taxi and 10 minibus drivers in two locations: (i) the municipality of Istanbul's bus terminal departure hub around the Bosporus crossing in Kadikoy (Asia) and (ii) a traditional large taxi driver hub in the business district of Besiktas (on the European side of Istanbul). The main form of data collection was through unstructured and semi-structured interviews with drivers while waiting for their next departure. Drivers were individually asked to give their informed consents before the interviews. Participant driver's names or surnames were not recorded and full confidentiality was provided as the act of using a mobile phone while driving is illegal in Turkey. A total of 20 interviews, in May-June 2009, were conducted - ten in each location (see Table 1). Each interview lasted approximately 20 minutes; some drivers were interviewed on multiple occasions to fit within their work pattern. Drivers were not fed back each other's answers. The researcher collected narratives in Turkish. All data were translated using the back-to-back translation method and recorded in a word document. The overarching purpose of the interviews was to reveal public discourses of commercial drivers' mobile phone use while driving.

Interviews were semi-structured in that the intention was to explore a framework of themes on participants (Kvale, 1996; Lindlof & Taylor, 2002). Both closed and open-ended questions were included in the semi-structured interviews. Three main aspects were analysed including - (i) drivers' mobile phone rituals, relationships and patterns of usage, (ii) triggers for using mobile phone while driving, (iii) market legitimacy or external pressures, (iv) identity legitimacy or internal pressures, (v) obsessive behaviour derived from the above or observed among other drivers.

Table 1. Characteristics of qualitative interview sample

	Minibus	**Taxi**
Age	35-66	27-65
Education		
Primary school dropout	1	3
Primary school graduate	6	4
High school graduate	2	1
University graduate	1	2
Own the vehicle?		
Yes	-	2
No	10	8
Number of years as a driver		
1-5 years	-	1
6-15 years	1	2
16-25 years	3	6
26-35 years	6	1
Self-declared mobile usage while driving		
For personal purpose	2	1
For business purpose	3	1
For both	5	8

* all male
* driver as the main job (all - 1 civil servant's part time job)
* all working in the Istanbul metropolitan area

Firstly, we started reading the transcripts and familiarising ourselves with the data (open coding, Strauss & Corbin, 1998). Reflecting the spirit of grounded theory as an inference method, this study allows subsequent interviews to benefit from earlier insights. Secondly, we coded the emerging salient themes in the data (each author with his own observations) and undertook an early comparison exercise linking to more abstract concepts reviewed in the literature (axial coding, Glaser 2005, p.105). The findings are classified by 'patterns and recurring organizations' arising from the analysis (Wetherall, Potter, & Antaki,

Figure 1. Legitimacy continuum

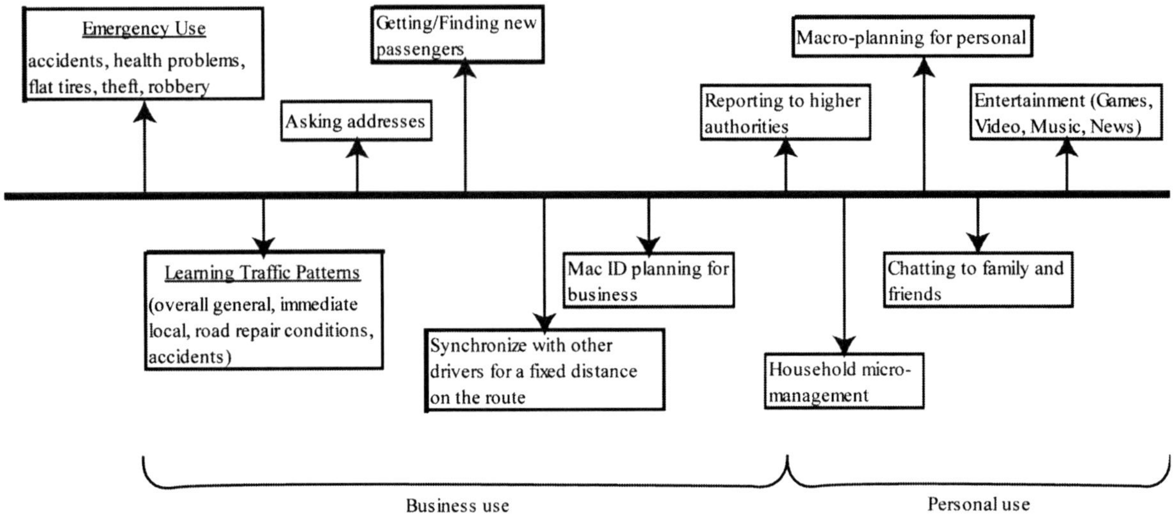

1988, p.177). A process of iteration was used to reach consensus among authors in order to construct a final thematic for analysis (Miles & Huberman, 1994). This follows the pattern matching strategy developed by Yin (1994). Thirdly, we started the indexing process in which the agreed thematic framework was systematically applied to the whole data (Gibbs, 2002). Furthermore, according to Henwood and Pidgeon (2003), the idea of theoretical agnosticism that investigates "things theoretically, with a range of theories, thinking about which one might speak to the data, considering overall 'fit' and 'workability'" was used leading to a larger number of categories that were subsequently refined over time. "M" represents quotes made by mini-bus drivers while, "T" represents quotes made by taxi drivers.

FINDINGS

The configuration allowing the identification of the different legitimacy and self-regulatory factors depends largely on a number of non-independent causes. To set the context of the findings' analysis, we first provide an overview of the frame of mind that our respondents were exhibiting. Subsequently, Figure 1 presents a continuum of legitimacy triggers emerging from the interviews. Legitimacy of use of a mobile phone by professional drivers, while driving, is justified under two main heading related to business and personal reflections. In the following sub-sections, we present firstly how the current frame of mind of drivers triggers events, followed by the four emerging salient dimensions of legitimacy and reasoning for some self-regulation namely (i) market legitimacy, (ii) identity legitimacy, (iii) obsessive legitimacy and (iv) self-regulation at work.

Framing Drivers' Minds and Triggers Within the Context of Emerging Market

At the outset every driver was keen to provide their perspective on using the mobile phone. It was described as a hotly debated topic, especially since it was made illegal to use one while driving in Turkey. Opinions were polarized. Drivers had found an emotional coping mechanism to comfort themselves that (i) they are better than other drivers to justify the usage and (ii) none ever get caught. Analysis of the responses suggested that drivers agreed that using mobile phone while driving is very dangerous, while at the same time believing that the dangers do not apply to them since they are careful, professional, and experienced drivers.

T6: Talking on the phone while driving is not dangerous for taxi drivers. But I am against the notion that normal citizens use mobiles while driving, especially the women drivers. T9: No, it is not dangerous for professional drivers. But if a normal driver tries to talk on the phone while driving, he may cause accidents and injuries...I drive carefully even if I am not on the phone.

The second source of legitimization arises from the lack of law enforcement and the fact that most drivers routinely use mobile phones. Findings demonstrate that the majority of professional drivers uses their mobile phone while driving because everybody else is witnessed doing it, even the police. Reflected justification based on the police's anti-normative behaviour offers a particularly strong signal. Those who do use their mobile phones while driving are not penalized and those who are not using are not rewarded but they miss out on the opportunities a mobile phone provides. Examples of reflected justification rationales are as follows;

M4: I have not seen anyone enforce the law. M9: I never saw anyone who was penalized with that law. Sometimes, the policeman takes 20 Turkish Liras and that is it...no tickets. M10: Even the traffic police use it while driving, and they are not punished, why should I be punished then? T1: I have not heard of any punished drivers. Under normal conditions, it is punished but everybody uses the mobiles while driving. Don't you think that police is not part of it? I think also now most of the drivers are using their mobiles but for professional drivers like us, using the mobile while driving has become an art...T2: I believe the number of minutes and the sequence of calls would decrease if other drivers would not use their mobiles

Following these, a number of triggers initiating or precipitating a chain of events, processes or reactions were described. In the legitimizing process of mobile phone use while driving, a number of factors acentuate the cognitive and behavioural mechanisms of an individual in respect of his decision to re-define the position of legitimacy borders. Significantly, some of these acts are interpreted as more legitimate than the other. Figure 1 proposes a legitimacy continuum listing the different triggers mentioned by our respondents. The continuum has to be considered dynamic with the intensity of each element and ranking changing significantly upon each individual, timing and road circumstances. Our respondents have classified the trigger events under three main categories related to emergency, business and personal characteristics.

i. Emergency legitimacy: Driving as a profession is a risky business with many unknown passengers carried to many different places for a cash reward. From this perspective, regular mobile phone contact with the outside world is described as legitimate allowing re-assuring news and progress reports to be made preventing attack and allowing quick reactions in case of robbery. As a secondary logic, drivers may also have to deal with vehicle malfunctions such as flat tyres or a damaged engine and they will ask for support. Thirdly, as professionals they witness many other accidents and incidents allowing them a voice in road traffic situations.
ii. Business legitimacy: Getting more business, especially in congested metropolitan roads has become a mobile phone game. Drivers seem to agree that a new kind of ritual/belief developed in the profession is encouraging

them to constantly update colleagues and co-drivers on road situations and opportunities. This mundane routine, making the mobile phone as important as 'good driving' allows better control of road traffic conditions. Phones are described as an empowerment tool for individuals working in the relative isolation of their car. Secondly, many drivers are still learning and re-learning the city map. With close to eleven million citizens, Istanbul is in constant development and roads change rapidly. New addresses appear everyday while little help is available from (i) technological tools such as GPS which is perceived as too expensive and diminishes the professional driver's prestige and (ii) road side individuals as they generally provide very unreliable information and may prove dangerous. Beyond the pure business functions, mutual agreements of the other drivers and passengers have set a platform for rationalization of the use of mobile phones on the road as a necessity rather than a choice-behaviour.

iii. Personal legitimacy: Mobile phone activities were described as more intermittent when deployed for personal use, often when passengers were not in the car or on return runs. This includes using the mobile phone on the road to chat with family and friends on subjects irrelevant to immediate security and traffic conditions. Driver's reported that the principal purpose of mobile phone usage was to overcome boredom. Entertainment devices such as radio and DVD players were also mentioned to be used while driving, sometime making the use of phone a less risky alternative in term of attention and concentration. Those occurrences are interpreted as the least legitimate purposes for the use of mobiles while driving. Other rationalizations were quoted ranging from (i) advertising relevant offers, (ii) micro-planning related to family life cycle, (iii) micro-management of other activities, (iv) second job related errands.

T7: I use mobile phone in real emergencies while driving, like if I see an accident in front of me, I call the ambulance or police. M10: I use it for work to know where I should take the minibus at the end of the working hours. I have to meet the owner of the minibus everyday. T7: I use it for my work. We get information on traffic, addresses, we talk to other drivers. T10: I talk to my family, usually they call me, to give the list of shopping, or to organize night events.

Market Legitimacy

For commercial vehicle drivers specifically, preserving or expanding their market place was described as a key factor of success and was translated as using any means to increase an individual driver's number of customers in any working day. In the past the only solution was to keep driving. Tiredness can kill and the mobile phone was described as a partial legitimate solution to satisfy both market requirements and effective driver management. In addition, road management technique described as covering as few kilometers as possible was also mentioned as good for business. Secondary types of market legitimacy were derived related to (i) time saving not having to stop or queue at taxi stations (ii) business management such as stopping the engine in blocked traffic or planning for fuel pitstops, (iii) decision to cross the Bosphorus or not following an evaluation of potential business on the other side, (iv) perceived competitive tension between drivers and (v) perceived prestige provided to drivers who allow others to get more business.

M1: We earn more money. If we learn which way to use to cross Bogazici bridge, we can make more rounds. T7: At least I am learning addresses that I do not know so I can take my customer. Most of the times, communicating is beneficial in order to learn how many taxies are waiting at the taxi stop, thus you decide for yourself to get in the line or look for customers on the road. M7: Otherwise, we lose so much time when the minibus stops in the middle of the road because of a problem in the engine.T1: You can be informed of the waiting list of customers and drive to the closest customer. If someone already has a customer, he forwards the potential customer to other drivers. So, we make more money instead of looking for customers all around. T2: We earn more money by using mobile phones. We inform each other of the waiting customer on the streets.... taxi stops prefer the drivers who have a mobile phone and use it. T9: Yes, it increases my competitive power. We get road traffic information. We find customers easily when we inform each other of the waiting customers.T10: I do more work because I learn the emptiest road from my friends, and I make more money.

Identity Legitimacy

Identity emerged as a differentiating concept between commercial drivers and other drivers. Most of the sample profesional drivers described talking on the mobile phone while driving without causing or having any accidents as contributing to their charismatic image. As taxis need to be clean at all the time, drivers need to have a phone described as part of their uniform. A certain element of fashion was present with some drivers based on the type of phone they used. This was sometime discussed with passengers. However, many had basic phones which they described as more suitable (less chance of theft and less worry when they fall). Many drivers also had more than one phone and other equipment such as an in-car charger, and they often connected their phone to a music system. The ability to multitask with the different devices was perceived as a positive advantage. Still, a majority admitted that passengers do not hesitate to ask them to stop using mobile. By default the ones not asking are described as "agreeing with the practice". Opinion was polarized between driving safely and reaching the destination in a minimum amount of time. Answers from respondents with regard to their identity are as follows:

T2: You appear charismatic.M1: Some people like it, some people are angry at our busy image. M7: Mobile phone is the accessory of an empty man. T10: It would not mean a thing if my colleagues would not be using it. Where would I get the traffic newsletter then? M2: Yes, my friends are pleasant when I tell them road conditions. M1: Some of the passengers get angry when I am chatting on the phone. But some of them want us to ask the other drivers on the phone which way they should go. The passengers tell me "Careful son, be careful, do not crash." T9: They get angry when I am chatting a friend on the phone. They tell me "You will cause an accident. Just drive. Do not take your eyes off the road. M5: The passengers their only matter is how many minutes I took them from somewhere to somewhere. They are not interested in the content of the drive.T10: They get annoyed when I am chatting with a friend and tell me to chat later. But they want me to ask the other drivers the road conditions.

Obsessive Legitimacy

The last aspect emerging from our exploratory work relate to the lack of self-regulation and the emergence of clearly obssessive behaviour. This particular set of factors relates to the frequency of usage and checking of the device. Not communicating or playing with the device for a long period of time (e.g. one hour) has become a source of anxiety for many drivers. Various functions such as the setting control or the volume control are checked out regularly. Other drivers rely on extra SMS services, mainly advertising, to keep them busy. A ritual of calling to be called is also in place. A certain order of call is put in place whereby key individuals in a certain order are contacted on a regular basis. This emerging obsession is possible because of the new pricing structure that has emerged in the Turkish mobile phone market with three competing providers. Many drivers have subscription contracts allowing them nearly unlimited free calls or texts between certain numbers or within certain times for a reasonable cost.

M7: Yes, sometimes I check the screen because I may not hear the beeps due to the engine noise. M9: If I do not have it with me, I feel a sense of emptiness. It is like a habit though, it is kind of automatic behaviour to open the phone whenever it rings. T4: Some people perceive it as a habit but in fact it is a requirement, it is a need. You must have a mobile phone for certain occasions, at the right time, with the right people, it shall be used, so owned. T9: It is an inseparable part of us and now it is quite cheap to run.

Emerging Forms of Self-Regulation at Work

Different forms of self-regulation have also emerged from the narratives. Firstly, functional and practical aspects were described whereby the usage, outgoing or incoming messages/calls were somehow disguised using the vibrating mode or a small beeping sound. Secondly, an emotional aspect, often derived from news in the media, whereby accidents are described with mobile usage as the root of the problem, was causing concern and reflective behaviour. Close calls with road accidents were also described as self-regulating behaviour but with a limited temporal effect. Thirdly, a certain realisation that even with mobile technology the traffic and road conditions do not improve dramatically was described. A new definition of urgency was being re-negotiated. Fourthly, an underlying fear that the law was eventually going to be applied justified some rationing of usage. Lastly, the main reason for self-regulation was emerging as an outside environmental factor - a secondary effect of the fact that the roads were becoming busier every month, increasing exponentially the risk taken while multitasking.

M2: Yes, I have a special mobile phone bag next to the wheel. It is easy to reach. M4: I switch the vibration mode on to prevent customers getting annoyed if my mobile phone rings. M5: Yes, I keep it on the door space on the left. It disturbs me in my pocket. M7: I wait for the red light. M8: I sometimes pull on the right, find a refuge or wait for a red light. M9: If I switch to vibration mode, I should keep it in my pocket, otherwise I would not hear it because of the noise of the engine. T1: If the call is urgent, I make the call while driving. But if I know the person calling is for chat, I do not always answer his phone. Sometimes not to be disturbed, I put it in the special closed case on my garnment and will check later if it has vibrated. T2: We don't open every incoming call. If we think the call is not urgent, we don't open while driving on a highway. Moreover, we don't answer the calls in the first few minutes when we've got a new customer and we make sure we are directed to the correct address......yes, I do switch to vibration mode. If my mobile is on the front console of the car, I can see it is ringing from the lighting of the screen. T5: If I am driving fast or the traffic flow is critical, I say "I will call you later" and hang up or I do not open the phone. T6: I carry it in my pocket to prevent theft. Because the police also get mad when they see it in the car, they make jokes such as "are you listening to the radio on your mobile phone?" T8: I send the busy signal to the

Figure 2. Dynamics of the legitimation process for Professional Mini-bus and Taxi drivers in Turkey

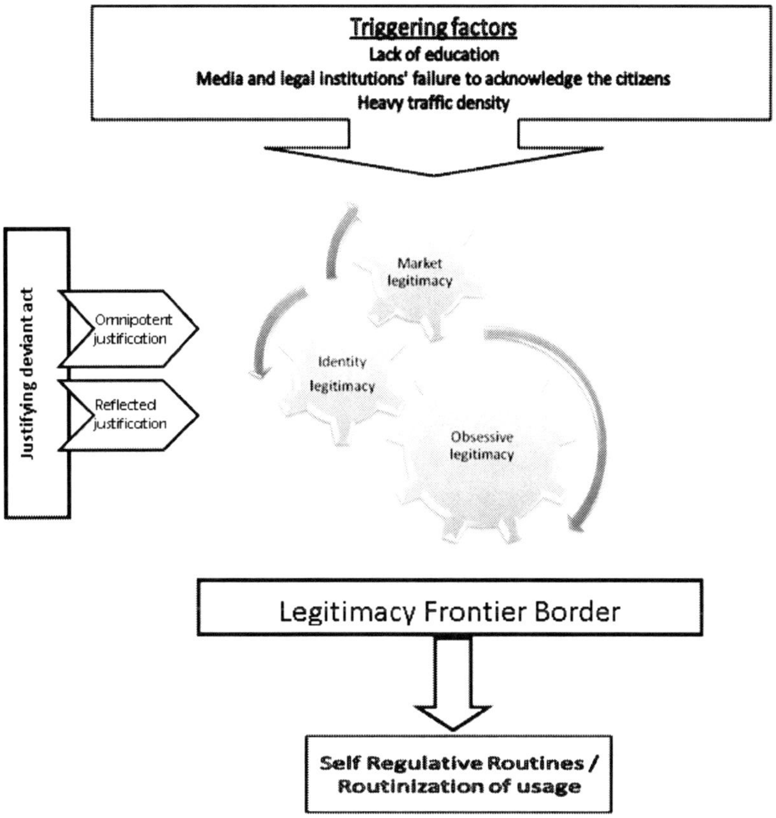

calling person if the road is not appropriate to make a phone call.

CONCLUSION

Based solely on personal short-term interests, people have reason to act in non-solidaristic manner, but when put in a situation where they must argue publicly and be held morally accountable for their actions, the significance of common norms becomes crucial to finding a solution to the tragedy of the commons (Rothstein, 2005 p.50)

Emerging countries have serious road accident problems that are increasingly linked to technological behaviour, including the usage of mobile phones. The findings emphasize the important role of attitude, subjective norms, legitimacy and self-regulation in developing intentions not to use mobile phone while driving. Findings also point out at a divergence between intentions not to use mobile phone and actual usage. In view of the fact that existing formal rules are largely irrelevant, call for greater enforcement of existing rules do not seem appropriate in the short-term. Penalties are portrayed as ineffective as they deal with the symptoms, not the causes of mobile phone usage while driving. The most noticeable aspect of the findings point out that individual's private matters seems to be strong justifications for usage rather than engaging with the institutional framework. While there is considerable pluralism between individuals in the values relevant to usage or not, social welfare is not mentioned by any of our respondents. As in many other welfare policies, a recent trend of new public management initiatives has been "to get individuals to take more responsibilities for the outcomes that affect them, often described as a transition from passive to an active welfare state" (Taylor-Gooby, 2008, p.169). This type of initiatives seem particularly adapted to emerging markets where the traditional threat rethoric has failed. Following Gintis et al (2005) and Rothstein' (2005), the view that individuals are 'often neither self-regarding nor altruistic but are strong reciprocators or conditional co-operators' seems to hold. Drivers behave altruistically as long as others are doing so as well. Following Taylor-Gooby (2008) "the implication is that it is social context and thus, in relation to modern welfare state societies, the way in which the design of social institutions frames choices, that is influencial in promoting a particular pattern of motivation"(p.174). Self-regarding reasons not to follow altruistic behaviour can, in turn, be controlled through within group sanctions against defectors. Indeed, professional drivers do justify their behaviour because other drivers engage in the same behaviour. Older drivers and managers, as those more experienced, following Ostrom's (2000) work, ought to encourage better behaviour and be targeted by educational and training policies.

Emerging mobile phones and their technologies have been found to have a greater impact than previously mentioned in the literature in encouraging negative practices. Our findings identify a series of triggers that create a set of new rituals. In turn, these new habits are justified by our sample of professional drivers through three main types of emerging mobile phone legitimacy justifying usage whilst driving. In addition, we also find that self-regulatory behaviours, in this particular context, are few and often non-sustainable within the current application of the legal framework. Legitimacy practices are described as dynamic and evolving rapidly. What is suprising from the findings is the special status that mini-bus and taxi drivers seem to have allocated to each other. Furthermore, whilst functional limits are appearing new emerging usage also seems to be regularly created. Indeed, the day to day context and apparently specific situations of our sample of professional drivers has yet to be located within the generic context of mobile phone and driving. The multiple ways in which legitimacy is portrayed leads many to think that rapid action is required. Following Head (1990), two types of intervention seem to be required to correct technological market failure through introducing more advanced technological means (e.g. in build car phone, ergonomic requirement preventing driving, GPS data etc) and preventing individual failure ranging from impulsiveness of behaviour to a lack of full information and preference endogeneity.

It has not been long since the government began regulating new technology usages to reduce potential negative social consequences in Turkey. While new standards are introduced on using new technologies, society as a dynamic unit also tries to resist certain norms and standards. The major contribution of this chapter lies in the fact that beyond regulation, civic discourses and relevant information on clearly demonstrated dangers has not been understood by the majority of the Turkish population. As found in other studies regarding seat beld use, policies ought to focus on the benefits of NOT using mobile phone while driving instead of emphasizing the negative outcomes. In doing so, this study contributes to the literature regarding policy making and self-regulation within the public space. It is not the technology per-se that is important but the daily usages and the impact of the practices. Our analysis supported the view that, raw information on presenting the dangers of mobile phone usage while driving or regulating their use is not sufficient to prevent the emergence of negative routines/behaviour. Legitimization processes obstruct the enforcement of the laws and generalize the negative use of mobile phones in a circular fashion rather than presenting benefits for the adoption of the appropriate behaviour. Today marginal non-uses of mobile phone while driving are no so much about the practices but the peer pressures felt while trying to reclaim and redefine higher standards of behaviour.

External social pressure through the ubiquitous 24/7 usage of m-technology, lack of alternative communication protocols, the real time need for action, personal boredoms and ignorance of clear and present dangers on a daily basis are blurring the boundaries between personal and organizational limits of suitable settings for mobile usage. As a secondary dynamic, factors such as education, drivers' work histories, impunity and a lack of strong consumer opposition appear central in shaping the development of the routines.

The research findings presented above must be interpreted in the light of the research method limitations, including a small sample size and limited geographical coverage which are preventing generalization. Future possible direction for such research includes a multi-sector analysis of professional drivers across countries. Secondly, further unpacking of the dynamic social processes at play is required. Thirdly, research should also concentrate on how to operationalize systematical change in behaviour in the context of emerging markets.

ACKNOWLEDGMENT

We are also very thankful to TUBITAK for funding this project (grant no 107K272).

REFERENCES

Aberg, L. (2001). Attitudes . In Barjonet, P. E. (Ed.), *Traffic psychology today* (pp. 119–137). USA: Kluwer Academic Publisher.

Adams, J. S. (1965). Inequity in social exchange . In Berkowitz, L. (Ed.), *Advances in experimental social psychology* (*Vol. 2*, pp. 267–299). New York: Academic Press. doi:10.1016/S0065-2601(08)60108-2

Ahuja, G. (2000). Collaboration networks, structural holes and innovation: A longitudinal study. *Administrative Science Quarterly*, *45*, 425–455. doi:10.2307/2667105

Ajzen, I. (1991). The theory of planned behaviour. *Organizational Behavior and Human Decision Processes*, *50*, 179–211. doi:10.1016/0749-5978(91)90020-T

Alm, H., & Nilsson, L. (1995). The effects of a mobile telephone task on driver behaviour in a car following situation. *Accident; Analysis and Prevention*, *27*(5), 707–715. doi:10.1016/0001-4575(95)00026-V

APS. (2004). *Psychological Aspects of Mobile Phone Use among Adolescents*. Australian Psychological Society.

Ashford, B. E., & Gibbs, B. W. (1990). The Double-Edge of Organizational Legitimation. *Organization Science*, *1*(2), 177–194. doi:10.1287/orsc.1.2.177

Ashforth, B. E., & Anand, V. (2003). The Normalization Of Corruption In Organizations. *Research in Organizational Behavior*, *25*, 1–52. doi:10.1016/S0191-3085(03)25001-2

Baharoglu, D., & Leitmann, J. (1998). Coping Strategies for Infrastructure: How Turkey's SpontaneousSettlements Operate in the Absence of Formal Rules. *Habitat International*, *22*(2), 115–135. doi:10.1016/S0197-3975(97)00313-5

Bandura, A. (1997). *Self-Efficacy: The Exercise of Control*. New York: Freeman.

Barney, J. B., & Hesterly, W. (1996). Organizational economics: Understanding the relationship betweenorganizations and economic analysis . In Clegg, S., Hardy, C., & Nord, W. R. (Eds.), *Handbook of Organization Studies*. Thousand Oaks, CA: Sage Publications.

Baumeister, R. F., & Heatherton, T. F. (1996). Self-Regulation Failure: An Overview. *Psychological Inquiry*, *7*(1), 1–15. doi:10.1207/s15327965pli0701_1

Baumeister, R. F., Heatherton, T. F., & Tice, D. M. (1993). When Ego Threats Lead to Self-Regulation Failure: Negative Consequences of High Self-Esteem. *Journal of Personality and Social Psychology*, *64*(1), 141–156. doi:10.1037/0022-3514.64.1.141

Becker, G. S. (1968). Crime and punishment: an economic approach. *The Journal of Political Economy*, *76*(2), 169–217. doi:10.1086/259394

Bianchi, A., & Phillips, J. G. (2005). Psychological predictors of problem mobile phone use. *Cyberpsychology & Behavior*, *8*(1), 39–51. doi:10.1089/cpb.2005.8.39

Brown, B., Green, N., & Harper, R. (2002). *Wireless World: Social, Cultural and Interactional Issues in Mobile Communications and Computing*. London: Springer.

Bruner, G. C., & Kumar, A. (2005). Explaining consumer acceptance of handheld Internet devices. *Journal of Business Research*, *58*(5), 553–558. doi:10.1016/j.jbusres.2003.08.002

Brusque, C., & Alauzet, A. (2008). Analysis of the individual factors affecting mobile phone use while driving in France: Socio-demographic characteristics, car and phone use in professional and private contexts. *Accident; Analysis and Prevention*, *40*(1), 35–44. doi:10.1016/j.aap.2007.04.004

Burkink, T. (2002). Cooperative and voluntary wholesale groups: channel coordination and interfirm knowledge transfer. *Supply Chain Management: An International Journal*, *7*(2), 60–70. doi:10.1108/13598540210425812

Calisir, F., & Lehto, M. R. (2002). Young drivers' decision making and safety belt use. *Accident; Analysis and Prevention*, *34*, 793–805. doi:10.1016/S0001-4575(01)00079-3

Carver, C. S., & Scheier, M. F. (1998), *On the Self-Regulation of Behavior*. Cambridge, UK: Cambridge.

Chapman, S., & Schofield, W. N. (1998). Lifesavers and Samaritans: emergency use of cellular (mobile) phones in Australia. *Accident; Analysis and Prevention*, *30*(6), 815–819. doi:10.1016/S0001-4575(98)00034-7

Chung, S., Singh, H., & Lee, K. (2000). Complementarity, status similarity, and social capital and drivers of alliance formation. *Strategic Management Journal*, *21*, 1–22. doi:10.1002/(SICI)1097-0266(200001)21:1<1::AID-SMJ63>3.0.CO;2-P

Clegg, S. (1975). *Power, Rule and Domination: A Critical and Empirical Understanding of Power in Sociological Theory and Organization Life*. London: Routledge and Kegan Paul.

Clegg, S. (1989). Radical revisions: Power, discipline and organizations. *Organization Studies*, *10*(1), 97–115. doi:10.1177/017084068901000106

Coleman, J. W., & Ramos, L. L. (1998). Subcultures and deviant behavior in the organisational context. *Research in the Sociology of Organizations*, *15*, 3–34.

COST269. (2003, 3-5 September 2003). *The good, the bad and the irrelevant: the user and the future of ICTs*. Helsinki (Finland).

Courpasson, D. (2000). Managerial Strategies of Domination: Power in SoftBureaucracies. *Organization Studies*, *21*(1), 141–162. doi:10.1177/0170840600211001

Crocker, J., Major, B., & Steele, C. (1998). Social stigma . In Gilbert, D., Fiske, S. T., & Lindzey, G. (Eds.), *Handbook of social psychology* (4th ed., pp. 504–553). Boston: McGraw Hill.

Dacin, M. T., Oliver, C., & Roy, J.-P. (2007). The legitimacy of strategic alliances: an Institutional perspective. *Strategic Management Journal*, *28*, 169–187. doi:10.1002/smj.577

Dedeoğlu, A. O. (2004). The Symbolic Use of Mobile Telephone Among Turkish Consumers. *Journal of Euromarketing*, *13*(2/3), 143–163. doi:10.1300/J037v13n02_08

Deegan, C., Rankin, M., & Voght, P. (2000). Firms' Disclosure Reactions to Major Social Incidents: Australian Evidence. *Accounting Forum*, *24*(1), 101–130. doi:10.1111/1467-6303.00031

Deutsch, M. (1985). *Distributive Justice*. New Haven, CT: Yale University Press.

Doring, N., & Gundolf, A. (2005). Your life in snapshots: Mobile weblogs (moblogs) . In Glotz, P., Bertschi, S., & Locke, C. (Eds.), *Thumb Culture: The Meaning of Mobile Phones for Society* (pp. 211–224). New Brunswick, NJ: Transaction Publishers.

Dowling, J., & Pfeffer, J. (1975). Organizational Legitimacy: Social Values and Organizational Behavior. *Pacific Sociological Review*, *18*(1), 122–1366.

Downing, A. J. (1985). Road Accidents in Pakistan and the need for improvements in driver training and traffic law enforcement. In: PTRC. Summer Annual Meeting, University of Sussex, July 15-18, 1985, Proceedings of Seminar H. London: PTRC Education and Research Services.

Downing, A. J., Baguley, C. J., & Hills, B. L. (1991). *Road safety in developingcountries: an overview*. Crowthome, UK: Transport and Road Research Laboratory.

Eisenhardt, K. M., & Schoonhoven, C. B. (1996). Resource-based view of strategic alliance formation: Strategic and social effects in entrepreneurial firms. *Organization Science*, *7*, 136–150. doi:10.1287/orsc.7.2.136

Elliott, R. (1994). Addictive Consumption: Function and Fragmentation in Postmodernity. *Journal of Consumer Policy*, *17*(2), 150–180. doi:10.1007/BF01016361

Elsbach, K. D. (2003). Organizational perception management. *Research in Organizational Behavior*, *25*, 297–332. doi:10.1016/S0191-3085(03)25007-3

Elsbach, K. D., & Sutton, R. I. (1992). Acquiring organizational legitimacy through illegitimate actions: A marriage of institutional and impression management theories. *Academy of Management Journal*, *35*(4), 699–738. doi:10.2307/256313

Esbjornsson, M., Juhlin, O., & Weilenmann, A. (2007). Drivers Using Mobile Phones in Traffic: An Ethnographic Study of Interactional Adaptation. *International Journal of Human-Computer Interaction*, *22*(1-2), 37. doi:10.1207/s15327590ijhc2201-02_3

Fishbein, M., & Ajzen, I. (1975). *Belief, Attitude, Intention, and Behavior: An Introduction to Theory and Research*. Reading, MA: Addison-Wesley.

Fombrun, C. J. (1996). *Reputation: Realizing Value from the Corporate Image*. Harvard Business School.

Fortunati, L. (1998). The ambigious image of the mobile phone . In Haddon, T. L. (Ed.), *Communications on the Move: The Experience of Mobile Telephony in the 1990s*.

Fortunati, L., Katz, J. E., & Riccini, R. (2003). *Mediating the Human Body. NJ*. Mahwah: LEA.

Gibbs, G. R. (2002). *Qualitative Data Analysis: Explorations with NVivo*. Buckingham: Open University Press.

Gintis, H., Bowles, S., Boyd, R., & Fehr, E. (2005). *Moral sentiments and material interests*. Cambridge, MA: MIT Press.

Glaser, B. G. (2005). *The Grounded Theory Perspective III: Theoretical coding*. Sociology Press.

Glaser, B. G., & Strauss, A. (1967). *The Discovery of Grounded Theory: Strategies for Qualitative Research*. Chicago, IL: Aldine Publishing Co.

Goffman, E. (1963). *Stigma: Notes on the Management of Spoiled Identity*. Prentice-Hall.

Gollwitzer, P. M., & Bayer, U. (1999). Deliberative Versus Implemental Mindsets in the Control of Action . In Chaiken, S., & Trope, Y. (Eds.), *Dual-Process Theories in Social* (pp. 403–422). New York: Guilford.

Goodman, M. J., Tijerina, L., Bents, F. D., & Wierwille, W. W. (1999). Using cellular phones in vehicles. *Transportation Human Factors*, *1*, 3–42. doi:10.1207/sthf0101_2

Goodman, P. S., & Friedman, A. (1971). An Examination of Adams' Theory of Inequity. *Administrative Science Quarterly*, *16*(3), 271–288. doi:10.2307/2391900

Greenberg, J. (1990). Employee theft as a reaction to underpayment inequity: The hidden cost of pay cuts. *The Journal of Applied Psychology*, *75*(5), 561–568. doi:10.1037/0021-9010.75.5.561

Gulati, R. (1999). Network location and learning: The influence of network resources and firm capabilities on alliance formation. *Strategic Management Journal*, *20*(5), 397–420. doi:10.1002/(SICI)1097-0266(199905)20:5<397::AID-SMJ35>3.0.CO;2-K

Ha, I., Yoon, Y., & Choi, M. (2007). Determinants of adoption of mobile games under mobile broadband wireless access environment. *Information & Management*, *44*, 276–286. doi:10.1016/j.im.2007.01.001

Harmer, J. A. (2003). Mobile Multimedia Services. *BT Technology Journal*, *21*(3), 169–180. doi:10.1023/A:1025175518841

Head, J. G. (1990). On merit wants: reflections on thr evolution, normative status and policy relevance of a controversial public finance concept. In G. Brennan & C. Walsh (Eds.), *Rationality individuals and public policy* (pp. 211-244). Centre for research on Federal Financial Relations, Australian National University.

Hearit, K. M. (1995). 'Mistakes Were Made': Organizations, Apologia, and Crises of Social Legitimacy. *Communication Studies*, *46*(1-2), 1–17.

Heckhausen, H., & Gollwitzer, P. M. (1987). Thought Contents and Cognitive Functioning in Motivational Versus Volitional States of Mind. *Motivation and Emotion*, *11*, 101–120. doi:10.1007/BF00992338

Henwood, K. L., & Pidgeon, N. F. (2003). Grounded theory in psychology . In Camic, P. M., Rhodes, J. E., & Yardley, L. (Eds.), *Qualitative Research in Psychology: Expanding Perspectives in Methodology and Design* (pp. 131–155). Washington, DC: American Psychological Association Press. doi:10.1037/10595-008

Hirschman, E. C. (1992). The consciousness of addiction: Toward a general theory of compulsive consumption. *The Journal of Consumer Research*, *19*, 155–179. doi:10.1086/209294

Hoch, S. J., & Loewenstein, G. (1991). Time-Inconsistent Preferences and Consumer Self-Control. *The Journal of Consumer Research*, *17*(March), 492–507. doi:10.1086/208573

Hollander, E. P. (1993). Legitimacy, power, and influence: a perspective on relational features of leadership . In Chemers, M. M., & Ayman, R. (Eds.), *Leadership Theory and Research: Perspectives and Directions* (pp. 29–47). Orlando, FL: Academic Pres.

Hoyer, W. D., & MacInnis, D. J. (2001). *Consumer Behaviour*. Boston: Houghton Mifflin Company.

Hulme, M., & Peters, S. (2003). *Me, My phone and I: The Role of The Mobile Phone*. 3GSM World Congress Cannes: Teleconomy Research House.

Hunt, J., & Manning, P. K. (1991). The Social Context of Police Lying. *Symbolic Interaction*, *14*, 51–70. doi:10.1525/si.1991.14.1.51

Hussain, K., Al-Shakarchi, J., Mahmoudi, A., Al-Mawlawi, A., & Marshall, T. (2006). Mobile phones and driving: a follow-up. *Journal of Public Health*, *28*(4), 395–396. doi:10.1093/pubmed/fdl057

Ireland, R. D., Hitt, M. A., & Vaidyanath, D. (2002). Alliance management as a source of competitive advantage. *Journal of Management*, *28*(3), 413–446. doi:10.1177/014920630202800308

Jacobs, H. H. (1976). Effects of alcohol on sustained attention in the presence of competing stimuli. *Dissertation Abstracts International*, 37, 3113.

Jacoby, J., Berning, C. K., & Dietvorst, T. F. (1977, April). What About Disposition? *Journal of Marketing*, 22–28. doi:10.2307/1250630

Janssen, W., & Tenkink, E. (1988). Risk Homeostatis theory and its critics: time for an agreement. *Ergonomics*, *31*(4), 429–433. doi:10.1080/00140138808966689

Jentoft, S. (2000). Legitimacy and disappointment in fisheries management. *Marine Policy*, *24*(2), 141–148. doi:10.1016/S0308-597X(99)00025-1

Jessop, G. (2008). Who's on the line? Policing and enforcing laws relating to mobile phone use while driving. *International Journal of Law, Crime and Justice*, *36*(3), 135–152. doi:10.1016/j.ijlcj.2008.03.001

Jost, J. T., Banaji, M. R., & Nosek, B. A. (2004). A decade of system justification theory: Accumulated evidence of conscious and unconscious bolstering of the status quo. *Political Psychology*, *25*, 881–919. doi:10.1111/j.1467-9221.2004.00402.x

Kale, P., Singh, H., & Perlmutter, H. (2000). Learning and protection of proprietary assets in strategic alliances: Building relational capital. *Strategic Management Journal*, *21*(3), 217–237. doi:10.1002/(SICI)1097-0266(200003)21:3<217::AID-SMJ95>3.0.CO;2-Y

Katz, J. E. (2003). *Machines that Become Us: The Social Context of Personal Communication Technology*. NJ: Transaction, New Brunswick.

Katz, J. E., & Sugiyama, S. (2006). Mobile phones as fashion statements: evidence from student surveys in the US and Japan. *New Media & Society*, *8*(2), 321–337. doi:10.1177/1461444806061950

Kemper, T. D. (1966). Representative Roles and the Legitimation of Deviance. *Social Problems*, *13*(3), 288–298. doi:10.1525/sp.1966.13.3.03a00040

Kenis, P., & Knoke, D. (2002). How organizational field networks shape interorganizationaltie-formation rates. *Academy of Management Review*, *27*(2), 275–293. doi:10.2307/4134355

Khanna, T., Gulati, R., & Nohria, N. (1998). The dynamics of learning alliances: Competition, cooperation, and relative scope. *Strategic Management Journal*, *19*(3), 193–210. doi:10.1002/(SICI)1097-0266(199803)19:3<193::AID-SMJ949>3.0.CO;2-C

Kosasih, D., Robinson, R. & Snell, J. (1987). *A review of some recent geometric road standards and their application to developing countries*. TRRL, Research Report 114. Crowthorne: Transport and Road Research Laboratory.

Kubey, R. W., Lavin, M. J., & Barrows, J. R. (2001). Internet use and collegiate academic performance decrements: early findings. *The Journal of Communication*, *51*(2), 366. doi:10.1111/j.1460-2466.2001.tb02885.x

Kvale, S. (1996). *Inter Views: An introduction to qualitative research interviewing*. Thousand Oaks, CA: Sage.

Laclau, E., & Mouffe, C. (1985). *Hegemony and socialist strategy*. London: Verso.

Laclau, E., & Mouffe, C. (2001). *Hegemony and socialist strategy: Towards a radical democratic politics* (2nd ed.). London: Verso.

Lajunen, T., & Räsänen, M. (2004). Can social psychological models be used to promote bicycle helmet use among teenagers? A comparison of the Health Belief Model, Theory of Planned Behaviour and the Locus of Control. *Journal of Safety Research*, *35*, 115–123. doi:10.1016/j.jsr.2003.09.020

Lehto, M. R. (1991). A proposed conceptual model of human behaviour and its implications for design of warning. *Perceptual and Motor Skills*, *73*, 595–611. doi:10.2466/PMS.73.5.595-611

Leopoldina, F., & Manganelli, A. M. (2002). *Young People and the Mobile Phone*. Retrieved December 4, 2006, from http://www.itu.int/osg/spu/ni/ubiquitous/Papers/Youth_and_mobile_2002.pdf

Lerner, M. J. (1980). *The belief in a just world: A fundamental delusion*. New York: Plenum Press.

Lin, A., & Tong, A. (2008). Text-messaging culture of college girls in Hong Kong: SMS as resources for achieving intimacy and gift-exchange with multiple functions . In Goggin, G. (Ed.), *Mobile Phone Cultures* (pp. 158–170). London: Routledge.

Lindlof, T. R., & Taylor, B. C. (2002). *Qualitative communication research methods*. Thousand Oaks, CA: Sage.

Ling, R. S. (2004). *Mobile connection: The cell phone's impact on society*. San Francisco: Morgan Kaufmann.

Malayapa, R. S., Ahern, E. W., Bi, C., Struabe, W. L., LaRegina, M., Pickard, W. F., & Roti Roti, J. L. (1998). DNA damage in rat brain cells after in vivo exposure to 2450 MHz electromagnetic radiation and various methods of euthanasia. *Radiation Research*, *149*, 637. doi:10.2307/3579911

McEvoy, S. P., Stevenson, M. R., McCartt, A. T., Woodward, M., Haworth, C., Palamara, P., et al. (2005). Role of mobile phones in motor vehicle crashes resulting in hospital attendance: a case-crossover study [Electronic Version]. *British Medical Journal*, bmj.38537.397512.397555. Retrieved July 12, 2005 from http://www.bmj.com/cgi/content/abstract/bmj.38537.397512.55v1

Merriam, S. B. (1998). *Qualitative Research and Case Studies Applications in Education*. San Francisco: Jossey-Bass Publications.

Miles, M. B., & Huberman, A. M. (1994). *Qualitative Data Analysis: An Expanded Source Book*. Thousand Oaks, CA: Sage.

Nielsen, J. R. (2003). An analytical framework for studying: compliance and legitimacy in fisheries management. *Marine Policy*, *27*, 425–432. doi:10.1016/S0308-597X(03)00022-8

Ostrom, E. (2000). Collective action and the evolution of social norms . *The Journal of Economic Perspectives*, *14*(3), 137–158. doi:10.1257/jep.14.3.137

Ozcan, Y. Z., & Koçak, A. (2003). Research Note: A Need or a Status Symbol? Use of Cellular Telephones in Turkey. *European Journal of Communication*, *18*(2), 241–254. doi:10.1177/0267323103018002004

Pagani, M. (2004). Determinant of Adoption of Third Generation Mobile Multimedia Services. *Journal of Interactive Marketing*, *18*(3), 46–59. doi:10.1002/dir.20011

Palen, L., Salzman, M., & Youngs, E. (2001). Discovery and integration of mobile communications in everyday life. *Personal and Ubiquitous Computing*, *5*(2), 109–122. doi:10.1007/s007790170014

Pieters, R. (1993). Consumers and their garbage a framework and some experiences from the Netherlands with garbage separation programs. *European Advance in Consumer Research*, *1*, 541–546.

Prochaska, J. O., DiClemente, C. C., & Norcross, J. C. (1992). In search of how people change: Applications to addictive behaviors. *The American Psychologist*, *47*(9), 1102–1114. doi:10.1037/0003-066X.47.9.1102

Redelmeier, D. A., & Tibshirani, R. J. (1997). Association between cellular telephone calls and motor vehicle collisions. *The New England Journal of Medicine*, *336*, 453–458. doi:10.1056/NEJM199702133360701

Rettie, R. (2007). Mobile Phones as Network Capital: Facilitating Connections. *Mobilities*, *3*(2), 291–311. doi:10.1080/17450100802095346

Rheingold, H. (2002). *Smart Mobs: the Next Social Revolution*. New York: Perseus.

Roster, C. A. (2001). Letting Go: The Process and Meaning of Dispossession in the Lives of Consumers . In Meyers-Levy, J., & Gilly, M. (Eds.), *Advances in Consumer Research* ((pp. 425–430). Provo, UT: Association for Consumer Research.

Rothengatter, T., & Manstead, A. S. R. (1997). The role of subjective norm in predicting the intention to commit traffic violations . In Rothengatter, T., & Carbonell Vaya, E. (Eds.), *Traffic & Transport Psychology, Theory and Application* (pp. 389–394). London: Pergamon.

Rothstein, B. (2005). *Social Traps and the problem of trust*. Cambridge: Cambridge University Press. doi:10.1017/CBO9780511490323

Sayer, L.A. & Downing, A.J. (1981). *Driver knowledge of road safety factors in three developing countries*. TRRL Supplementary Report 713. Crowthorne: Transport and Road Research Laboratory.

Shade, L. R. (2008). Feminizing the mobile: Gender scripting of mobiles in North America . In Goggin, G. (Ed.), *Mobile Phone Cultures* (pp. 42–52). London: Routledge.

Sharma, G., & Sturges, P. (2007). Using ict to help the poor access public services: An action research programme. *Information Development*, *23*(1), 15–24. doi:10.1177/0266666907075623

Simsekoglu, O., & Lajunen, T. (2008). Social psychology of seat belt use: A comparison of theory of planned behavior and health belief model. *Transportation Research Part F: Traffic Psychology and Behaviour*, *11*, 181–191. doi:10.1016/j.trf.2007.10.001

Singh, J., & Pandya, S. (1991). Exploring the effects of Consumers' dissatisfaction level to complaint behaviours . *European Journal of Marketing*, *25*(7), 7–21. doi:10.1108/EUM0000000000621

Strauss, A., & Corbin, J. (1998). *Basics of Qualitative Research Techniques and Procedures for Developing Grounded Theory*. Thousand Oaks, CA: Sage.

Stroebe, V. (2000). *Social psychology and health*. Buckingham: Open University.

Suchman, M. C. (1995). Managing Legitimacy: Strategic and Institutional Approaches. *Academy of Management Journal*, *20*(3), 571–610. doi:10.2307/258788

Sykes, G. M., & Matza, D. (1957). Techniques of Neutralization: A Theory of Delinquency. *American Sociological Review*, *22*, 664–670. doi:10.2307/2089195

Tajfel, H. (1974). Social identity and intergroup behaviour. *Social Sciences Information. Information Sur les Sciences Sociales*, *13*(2), 65–93. doi:10.1177/053901847401300204

Taylor-Gooby, P. (2008). Choice and Values: Individualised Rational Action and Social Goals. *Journal of Social Policy*, *37*(2), 167–185. doi:10.1017/S0047279407001699

Tyler, T. R. (1990). *Why people obey the law*. New Haven: Yale University Press.

Vrechopoulos, A., Constantiou, I., Sideris, S., & Doukidis, G. (2003). The critical role of consumer behaviour in mobile commerce. *International Journal of Mobile Communications*, *1*(3), 329–340. doi:10.1504/IJMC.2003.003498

Walker, L., Williams, J., & Jamrozik, K. (2006). Unsafe driving behaviour and four wheel drive vehicles: observational study. *British Medical Journal*, *333*(7558), 71–74. doi:10.1136/bmj.38848.627731.2F

Walsh, S. P., White, K. M., Hyde, M. K., & Watson, B. (2008). Dialling and driving: Factors influencing intentions to use a mobile phone while driving. *Accident; Analysis and Prevention*, *40*(6), 1893–1900. doi:10.1016/j.aap.2008.07.005

Weilenmann, A. (2001). Negotiating use: making sense of mobile technology. *Personal and Ubiquitous Computing*, *5*(2), 109–122. doi:10.1007/PL00000015

Whetherall, M., Potter, J., & Antaki, C. (1988). *Analyzing Everyday Explanation*. Thousand Oaks, CA: Sage.

WHO. (2000). *Fact Sheet on Electromagnetic Fields and Public Health: Mobile Phones and Their Base Stations*. Geneva: World Health Organization.

Wittenbraker, Gibbs, & Kahle. (1983). Seat belt attitudes, habits and behaviors: an adaptive amendment to the Fishbein model . *Journal of Applied Social Psychology*, *13*, 406–421. doi:10.1111/j.1559-1816.1983.tb01748.x

Woodruffe, H. R. (1997). Compensatory Consumption: Why women go shopping when they're fed up and other stories . *Marketing Intelligence & Planning*, *15*(7), 325–334. doi:10.1108/02634509710193172

Yin, R. (1994). *Case Study Research: Design and Methods*. Thousand Oaks, CA: Sage.

Zelditch, M. (2001). Theories of Legitimacy. In J. T. Jost & B. Major (Eds.), *The psychology of legitimacy: emerging perspectives on ideology, justice, and intergroup relations* (Vol. 2, illustrated, pp. 33-53). Cambridge University Press.

Chapter 13
Mobile Technology and the Gulf Society:
Perception and Attitude

Khaled A. Sabry
Hamdan Bin Mohammed e-University, UAE

Ahmed Al-Nakeeb
Al-Ain University, UAE

Khalid W. Alrawi
Al-Ain University, UAE

ABSTRACT

Globally, mobile technology alters the way individuals and groups conduct their day-to-day activities. It impacts on business, culture and society, as it changes the way people communicate, do business, socialise, and has considerable effect on its user's perception and attitude. This chapter reviews and explores mobile technology growth in the Gulf region with particular focus on the use of mobile phones in the UAE. It further explores, through a snapshot survey, people's perception, attitude, and possible implications of the technology on their behaviour. Based on the review and results of the survey we conclude with a discussion, recommendations and suggestions for future research.

INTRODUCTION

Mobile phones were originally created for business use for adults (Aoki & Downes, 2003). The growth of mobile phone technology was realised in 2002 through the increase of the number of mobile phone users worldwide, over those of land-line phone users (Srivastava, 2005). Mobile phones alter the way people conduct their daily activities with extensive implications on different cultures and societies, due to changes in the nature of communication, implications on people's relationships, and effects on economic activities as well as users' perceptions of themselves and the world (Plant, 2002).

Gender difference is another factor that may affect mobile phone's usage. For example, older female teens lag behind males in the ownership of mobile phones (Ling 1999). Also, males differ from females in terms of mobile phones' usage. For example, males are more interested in down-

DOI: 10.4018/978-1-61692-818-6.ch013

loading games, while females are more interested in downloading ringtones (BBC News, 2004a). According to a survey by market research firm JD Power and Associates, spending on mobile phones has increased, reaching £45 (USD 72) a month, including for example text, voice and other phone services (BBC News, 2004b). Mobile technology impacts on young people's peer groups making possible a "truly networked society" as well as on family relationships (Campbell, 2005). Further, the ubiquity of mobile phones, and integrated features, such as Web browsers and the Java platform, represent important basis for mobile learning (Kreutzer, 2008) and mobile marketing (Leppaniemi and Karjaluoto, 2008).

However, some schools and educational institutions argue that the use of mobile phones can have negative effects in terms of disrupting teaching and reducing students' attention in class (Campbell, 2005). This is in addition to its negative effect as a plagiarism tool for cheating in examinations (The Guardian, 2006). Furthermore, increasingly, members of modern societies are assuming the role of *technoverts*, "a method of using mobile technology to insulate themselves from human interaction" (Randall, 2008). Traditionally, for example, at universities, students are commonly seen in pairs or groups discussing various issues and interacting with their peers and lecturers, but, this is gradually disappearing. Students are increasingly occupied with their mobiles, either texting others, checking messages, listening to music or calling others. The lack of face-to-face interaction in a public environment, lack of eye contact and weaker social bonding seem to have become the norm. This is in addition to health effects (Telegraph, 2008), road safety issues (ROSPA, 2007), security and culturally related concerns, for example, possible implications and consequence of using the Blackberry technology (The National, 2010; Gulf News, 2010).

There is no doubt that mobile technology has its pros and cons, as with other types of technology. The challenge here is the extent to which the technology can be geared towards benefiting society and increasing its advantages and benefits. Despite much research conducted in area of mobile technology, there is a noticeable shortage of research investigating the Gulf region perspective. The following section attempts to investigate mobile technology in the Gulf area, focusing mainly on the most popular one, mobile phones. It reviews available literature, presenting available statistics, analysis, and critiques. The purpose is to offer an overview of literature and up-to date articles published on the topic as an initial investigation that may lead to further research and contribution to this field. The review aims to provide an understanding of issues related to mobile technology, and highlight problems and opportunities that may contribute to its success in the region. It attempts to answer the question: What are the problems caused by, or opportunities offered by mobile technology? What are the factors that may drive its success in the region?

ADVANCEMENTS OF MOBILE TECHNOLOGY AND THE GULF REGION

The use of mobile phones for buying goods and services (m-commerce) is steadily growing (Laudon & Laudon, 2007). Despite the world financial meltdown, the Middle East region is considered one of the world's fastest growing mobile subscription markets with a 47% yearly increase in 2008 (WFed, 2008). According to WFed (2008), mobile penetration in Arab countries reached a 56% high at end of the second half of 2008, with an expected surge in wireless broadband usage as networks are developed; driving an increase in data traffic levels in 2009. According to the Arab Advisors Group (AAG), 48.36% of Internet users in Saudi Arabia have been reported purchasing products and services online and through their mobile handsets during 2007 (AmeInfo, 2007). Also AAG revealed that 51.2% of UAE Internet users have purchased products and services online and through their mobile handsets over a 12 months period of 2007. Based on Madar Research Group and Orient Planet, the UAE topped the 2007 Arab ICT Use Index with a score of 2.19, breaking the 2.00 index barrier (Gulf News, 2009c). Mobile phone subscriptions have "remained the strongest driving force behind the widespread use of ICT tools across the Arab World", registering a 40.51% growth in 2007, with the UAE achieving a 42.61% growth in mobile phone subscriptions in 2007 while mobile phone penetration rate rose to 131.64%, the highest in the Arab World in 2007(Gulf News, 2009c).

Internationally, many developments have occurred in the last 10 years in the world of e-commerce (IGW, 2007) that impacted on business, government and personal computing, one of which is BlackBerry communications device which enables new mobile business culture. Research in Motion, the designer, manufacturer and marketer of innovative wireless solutions for the worldwide mobile communications market participated in GITEX 2007 in Dubai and promoted one of their products, BlackBerry solution in the region (Zawya, 2008). There is an extensive effort being made in the Gulf region to capitalize on the cyber technologies for the purpose of enhancing services provided to the citizens through e-government initiatives (Kostopoulos, 2003). iPhone sales raised to more than twice last year's sales as it entered more countries (Gulf News, 2009b; The National, 2009). The launch of Apple iPhone 3G and Blackberry in the gulf region with features such as user friendly interfaces, that provide many communication and wireless features such as GPS, internet, web, messaging and e-mail facilities. Mobile technology may boost the adoption and growth of mobile commerce (GulfNews, 2009). Mashreq Bank offered their customers a comprehensive mobile banking service including access to account details, money transfer and payment of bills (ITP.NET, 2008). Mobile technology may also transform education globally (GulfTimes, 2008). There are more mobile subscribers per person in the UAE than anywhere else in the world. According to the UAE Telecomm Regulatory Authority (TRA) there are 186 cell phone lines per 100 people (The National, 2009a); however there is no indication of gender or individual differences.

However, there are many issues that can result in negative outcomes of mobile technology use, including those related to safety. Using mobile phones while driving seems a common practice in the UAE, despite a continuous effort from the authorities to curb such potentially fatal practices. The Road and Transport Authority (RTA) indicated, according to a report by The National (2009b) that this was one of the main causes for road accidents. A campaign was launched in March 2009 with the slogan 'Don't Call Till You Arrive' to help cut the number of road fatalities which reached 293 death cases on Dubai roads in 2008. The use of mobile phones while driving raises the risk of accidents to 3-4 times higher, according to the World Health Organisation (The National, 2009b). According to The National (2009c), the rate of drivers fined for such offence in Dubai increased in February 2009 to 1,139 compared with 866 offences in January 2009, although there were no details mentioned of gender and/or individual differences.

The following section describes research we conducted to investigate people's perception of mobile phones, looking at issues related to individual differences, and the extent to which mobile phones are used for different purposes.

RESEARCH OBJECTIVES AND METHODOLOGY

This research aims to explore and highlight further issues related to the use of mobile technology in the Gulf region in relation to social and cultural aspects as part of a research on the potential and

Figure 1.

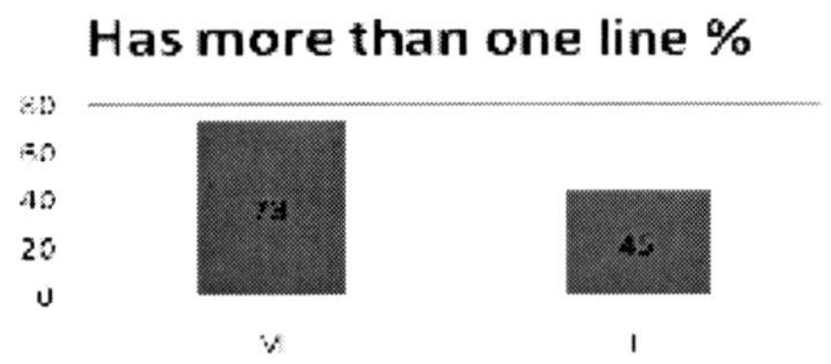

Figure 2.

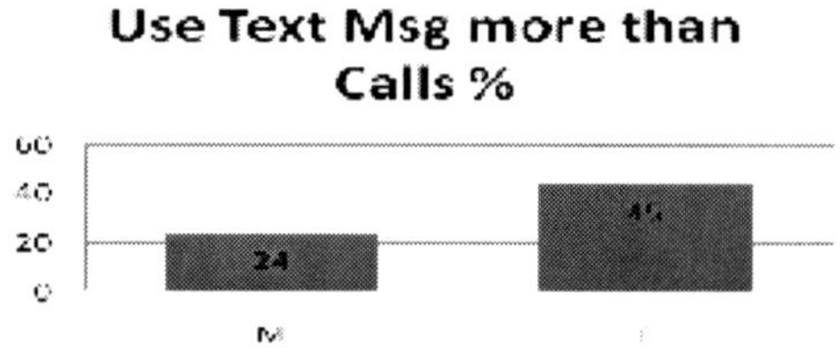

implications of mobile technology in the area. It is of an exploratory nature and the methodology described below is deemed appropriate for the purpose of this chapter. The research will seek to obtain and collect data as a snapshot and an initial exploration through conducting a survey to examine people's perception and attitude towards mobile technology. It represents a snapshot of current mobile phone use. The investigation took place at a UAE university involving undergraduate and post graduate students. The study took place at the beginning of the Spring Semester 2009. Out of 175 randomly selected students, 149 completed the questionnaire, a response rate of 85%. The sample included 74 female and 75 male students. The students spanned a range of abilities and backgrounds, with high number of working students and UAE nationals.

Students were informed of the research objectives and were guided through the questionnaire. The questionnaire was anonymous in order to encourage them to supply such details as freely and accurately as possible. The conducted investigation is based on self-reporting preferences to explore students' perspectives and possible relationships with other individual differences, rather than to obtain statistically significant results. The implications can have some generalisation of the results (Yin, 1994) which can be validated by further research. The following section displays the survey results followed by discussion and analysis of the findings in relation to the above literature review, with suggestions and recommendations for future research.

SURVEY RESULTS

Possession of Mobile Phones

The results showed that a high percentage of the population (40%) had possessed mobile phone/s for over 10 years. In terms of gender difference, however, a higher percentage of male population (59%) had possessed mobile phones for more than 10 years compared with female population (22%). The results also show that a high percentage of the population (59%) possessed more than one mobile phone. In terms of gender difference, however, a higher percentage of male population (73%) possessed more than one mobile phone (Figure 1) compared with female population (45%).

Use of Text Messaging and Calls

Overall, the majority of the population use calls more than text messaging. However, 34% of the population use text messages more than calls. In terms of gender difference (Figure 2), a higher percentage of females (45%) uses text messages more than calls in comparison with males (24%).

The Importance of the Internet Feature

The population were asked to indicate their perception towards the increasing features of mobile phones. Overall, a high percentage of the population (62%) indicated that the internet feature was either important or very important. However, in terms of gender difference (Figure 3), a higher

Figure 3.

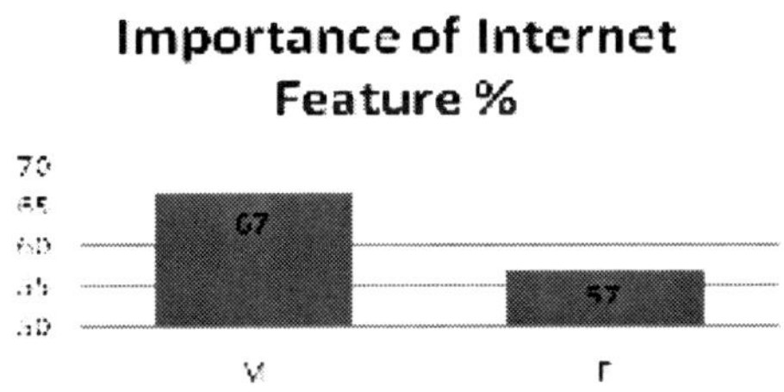

Figure 4.

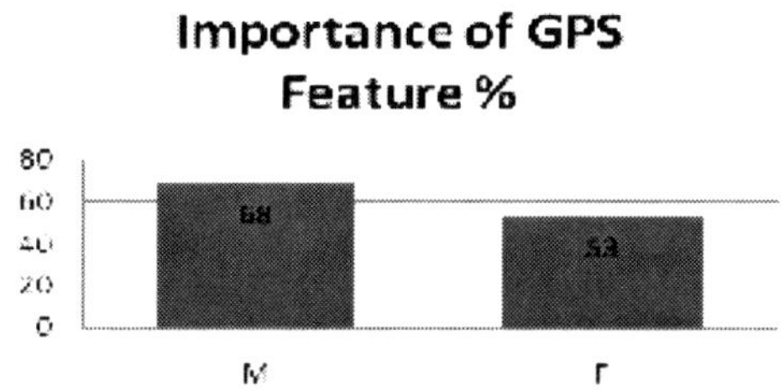

percentage of males (67%) seemed to appreciate this feature in comparison with females (57%).

The Importance of the GPS Feature

The population were also asked to indicate their perception towards the Global Positioning System (GPS) feature of mobile phones. Overall, a high percentage of the population (60%) indicated that the GPS feature was either important or very important. However, in terms of gender difference (Figure 4), a higher percentage of males (68%) seemed to appreciate this feature in comparison with females (53%).

Frequency of Changing Mobile Phones

Overall, a high percentage of the population change their mobile more than once a year (48%) However, in terms of gender difference (Figure 5), a higher percentage of females (55%) seemed to do so in comparison with males (40%).

Further, a high percentage of the population (58%) indicated that the reason for changing their mobile was generally to either get new features or obtain the latest new models available rather than changing the equipment for becoming old. However, in terms of gender difference, a higher percentage of females (62%) seemed to do so in comparison with males (49%).

Use of Mobile Phones While Driving

Overall, a high percentage of the population (77%) considered that using mobile phones while driving (without hand free or wireless kit) as a negative aspect. However, in terms of gender difference (Figure 6), a higher percentage of females (84%) seemed to realise the danger in comparison with males (71%).

The population was also asked to indicate whether they use wireless or hand free kits while driving. 33% reported that they wore it at all times. In terms of gender difference (Figure 7), only a small difference existed between males and females. 32% of the males wear it at all times, in comparison with the females (34%). However, 48% indicated that they wore it some of the time. In terms of gender difference, 56% of males wear

Figure 5.

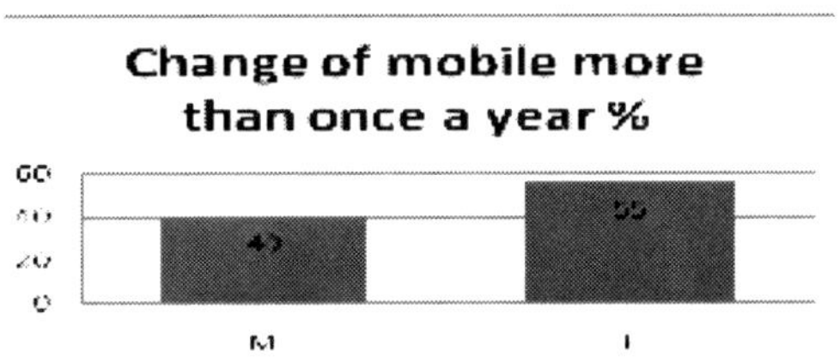

Figure 6.

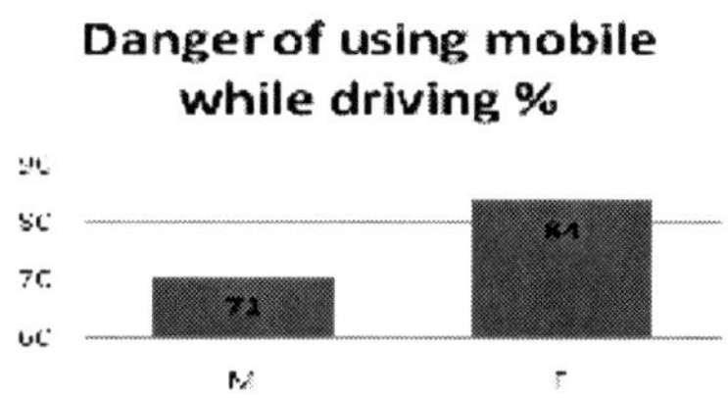

Figure 7.

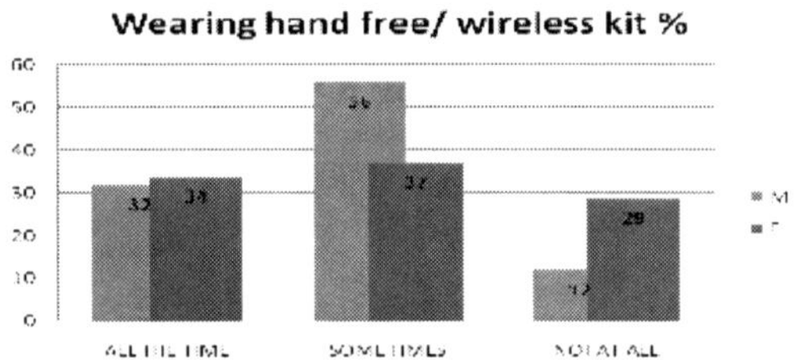

Figure 8.

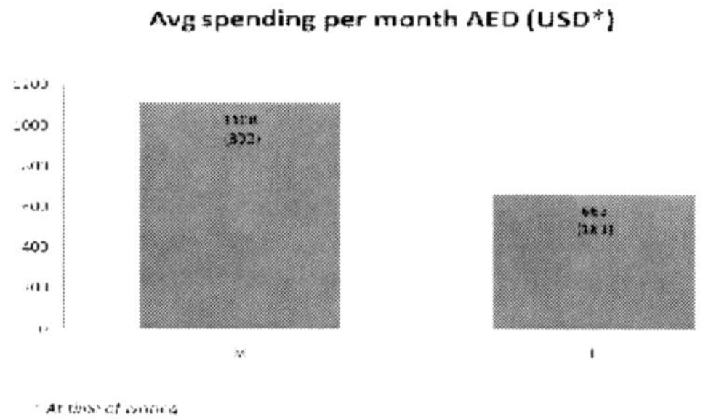

it some of the time, in comparison with females (37%). However, 19% indicated that they did not wear it at all. In terms of gender difference, 12% of males do not wear it at all, in comparison with females (29%).

Other Forms of Mobile Technology

Overall a high percentage of students (74%) uses other forms of mobile technology such as wireless laptops and car GPS. However, in terms of gender difference, a higher percentage of males (83%) seems to do so in comparison with females (66%).

Average Spending

The subjects were asked to indicate their average spending per person per month on mobile phones. Overall, the average individual spending was AED 887 (USD 241). In terms of gender difference (Figure 8), males spend far higher (AED 1108/ USD 302), nearly double the amount, in comparison with females (AED 663/ USD 181). The results also show that in some cases the spending reaches up to AED 5000/month (USD 1361/month) per person for males and AED 3000 (USD 817) for females.

Usage Level and Purpose

The subjects were then asked to indicate their usage level in terms of work, family, and other social activities. For work related calls, a low percentage of the population (8%) uses mobile phones ½ or more of their total calls. In terms of gender difference, only a small difference exists between them. For family related calls, 21% of the population use mobile phones ½ or more of their total call time. In terms of gender difference, a higher percentage of females (28%) uses mobile phones for half or more of their total calls, in comparison with males (13%). For other socially related calls, 59% of the population use mobile phones for half or more of their total calls. In terms of gender difference, only small differences exist between them.

Negative Aspects

They were also asked to indicate their perception of mobile phones' negative implications in terms of work, family and study. A low percentage of the population (13%) considers that mobile phones have negative effects on work related activities. In terms of gender difference, only 8% of males consider that, in comparison with the female population (18%). For family related calls, a very low percentage of the population considers negative effects (1%). Again, a low percentage of the population (11%) considers that mobile phones have negative effects on study related activities. In terms of gender difference, only 5% of the male population think so, in comparison with the female population (18%).

Children's Use of Mobile Phones

The population was asked to indicate the age at which they may allow their children to use mobile phones. 8% indicated that they will allow their children to have a mobile phone when they are less than 10 years old. In terms of gender difference, 5% males will allow it at less than 10, in comparison with females (13%). However, 41% indicated that they will allow their children to have a mobile phone when they are between 10-15 years old. In terms of gender difference, 47% of males will allow it between 10-15, in comparison with the female population (35%). 50% indicated that they will allow their children to have a mobile phone when they reach 15 years or over. In terms of gender difference, 48% of males will allow it at 15 or over, in comparison with females (53%).

DISCUSSION AND FUTURE WORK

Overall, the results have shown important findings in terms of business and marketing aspects as well as individual differences usage of mobile phones, particularly in terms of gender differences. The results show that the population generally has a different level of preference towards the technology and its features. It is suggested, therefore, that the technology should take into account such variation in relation to gender issues and users' perception. Mobile phones seemed to be a highly favoured communication technology in the UAE. The results highlighted the UAE market as one of the potential markets of this type of communication technology, taking into account the importance of gender difference aspects. Based on the above results, a reasonably high percentage of the population has been using mobiles for over 10 years, with a higher percentage of males in comparison with females. This is in conjunction with a high individual monthly spending, reaching in some cases USD 1361 for male and USD 817 per month for female, a higher figure than the one previously mentioned in the above literature (USD 72), representing an important finding in terms of business and marketing aspects.

A high percentage of the population possessed more than one mobile phone. This supports the literature regarding the high number of mobiles per person in the UAE, representing important business and marketing aspects as well as individual differences, as the results indicated, a higher percentage of males possessed mobile phones earlier than females (this supports the literature described above), and that a higher percentage of males possessed more than one mobile phone in comparison with females. The results have also highlighted some important marketing drivers represented in the significant percentage of subjects indicating that they change mobile more than once a year, with slightly higher percentage of females than males. Similarly, a reasonably high percentage of the population indicated that the reason for changing their mobile was generally for obtaining new features or the latest models available rather than changing the equipment when old, with a higher percentage of females than males.

The results have also shown that the majority of the population indicated that they made more calls than sent text messages. However, a significant percentage of the population indicated that they used text messages more than calls. In terms of gender difference, a higher percentage of females (nearly double) uses text messages more than calls in comparison with males. This finding is important in terms of design issues, business and marketing aspects as well as safety implications. Similarly, with regards to smart mobile phones and the increasing features of mobile phones, a high percentage of the population indicated that the internet feature was either important or very important, with a slightly higher percentage of males than females. Similarly the population's perception of the GPS feature was highly favoured, with a higher percentage of males than females. Further, a high percentage of the population used other forms of mobile technologies such as wireless laptops and car GPS, with higher percentage of males than females.

The results have also highlighted some safety concerns that need to be addressed in order to combat possible negative implications of the technology. In terms of safety awareness, a considerably high percentage of the population uses mobile phones while driving despite their awareness of the danger. This issue was highlighted in the literature in terms of high number of road accidents and rising number of offences. This will require mobile technology companies in cooperation with car manufactures to develop safer and innovative mobile systems that help to minimise in-car distractions and maximise drivers' concentration and focus on controls and the road in order to reduce possible negative implications. This is in addition to educational workshops/ seminars, awareness advertisements and enforcements of necessary legislations. In terms of mobile phones' negative implication towards work, family, study and other social activities, the majority of the population feel positive about its use. Most of the population were positive about the use of the mobile for family related communication. With reference to gender difference, however, more females compared to males seem to consider that mobile phones have negative implications on work and study matters. With reference to children's use of mobile phones, the majority of the population indicated that they prefer children to be 10 years old or over before starting to use mobile phones.

This chapter focused particularly on mobile phone technology in the Gulf region. It reviewed available literature and explored people's perception of the technology though a survey as part of a research in mobile technology. It looked at issues related to individual differences, and the extent to which mobile phones were used. Further research is needed to examine a larger number of people with a particular in-depth focus on marketing and safety issues as well as investigating the use of different types of mobile-based technologies. Research will be needed to compare between different cultures and the extent to which some of the findings may be generalised. Future work will also include the investigation of other individual differences (such as special needs, age and prior knowledge) as well as the use of mobile technology for educational and medical purposes.

ACKNOWLEDGMENT

We would like to thank everyone participated in the survey and made this research possible.

REFERENCES

AmeInfo. (2007). Saudi Arabia's Internet users spend over $3.28bn in B2C e-commerce during 2007. Retrieved November 2008 from http://www.ameinfo.com/143272.html

Aoki, K., & Downes, E. J. (2003). An analysis of young people's use of and attitudes toward cell phones. *Telematics and Informatics, 20,* 349-364. In M. Campbell (2005). The impact of the mobile phone on young people's social life. Paper presented to the Social Change in the 21st Century Conference. Retrieved April 2009 from http://eprints.qut.edu.au/3492/1/3492.pdf

Campbell, M. (2005). *The impact of the mobile phone on young people's social life*. Paper presented to the Social Change in the 21st Century Conference. Retrieved April 2009 from http://eprints.qut.edu.au/3492/1/3492.pdf

Gulf News (2009, February 22). iPhone launch could spur mobile commerce growth in Gulf.

Gulf News (2009b, April 24). Apple and eBay beat analysts' expectations.

Gulf News. (2009c, January 17). UAE at top of Arab ICT Index, by Naushad K. Cherrayil, January 17. Retrieved April 2009 from http://www.gulfnews.com/business/Technology/10276038.html

Gulf News (2010, August 2). BlackBerry to go offline.

GulfTimes (2008, August 21). Internet and Mobile phones spur economic development.

IGW. (2007, December 7) InterGovWorld. The top 10 developments in the world of e-commerce. Retrieved April 2009 from http://www.intergov-world.com/article/badbf45b0a0104080180bfe-1b7c9af9b/pg1.htm

ITP.NET. (2008). Mashreq Bank today launched what it says is the most rich mobile banking service in the region, November 10. Retrieved September 2009 from http://www.itp.net/537746-mashreq-launches-mobile-banking

Kostopoulos, G. (2003). E-government in the Arabian Gulf: a vision toward reality. In *ACM International Conference Proceeding Series; Vol. 130, Proceedings of the 2003 annual national conference on Digital government research* (pp. 1-7). Boston, MA.

Kreutzer, T. (2008). Assessing Cell Phone Usage in a South African Township School. Paper presented at e/merge 2008, July 7 - 18, 2008 in Cape Town, South Africa Retrieved April 2009 from http://emerge2008.net/access/metaobj/content/group/emerge2008/Presentations/opening%20worlds/Mobile%20Technology%20Usage

Laudon, K., & Laudon, J. (2007). *Management Information Systems: Managing the digital firm*. NJ: Prentice Hall.

Leppaniemi, M., & Karjaluoto, H. (2008). Mobile marketing: from marketing strategy to mobile marketing campaign implementation. *International Journal of Mobile Marketing*, *3*(1), 50–61.

Ling, R. (1999). We release them little by little: maturation and gender identity as seen in the use of mobile telephone. *1999 International Symposium on Technology and Society (ISTAS`99,* July 29-31, 1999 Rutgers University, New Brunswick, New Jersey. Retrieved September 2009 from http://www.richardling.com/papers/1999_we_release_them.pdf

News, B. B. C. (2004a, June). Mobile usage shows gender split. Retrieved September 2009 from http://news.bbc.co.uk/2/hi/technology/3766643.stm

News, B. B. C. (2004b, May). Monthly spend on mobiles rises. Retrieved September 2009 from http://news.bbc.co.uk/2/hi/technology/3749739.stm

Plant, S. (2002). On the mobile, the effects of mobile telephones on social and individual life. Report for Motorola. Retrieved April 2009 from http://www.motorola.com/mot/doc/0/267_MotDoc.pdf

Randall, E. (2008). Mobile Technology Changes the Social Hierarchy. Is the text message more important than the messenger? Retrieved April 2009 from http://educationalissues.suite101.com/article.cfm/florida_can_be_flexible_about_education_outcomes

Srivastava, L. (2005). Mobile phones and the evolution of social behaviour. *Behaviour & Information Technology*, *24*, 111–129. doi:10.1080/01449290512331321910

Telegraph (2008, April 19). New doubts raised over mobile phone safety, by Nic Fleming, Science Correspondent. Retrieved from http://www.telegraph.co.uk/news/uknews/1561587/New-doubts-raised-over-mobile-phone-safety.html

The Guardian. (2006, December 5). Jam mobile phones to combat exam cheating, report urges.

The National (2009, April 24). Apple underplays its success.

The National (2009a, February 5). UAE leads world in mobiles per person.

The National (2009b, March 15). Drive to cut road deaths by 40%.

The National (2009c, March 16). Far more drivers fined for phone calls.

The National (2010, August 11). Security scenario justifies concerns about BlackBerry.

WFed [Wireless Federation] (2008): Middle East mobile subscription rates set to hit 15% growth in 2009. Retrieved April 2009 from http://wirelessfederation.com/news/category/mobile-subscription

Yin, R. K. (1994). *Case study research: design and methods*. Thousand Oaks, CA: Sage.

Zawya (2008). RIM focused on Blackberry growth in the Gulf region. Retrieved January 2008 from www.zawya.com

Chapter 14
BlogWall:
Promoting Social Communication through Mobile Technology in Sri Lanka

Adrian David Cheok
National University of Singapore, Singapore

Owen Noel Newton Fernando
National University of Singapore, Singapore

Nimesha Ranasinghe
National University of Singapore, Singapore

Kening Zhu
National University of Singapore, Singapore

Chamari Edirisinghe
National University of Singapore, Singapore

ABSTRACT

Since the invention, the mobile phone is becoming more and more popular to lead the communication sector and it has been the spirit of personal communication from the beginning of 80s. Furthermore, developing countries which have always been on the search for affordable communication solutions found mobile communication the most popular method. Low cost communication, infrastructure, and maintenance are some of the key reasons that make mobile technologies popular in developing countries. Statistics depict that the usage of short messages is one of the main communication method in developing countries and most of the organizations are using SMS (Short Message Service) as a tool to assist people, especially in African and Asian continents. Sri Lanka, as a developing country, has a unique culture which has emerged scaling many centuries, mixing with various neighboring cultures. Recently the western cultural influence has dramatically changed the various cultural aspects of the urban population. The rapid economic growth, the changes in agriculture based economic environment, advances in communication and media, and globalization trends has transformed the cultural experiences of Sri Lankans. Taking into account the speedy progress of the mobile technology, especially the SMS, the evolution of the Sri Lankan way of living which has absorbed the culture that has developed with the use of mobiles, and the long literary history where poetry had played a major role in communication,

DOI: 10.4018/978-1-61692-818-6.ch014

we are observing the suitability of the Blogwall system, an interactive system which operates on user SMS and provides opportunities for creative poetry by combining visual art and poetry.

INTRODUCTION

The rapid developments of science and technology during the 20th century transformed the world more widely than at any time in the past and the mobile industry has gone through successive waves of innovation later in the century (Browne, 2003). The trend towards low power consumption and hand held devices have increased the challenges and possibilities in the mobile industry. However, only the voice service was primarily available through mobile devices. After coupling short messaging service with voice service, subscribers started to acquire novel ideas for new mobile services (Siau & Shen, 2003).

At present, mobile phones have become the most vital part of our daily lives making it very hard to imagine a life without a mobile phone. They have become inexpensive, user-friendly, comfortable, and emerging with almost every latest elements desired by people. From children to senior citizens, mobile phones have become a fashion as well as a way of staying ahead with the technology. The express lifestyles of the modern public are making the communications with the loved ones intricate. Mobile phone technology, which connected people wherever and whenever, is allowing people to overcome this dilemma.

Even though there are many technologies available for connecting people nowadays, most of them are contrastingly expensive than the mobile communication (Grinter & Eldridge, 2001). Furthermore, in developing countries the access to technology is distributed unevenly. Demography, economical imbalance, political environment and social conditions are the main factors behind this irregularity. The unequal access to the technology has created a technological divide in developing countries which is being positively answered by the mobile technology. The use of cellular technology instead of copper wire, the inexpensiveness, the user friendliness, and the mobility have provide a greater flexibility in accessing area where the minimum infrastructure prevails, thus reducing the communication and technological divide.

There are numerous instances where mobile technology was applied to reduce poverty in developing nations by opening up new opportunities (Slater & Kwami, 2005). In most occasions, the mobile phone allows users to conduct businesses, exchange information, and many more activities. Developing countries such as Sri Lanka are using mobile phones as the support for services (Samarajiva, 2000; Designing useful mobile services for Africa, 2009). Health, banking, government, education, business, and many other sectors are using mobile services which would qualitatively improve the life of under-privileged populations. As of today, the most widespread way to provide such services on mobile phones is with SMS-based applications. The reasons for that are identified as high availability, easy to use, and low and predictable costs when using short messages. However, there should be a well structured preparation to ensure the increase on positive impacts when the mobile technologies are introduced to an already established culture.

Sri Lanka has an illustrious culture based primarily on Buddhism and agriculture which has developed over centuries. Social connections and collective behavior are two main elements of this island nation. The introduction of mobile phones during the 1990s has tremendously transformed the everyday activities of the Sri Lankans. There has been a unique communication using poetry which evolved from the ancient times in Sri Lanka (Pollock, 2003). In addition, Sri Lankans are mainly using mobile phones for communicating through short messages since voice calls are more expensive than the short messages. This phenomenon is common in most of the developing countries since people could not tolerate the voice

Figure 1. A usage scenario of the BlogWall system

charges compared to the charges of short messages (Lacoh´ee & Wakeford, 2003). As a developing country, Sri Lankan society is becoming more and more industrialized and moving away from the arts and literary interests which were more essential part of the society throughout the history.

The nation which is celebrated for its social communications through poetry (Wijesekara, 1990) and prose from ancient times, has embraced the text messaging that associated with mobile technology, to express extensively. The integration of mobile communication to the already well established and highly connected culture of Sri Lanka is efficiently and effectively transpired thus making it an integral part of the lives of modern populace.

Taking into account the mobile users, the amount of text messaging that transferred and the traditional social communication roots related to poetic literature, we are introducing a system called Blogwall. BlogWall introduces a new form of communication in the networked digital era representing a combination of digital bits and cultural bits. "Mixing" or "mash-up" using poetry is the major element of this system which transforms the users into experiencing the state of being a poet by mixing short messages into poems. "Mixing" allows creation of unique and personal art work on top of bit literature and a distinctive poetry mixer can be created sans any prior experience or practice. As in figure 1, the user sends a short message to the system which is preconfigured with a mobile number. Then the extracted text from the SMS will be transferred to the processing unit, exempting any inappropriate words, will process and mix to generate a new poetry and the end result will be displayed.

MOBILE CULTURE IN DEVELOPING COUNTRIES

Since its introduction, telephone is involved in the bilateral interactions. According to Hans Geser (2005) phone is a medium of organization. "It enables the real-time integration of highly complex organizations ... coordination processes within cities ... could not be realized on the basis of face to face interaction". The phone supports the most ancient mode of exchange: the oral communication which scales the social boundaries. Mobile phone culture which Genevieve Bell (2005) aptly describe as 'thumb culture', which connects people who are on the move, who does not take time off, go to a certain place to reach to people. Every moment in the mobile world is spent on planning and interacting with others, and the use of the device provides them with a sense of relaxation regarding the change of schedules and altering of plans so long as the users are coordinating over the mobile phones (Katz, 2005). The mobile environment provides the user with the option of concealing their interactions and limiting them to a certain circle.

The Asian cell phone market has seen a dynamic growth in the late 1990s (Bell, 2005). It has become a social object that is ever present in every aspect of one's daily life; be it at the place of work or at home. A research conducted in African community concluded that monthly income was not a limiting factor to have a mobile phone (Samuel, Shah, & Hadingham, 2005) and the people tend to use short message service for their day to day activities ranging from social interactions to the crop prices. In addition, popularity of short message service in developing countries is higher than the popularity of voice messages. For example, Sri Lanka has a rapidly growing mobile communication subscriber, who is overcoming the basic transportation problems and communications inconveniences of domestic landlines by using the mobile phone technology to its maximum.

Mobile Culture in Sri Lanka

As a developing country during the previous decade number of subscribers is enormously increased in the mobile communication sector in Sri Lanka (Statistical overview of the telecommunication sector as at end of 4th quarter 2008, 2009). The traditional Sri Lankan culture is now totally mixed with the new mobile communications. Approximately, all the people living in urban areas and greater part from the rural areas are using mobile phones for their daily activities. Recently, the mobile infrastructure was advanced by enabling third generation (3G) communication facilities by all the service providers in the island. Peer to peer communication, entertainment, find daily information on weather, politics, trading, and stock market are some of the primary usages in Sri Lankan mobile communication industry.

According to the statistics from Telecommunications Regulatory Commission of Sri Lanka, about 11,082,454 mobile handsets are circulating among Sri Lankans by the end of 2008 (Statistical overview of the telecommunication sector as at end of 4th quarter 2008, 2009) and about 29, 166, 409 short messages are sent during the year 2003 (Bandusiri, 2009). This is becoming a universal phenomenon in most of the Asian and African region countries, commonly developing countries, since the limited wired or fixed telecommunication services. For instance, in Sri Lanka, the fraction of text messages sent (29166409) to total outgoing mobile minutes (154937) calculated as 188 to 1 (Bandusiri, 2009). Owing to the above facts, it is well known that usage of text/short messages is more popular than voice calls in most of the developing countries.

Akin to most of the modern technological inventions, the mobile technology has developed and established its niche in Sri Lanka's culture. It is changing the local customs making mobile conversation a revered part of their daily interactions; a mobile conversation, even if it is inconsequential, is considered important above all else. As mentioned by Genevieve Bell (Katz, 2005), mobile phones are re-charting social relationships. Sri Lankans are making friends, introducing new friends, beginning, renewing and breaking up relationships and meeting their life partners by the means of mobile phones. Moreover, the express flow of daily news through SMS keeps Sri Lankans connected to situations and issues. Soon after the 2004 Tsunami which devastated coastal areas of Sri Lanka, total chaos regained in the coastal areas of Sri Lanka due to extensive inter-connectivity of SMS, which relayed an inaccurate Tsunami warning.

The rapid escalation signifies the high social mobility outside of home which encourages fresh relationships and inter-connectivity of people who are using the same mobile subscription. Since Sri Lankan mobile telephone companies are offering extremely attractive and groundbreaking connectivity, the subscribers are rallying round to certain connection providers for the best possible advantages thus creating new circles of friends. The mobile coverage in Sri Lanka spans round the island covers almost all the country, with a few local and international service providers competing for the subscribers. It is not only the urban population who uses mobile phones excessively but also the rural population, who conduct their day to day transactions with the cities through mobile phones. One example is that a merchant who is living 205km out of the city of Colombo ordering his goods with the use of mobile phone with the intention of getting them delivered the same day. These type of business transactions are a way of life in Sri Lanka where the transportation of all types are time-consuming. Similar to most countries in Asia, mobile telephones and connectivity is a fashion statement. Teenagers and Young consider the ownership of a state of the art mobile phone as a statement of their wealth and their fashionable affiliation to the modern technology.

POETRY CULTURE IN SRI LANKA

Poetry has been a major part of Sri Lankan tradition. There had been numerous efforts at communication using poetry throughout the history. There have been various poets, notably monks, rulers and nobles in Sri Lankan literary history who has written extensively to communicate emotions, satire and joy. The Sigiriya Graffiti, which dated around 8th century AD, that are well preserved on the Mirror Wall of Sigriya, are the poems of the travelers who had visited and admired the well structured gardens and exquisite frescoes (Paranavitana, 1956). They have expressed their feelings and experiences in poetic form. The Sandesha Kavyas (Wijesekara, 1990) (Poetic messages) which were written during the 12th to 16th centuries are some of the most famous example of Sri Lankan affiliation to literature.

Creating poetry and communicating the emotions were elements of the Sri Lankan society until the emergence of electronic media as a popular cultural pattern. The culture associated with agriculture promoted a tradition of poetry. Since the majority of Sri Lankans during the early half of the 20th century were paddy cultivators, since the cultivation of paddy is a process that continued year round, people developed a literature associated with every event. They created poetry and communicated their passion, despair, joy, love, and even loneliness while vigilant over their harvest, protecting it from wild animals. Those are the folk verses, which were rarely documented yet celebrated for their eloquence and charm. The similar kind of poetic culture is associated with miners, who were in danger constantly and lonely since the graphite mining introduced during the colonial period was treacherous, requiring the labor force, which came from various parts of Sri Lanka searching for jobs, leaving their loved ones, to reside onsite. The transporters of the early 20th century, who delivered goods to the city from various parts of Sri Lanka, who travelled on days in their bullock carts and loggers who raft logs along rivers, constructing a boat out of logs, had developed a similar poetic culture. There are poetry councils, still famous in Sri Lanka, where poets from all over the country joined in, to debate in poetry. They construct a special abode decorating it with palm leaves, where poets would assemble and start a poetic debate to an audience. One of the famous female poets of the Dutch occupational period by the name of 'Gajaman Nona' (Gunavardhana, 1991), is immensely popular for her skilled wit in creating poetry instantly in reply to another poem. The poetic discussion she had with one of the aristocrats who was pursuing her is legendry in its veiled sensuality and contempt.

This historical attachment Sri Lankans have for poetic literature and the present attachment for mobile technology, especially the SMS, we found as an appropriate background for the BlogWall.

THE BLOGWALL

Blogwall is a system where poetry is created by the users' SMS and the users are encouraged to interact with every participant rather than limiting their connectivity to peer to peer. The general setup of BlogWall requires a high-end computer with a good graphics card, projector, and a screen. Dedicated GSM / GPRS modem is used to receive SMS messages. The application issues AT commands to the modem to locate SMS. All the messages received by the server will be written to a log file along with originator phone number and the date / time. The application consists of several modes of operation. Based on the enabled

Figure 2. The system architecture

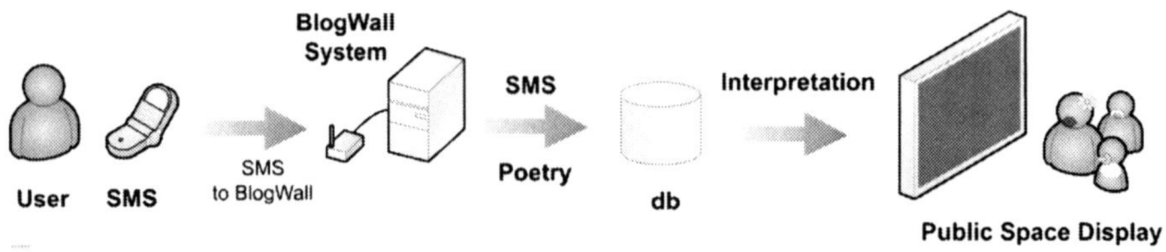

modes, it offers different services to the user. The complete system architecture is shown in Figure 2.

There are several other ways it can be used. The additional modes of BlogWall can be activated based on user's requirements, although the main focus of BlogWall is poetry. These modes are mainly used as value addition to the overall system. If the application is in the standard display mode, it will immediately display the text message with random animation. In the polling mode the system displays polling question and available answers which are indexed by single alphabetical letter within parenthesis. To vote, users send SMS with appropriate indexed letter of the answer to the system. BlogWall also has the capability to provide statistical data to system administrators. Keyword triggering mode enables the application to trigger an internal function based on a word in the SMS. This feature is somewhat similar to the features found in popular chat programs like Windows Messenger. For example, if the SMS contains word "love" the application may replace the word "love" with an image of a heart. The keyword triggering mode can also display a small verse based on the words found in the SMS. The images as well as verses are selected from an internal database.

The most prominent feature of BlogWall is its ability to mix poetry. The technique used in BlogWall integrates a number of ideas from different disciplines such as information retrieval and natural language understanding, specifically word sense disambiguation (WSD) and topic summarizing, and augments the system with genetic algorithm based methods to create a model for coherent poetry output. In the procedure of poetry generation after receiving the message, as shown in Figure 3, the input message will firstly be analyzed by the Message Filter module in which SMS language and emoticon will be translated into normal English. The translated message will be passed to the Word Sense Disambiguation (WSD) layer which is used to understand the specific meaning or sense of each word in the message. In WSD layer, WordNet (Fellbaum, et al., 1998) is used as the lexical database of English words. It is modified by adding the analysis of determiners, prepositions, pronouns, conjunctions, and particles, in order to increase the accuracy of word sense disambiguation.

After the processing has done on WSD layer, the results are transferred to the topic summarization layer. In this attempt the salience measure (Boguraev & Kennedy, 1997) of a particular topic is calculated and the topics with highest salience measure are selected. By matching the specific words and the related topics in input message with the existing poem lines system will generate a preliminary list of probable poem lines in the topic summarizing layer.

In the stage of final poem mixing, 20 combinations of three poem lines will be randomly and iteratively generated from the shortlisted probable poem lines. The salience measure of each combination is calculated by the fitness function. The combination with highest salience measure will be output as the final poem.

The unique mixing system enables the system to borrow random but relative lines of poetry from different poets. Therefore, the final outcome

Figure 3. Overview of Poetry Generation

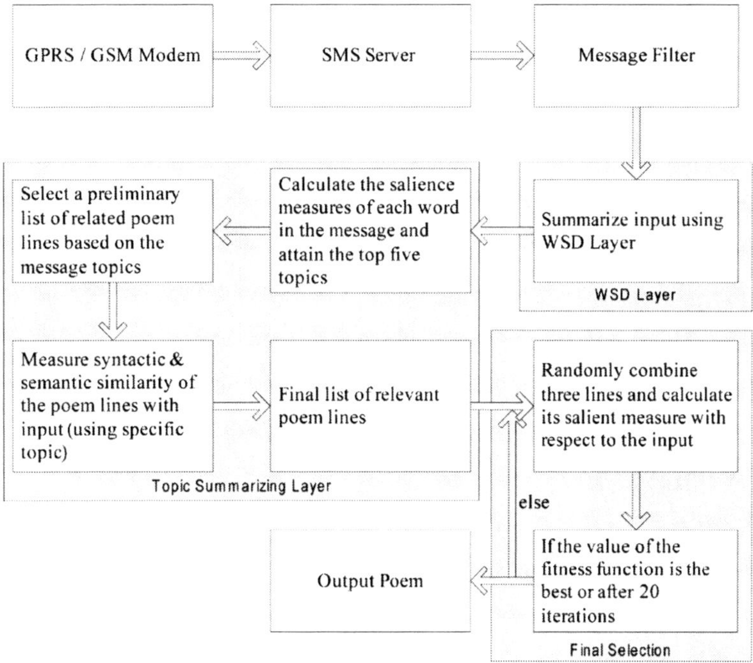

of the system could be unusual, surprising, or maybe amusing. As an example, the user sends the SMS"Happiness is made of good health and bad memory". Firstly the system will pass into the WSD layer to summarize the input message. The analyzed output of the WSD layer is as shown in Table 1.

Based on the result of WSD, the system further analyzes the topics of the input message. As shown in Table 2, the specific topic of each word will be obtained and the top five topics with highest salience measure will be selected as the topics of the input message.

According to the topics of the input message, poem lines with the specific word under each topic are shown in Table 3, Table 4, Table 5, Table 6, and Table 7.

Table 1. Result of the WSD layer

Word	Type	Sense
happiness	noun	emotions experienced when in a state of well-being
is	adverb	have the quality of being
made	verb	make or cause to be or to become
of	preposition	A preposition
good	adjective	having desirable or positive qualities especially those suitable for a thing specified
health	noun	a healthy state of wellbeing free from disease
and	conjunction	coordinating connector
bad	adjective	having undesirable or negative qualities
memory	pronoun	something that is remembered

Table 2. Result of the Topic Modeling layer

Word	Category	Salience Measure
happiness	feeling	25.80
health	state	22.28
bad	attribute	21.10
good	attribute	21.09
memory	cognition	17.66

Table 3. Shortlisted Poem lines about "happiness" and "feeling"

No	Poem Line
1	But being too happy in thine happiness,
2	A success brings endless happiness,
3	Eternal souls eternally glow
4	In life we think we pursue riches, success or even happiness
5	A rose can say I wish you happiness,
...	...

Table 4. Shortlisted Poem lines about "health" and "state"

No	Poem Line
1	Spontaneous wisdom breathed by health,
2	So while things are good and going real well,
3	The best lack all conviction, while the worst
4	A success brings endless happiness,
5	Dressing and really good turkey meat.
...	...

Table 5. Shortlisted Poem lines about "bad" and "attribute"

No	Poem Line
1	Poverty is thing bad
2	to substitute bad days, for the best ones you can remember,
3	Of moral evil and of good,
4	we went thru moments that were good and bad,
5	The best lack all conviction, while the worst
...	...

Table 6. Shortlisted Poem lines about "good" and "attribute"

No	Poem Line
1	Love is its own best law
2	to substitute bad days, for the best ones you can remember
3	Of moral evil and of good,
4	Faith of flower was right and better
5	So while things are good and going real well,
...	...

Table 7. Shortlisted Poem lines about "memory" and "cognition"

No	Poem Line
1	to substitute bad days, for the best ones you can remember,
2	The bad news: heaven is one endless day;
3	Dressing and really good turkey meat.
4	Visions of glory, spare my aching sight,
5	Love is its own best law
...	...

For the purpose of providing different entertaining results for the same input message, the system will generate random combinations of three poem lines from the shortlisted list iteratively. The final output which passes the fitness function or has the highest salience measure in the fitness function is as follows:

In life we think we pursue riches, success or even happiness
Faith of flower was right and better'
The bad news: heaven is one endless day.

The final output of BlogWall is shown Figure 4, and Table 8 shows more examples generated from the system.

Figure 4. The output from the BlogWall system

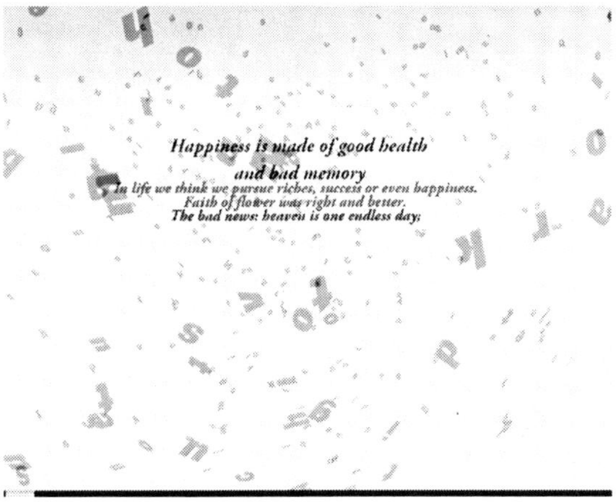

Preliminary results of the algorithm are very encouraging, and definite qualitative. People find the system entertaining, and are curious to experiment. The output of the system generally appealed to the user. The key novelties in the algorithm that makes it more efficient are listed as follows,

- **Information Summarization:** This is a direct result of the Topic Summarization method. Understanding of the context gives rise to a more efficient summary of the poem.
- **Information Retrieval: Quality wise:** By understanding intra document links and understanding of the sense of the words, the quality of poem lines selected is achieved. Quantity wise: By understanding the context, the algorithm is able to retrieve more relevant poem lines.
- **Remixing Quality (Especially Coherence):** By using inter line links to reinforce the input message's theme the remixed poem is of good standards.
- **Limited information Loss:** The top three synonyms were used to find the relevant lines.

Table 8. Three examples of input messages and output poems

Input Message	Poem
I want to realize my dreams, not just record them.	Records of great wishes slept with, Dream moon, cream moon, first he kissed me. He'd have God for his father & never want joy.
All I want is someone to listen	At starting, is my object? Nay we 'll go Someone had better be prepared for rage I want to be someone
Sometimes we should slow down and enjoy the lives	From the wisdom of many life-times in my heart, I love him dear Draining our life energy, its role

USER EVALUATION

Design of the User Evaluation

In this user evaluation, 15 participants were selected from Sri Lanka with the average age of 26. Even though there are several modes in the BlogWall system, the main emphasis was given into the poetry generating system which is the main module in the system. To understand the users' engagement and social experience, we have focused on three main directions on BlogWall system in this user evaluation. The usability and acceptance of the system, interaction and social communication using the system, and the possibilities of using the system to preserve the demolishing poetry culture in Sri Lanka are the three main sections focused on this study.

1. Acceptance of BlogWall
   - Question: Are users comfortable with BlogWall?
   - Hypothesis: Users would satisfy with the system regarding to the experience and the related outputs.
   - Example Statement 1.1: I like the system

- Example Statement 1.2: Generated poetry is related to the input SMS
- Example Statement 1.3: I prefer to see the SMS directly on the screen
- Example Statement 1.4: I prefer generating poetry from the SMS
- Example Statement 1.5: When using the system I was felt like I am a poet

2. Social Communication using BlogWall
    - Question: Could BlogWall be used as a tool for social communication?
    - Hypothesis: The system could be used for indirect social communication through poetry.
    - Example Statement 2.1: I think this system could use to communicate with others
    - Example Statement 2.2: I think this system could use as an indirect communication tool with my friends (for example, blogs, facebook, and twitter are used as indirect communication tools)
    - Example Statement 2.3: Poetry could use as a medium for communicating at present
3. Preserve Poetry Cultural in Sri Lanka
    - Question: Could BlogWall be used preserve the demolishing poetry culture in Sri Lanka?
    - Hypothesis: The system could preserve the poetry culture in Sri Lanka.
    - Example Statement 3.1: I think the traditional poetry culture in Sri Lanka is demolishing from the new generation
    - Example Statement 3.2: I think this system could help to preserve the poetry culture in Sri Lanka
    - Example Statement 3.3: I would like to communicate with others through poetry

Table 9. Summarized Results of user evaluation based on three main sections

	Negative	Neutral	Positive
1.Usability	18%	29%	53%
2.Social Communication in Sri Lanka	16%	32%	52%
3.Preserve poetry culture in Sri Lanka	15%	32%	53%

Table 10. Detailed results of user evaluation based on individual statements

Statement	Options		
	No	Neutral	Yes
1.1	00%	28%	72%
1.2	36%	32%	32%
1.3	20%	28%	52%
1.4	16%	24%	60%
1.5	20%	32%	48%
2.1	20%	36%	44%
2.2	16%	16%	68%
2.3	12%	44%	44%
3.1	04%	32%	64%
3.2	24%	28%	48%
3.3	16%	36%	48%

Results of the User Evaluation

Data collected from the survey after using the system are expressed as mean unless otherwise specified. Results of the survey are presented in Table 9 and Table 10. Table 9 presented the high level results according to each section and Table 10 presented the results according to the Statements. Of all elements explored with this survey, most of them performed positively since more than 50% selected the favorable choice to the Statements. Only the statement number 1.2 is based on generated poetry and the input SMS provided a balanced result. A detailed analysis of results described further.

In the aspect of acceptance of BlogWall, none of the users commented negatively on statement 1.1 which denoted that all the participants are partial to the system. Although the system could not accurately generate poems according to the meaning of the input message, some of the participants still provided positive feedbacks for this part. One participant said it was really nice to see the response corresponding to the SMS. Another user commented that the poems generated by the system are meaningful to some extent and observed the various possibilities for further improvements. For example, the system still needs more accurate result in understanding the input SMS. One user suggested that it would be better if this system can input and output in other languages

In the section of social communication using BlogWall, more than 60% of the participants agreed that poetry could be used as a medium for communicating at present and they could use the system to communicate with their friends indirectly. One other suggestion for improvement was that as a communication media it would be better if the system could give related images which will provide visuals for the poem.

In the final section in the questionnaire about preserving poetry culture in Sri Lanka using the BlogWall system, as can be seen in Table 9, 58% of users do agree that the system provides prospects to preserve poetry culture in Sri Lanka. It is observed that most of the participants agreed that even though the younger generation of Sri Lanka is distancing themselves from poetic literature, they would interact positively with a system operates using the popular applications they are familiar with. Given below are some of the positive comments from the participants.

I strongly believe that this system could be effective in immersing young people in poetry

This is so fascinating.. I am animated and stimulated by the system's response to my SMS

In addition, another participant commented that the system should understand the different cultural elements in order to generate better poems. However, these excellent comments prove that the introduction of BlogWall system into the modern society of Sri Lankan would be a rewarding experience while to it helps to create a discourse on preserving of traditional poetry and integration of it to modern cultural trends. Since the vision behind the creation of BlogWall is remixing the traditional culture with modern applications thus promoting robust social interactions and to help to preserve and originate a discourse on poetry. Considering the user responses, BlogWall has successfully achieved it.

DISCUSSION

Examining the user experience carefully and allowing for the availability of mobile technology and use of SMS as a communication method, we are hopeful that BlogWall could be used to enhance the social interactivity of people of developing countries like Sri Lanka in a positive manner, and develop the society into a caring and sharing community, where the communication intensity is humane and modern. For example, most of the students in Sri Lanka, both male and female, are subdued in schools and their interaction with the teacher is at best tentative (Gunawardena, 2008). The BlogWall system could be an appropriate platform for them to express their views without directly refereeing to the teacher. By this method they could overcome their social quietness of expressing themselves in public which will be beneficial to their self confidence. This indirect way of communication in the classroom could encourage them to participate in discussions directly in the future. In addition, BlogWall could be installed in public spaces, such as train stations, shopping malls, hospitals, etc., for people to express themselves freely.

Furthermore, it is well known that language difference is one of the main problems that cause misunderstandings between various cultures. Therefore, by translating poetry from local language into English, which is widely used in the world, and from English into local language, we hope BlogWall could enhance the cultural communication and understanding between different races.

CONCLUSION

Sri Lanka has a rich, traditional culture that had developed over centuries. The cultural practices and traditions that unite them are part of their lifestyles to the present. The agricultural, industrial and economic changes, political dissensions, modern technology and globalization have transformed the communities, accustoming them to new life patterns which are a hybrid of old and new. Our intention is to introduce a social communication method that is operating within the familiar life patterns, a system that would re-familiarized them with cultural elements such as poetry which has always been part of their lives thus the development of Blogwall.

Since Sri Lanka as a developing country has achieved tremendous maturity in mobile technology, we envisage the application of the technology of BlogWall would attain a remarkable success. It is noted that most people of Sri Lanka are culturally subdued in expressing their feelings in public. Therefore, the possibility to use mobile technology, especially SMS, to promote social interactivity in Sri Lankan community is explored using Blogwall. By combining traditional culture and modern mobile technology, BlogWall provides a novel mobile artistic platform for indirect self-expression and public communication. In view of the fact that high accessibility of mobile technology and inexpensive SMS, BlogWall could be launched to create a new era of public communication in developing countries.

REFERENCES

Bandusiri, A. S. W. (2009). Statistical overview of the telecommunications sector in Sri Lanka. Retrieved from http://www.itu.int/

Bell, G. (2005). The age of the thumb: A cultural reading of mobile technologies from Asia. In P. Glotz, S. Bertschi, & C. Locke (Eds.), *Thumb Culture: The Meaning of Mobile Phones for Society* (pp. 67-88). Transcript, Verlag, Bielefeld.

Boguraev, B., & Kennedy, C. (1997). Salience-based content characterization of text documents. In *Proceedings of a Workshop on Intelligent Scalable Text Summarization Sponsored by the Association for Computational Linguistics*, Madrid, Spain.

Browne, L. (2003). *Technology explosion, early 20th century*. Retrieved from http://www.economicadventure.org/gazette/ch3.pdf

Designing useful mobile services for Africa (2009). Retrieved from http://googleblog.blogspot.com/2009/07/designing-useful-mobile-services-for.html

Fellbaum, C. (1998). *WordNet: An electronic lexical database*. Cambridge, MA: MIT Press.

Geser, H. (2005). Is the Cell Phone Undermining the Social Order? Understanding Mobile Technology from a Sociological Perspective. In P. Glotz, S. Bertschi, & C. Locke (Eds.), *Thumb Culture: The Meaning of Mobile Phones for Society* (pp. 23-37). Transcript, Verlag, Bielefeld.

Grinter, R., & Eldridge, M. (2001). y do tngrs luv 2 txt msg. In *Proceedings of the Seventh European Conference on Computer-Supported Cooperative Work ECSCW* (vol. 1, pp. 219-238).

Gunavardhana, D. (1991). *The Gajaman story: Gajaman puvatha*. Sridevi Printers.

Gunawardena, C. (2008). Gender equity in higher education in Sri Lanka. *International Studies in Education*. Retrieved from http://iise.pitt.edu/publications/ncontents/v9/n1/ise200904.pdf

Katz, J. (2005). Mobile Communication and the Transformation of Daily Life. In P. Glotz, S. Bertschi, & C. Locke (Eds.), *Thumb Culture: The Meaning of Mobile Phones for Society* (pp. 171-184). Transcript, Verlag, Bielefeld.

Lacoh'ee, H., & Wakeford, N., I., P. (2003). A social history of the mobile telephone with a view of its future. *BT Technology Journal*, *21*(3), 203–211. doi:10.1023/A:1025187821567

Paranavitana, S. (1956). *Sigiri graffiti: Being Sinhalese verses of the Eighth, Ninth and Tenth centuries (archaeological survey of Ceylon)*. Oxford University Press.

Pollock, S. (2003). *Literary cultures in history: reconstructions from South Asia*. University of California Press.

Samarajiva, R. (2000). The role of competition in institutional reform of telecommunications: Lessons from Sri Lanka. *Telecommunications Policy*, *24*(8-9), 699–717. doi:10.1016/S0308-5961(00)00057-4

Samuel, J., Shah, N., & Hadingham, W. (2005). *Mobile communications in South Africa, Tanzania, and Egypt: Results from community and business surveys*.

Siau, K., & Shen, Z. (2003). Mobile communications and mobile services. *International Journal of Mobile Communications*, *1*(1), 3–14. doi:10.1504/IJMC.2003.002457

Slater, D., & Kwami, J. (2005). *Embeddedness and escape: Internet and mobile use as poverty reduction strategies in Ghana* (Working Paper 4).

Statistical overview of the telecommunication sector as at end of 4th quarter 2008 (2009). Retrieved from http://www.trc.gov.lk/images/pdf/stodec1.pdf

Wijesekara, N. (1990). *Archaeological department centenary (1890-1990): Inscriptions*. Department of archaeology, Sri Lanka, Colombo.

Chapter 15
The Role of Information Communication Technologies within the Field of Communication for Social Change

Jan Servaes
University of Massachusetts, USA.

ABSTRACT

There is a lot of talk about the 'newness' of mobile and wireless Information and Communication Technologies (ICTs) these days. What is so 'new' about them? And in what way will they solve the still unresolved problems of poverty, inequality and information divides in the world? This chapter takes a bird's eye perspective and presents a number of observations regarding the role of ICTs within the field of Communication for Development and Social Change (CDSC). All those involved in the analysis and application of Communication for Development and Social Change would probably agree that in essence communication for social change is the sharing of knowledge aimed at reaching a consensus for action that takes into account the interests, needs and capacities of all concerned. It is thus a social process, which has as its ultimate objective sustainable development at distinct levels of society. Communication media and Information and Communication Technologies (ICTs) are important tools in achieving social change but their use is not an aim in itself—interpersonal communication and traditional group media must also play a fundamental role. This basic consensus on development communication has been interpreted and applied in different ways throughout the past century. Both at theory and research levels, as well as at the levels of policy and planning-making and implementation, divergent perspectives are on offer. In this chapter, the author presents a brief overview of the field of Communication for Development and Social Change (CDSC) and elaborates on the role and impact of Information and Communication Technologies (ICTs) for Development and Social Change.

DOI: 10.4018/978-1-61692-818-6.ch015

SUMMARIZING THE FIELD OF COMMUNICATION FOR DEVELOPMENT AND SOCIAL CHANGE

There are at least three levels at which we can summarize the field of Communication for Development and Social Change (CDSC): by identifying the different theoretical (1) development and (2) communication paradigms, and (3) by looking at the research priorities in different time periods:

1. Development Paradigms

1. After the Second World War, the founding of the United Nations stimulated relations among sovereign states, especially the North Atlantic Nations and the developing nations, including the new states emerging out of a colonial past. During the cold war period the superpowers—the United States and the former Soviet Union—tried to expand their own interests to the so-called Third World or developing countries. In fact, the USA was defining development and social change as the replica of its own political-economic system and opening the way for the transnational corporations. At the same time, the developing countries saw the 'welfare state' of the North Atlantic Nations as the ultimate goal of development. These nations were attracted by the new technology transfer and the model of a centralized state with careful economic planning and centrally directed development bureaucracies for agriculture, education and health as the most effective strategies to catch up with those industrialized countries. This mainly economic-oriented view, characterized by endogenism and evolutionism, ultimately resulted in the *modernization and growth* theory. It sees development as an unilinear, evolutionary process and defines the state of underdevelopment in terms of observable quantitative differences between so-called poor and rich countries on the one hand, and traditional and modern societies on the other hand (for more details, see Servaes, 1999, 2003, 2008).

2. As a result of the general intellectual 'revolution' that took place in the mid 60s, this Euro- or ethnocentric perspective on development was challenged by Latin American social scientists, and a theory dealing with *dependency and underdevelopment* was born. The dependency approach formed part of a general structuralist re-orientation in the social sciences. The 'dependistas' were primarily concerned with the effects of dependency in peripheral countries, but implicit in their analysis was the idea that development and underdevelopment must be understood in the context of the world system. This dependency paradigm played an important role in the movement for a New World Information and Communication Order from the late 1960s to the early 1980s. At that time, the new states in Africa, Asia and the success of socialist and popular movements in Cuba, China, Chile and other countries provided the goals for political, economic and cultural self-determination within the international community of nations. These new nations shared the ideas of being independent from the superpowers and moved to form the Non-Aligned Nations. The Non-Aligned Movement defined development as political struggle.

3. Since the demarcation of the First, Second and Third Worlds has broken down and the cross-over centre-periphery can be found in every region, there is a need for a new concept of development which emphasizes *cultural identity and multidimensionality* (further discussed in De Cuellar, 1995; Robertson, 1992; Sen, 2004; Servaes, 1999, 2008, Tomlinson, 1999). The present-day 'global' world, in general as well as in its distinct regional and national entities, is confronted with multifaceted crises. In other words, the previously held dependency perspective has become more difficult to support because of the growing interdependency of regions, nations and communities in our globalized world.

From the criticism of the two paradigms above, particularly that of the dependency approach, a new viewpoint on development and social change has come to the forefront. The common starting point here is the examination of the changes from 'bottom-up', from the self-development of the local community. The basic assumption is that there are no countries or communities that function completely autonomously and that are completely self-sufficient, nor are there any nations whose development is exclusively determined by external factors. Every society is dependent in one way or another, both in form and in degree. Thus, a framework was sought within which both the Centre and the Periphery could be studied separately and in their mutual relationship, both at global, national and local levels.

More attention is also being paid to the *content of development*, which implies a more normative, holistic and ecological approach. 'Another development' questions whether 'developed' countries are in fact developed and whether this genre of progress is sustainable or desirable. It favors a multiplicity of approaches based on the context and the basic, felt needs, and the empowerment of the most oppressed sectors of various societies at divergent levels. A main thesis is that change must be structural and occur at multiple levels in order to achieve sustainable ends.

2. Communication Paradigms

1. The above more general typology of the so-called development paradigms can also be found at the communication and culture level. The communication media are, in the context of development, generally used to support development initiatives by the dissemination of messages that encourage the public to support development-oriented projects. Although development strategies in developing countries diverge widely, the usual pattern for broadcasting and the press has been predominantly the same: informing the population about projects, illustrating the advantages of these projects, and recommending that they be supported. A typical example of such a strategy is situated in the area of family planning, where communication means like posters, pamphlets, radio, and television attempt to persuade the public to accept birth control methods. Similar strategies are used on campaigns regarding health and nutrition, agricultural projects, education, HIV/Aids prevention and so on.

This model sees the communication process mainly as a message going from a sender to a receiver. This hierarchic view on communication can be summarized in Laswell's classic formula, -- 'Who says What through Which channel to Whom with What effect?' --, and dates back to (mainly American) research on campaigns and diffusions in the late 40s and 50s (Lerner, 1958; Lerner & Schramm, 1967; Schramm, 1954, 1964; Schramm & Lerner, 1976).

The American scholar Everett Rogers (1983, 1986) is said to be the person who introduced this diffusion theory in the context of development. Modernization is here conceived as a process of diffusion whereby individuals move from a traditional way of life to a different, more technically developed and more rapidly changing way of life. Building primarily on sociological research in agrarian societies, Rogers stressed the adoption and diffusion processes of cultural innovation. This approach is therefore concerned with the *process of diffusion and adoption of innovations* in a systematic and planned way. Mass media are important in spreading awareness of new possibilities and practices, but at the stage where decisions are being made about whether to adopt or not to adopt, personal communication is far more likely to be influential. Therefore, the general conclusion of this line of thought is that *mass communication is less likely than personal influence to have a direct effect on social behaviour*.

2. Newer perspectives on development communication claim that this is a limited view of development communication. They argue that this diffusion model is a vertical or one-way perspec-

tive on communication, and that development will accelerate mainly through active involvement in the process of the communication itself. Research has shown that, while groups of the public can obtain information from impersonal sources like radio, television, and nowadays the Internet, this information has relatively little effect on behavioral changes. And development envisions precisely such change. Similar research has led to the conclusion that more is learned from interpersonal contacts and from mass communication techniques that are based on them. On the lowest level, before people can discuss and resolve problems, they must be informed of the facts, information that the media provide nationally as well as regionally and locally. At the same time, the public, if the media are sufficiently accessible, can make its information needs known.

Communication theories such as the 'diffusion of innovations', the 'two-step-flow', or the 'extension' approaches are quite congruent with the above modernization theory. The elitist, *vertical or top-down orientation* of the diffusion model is obvious (for more details, see Fraser & Restrepo-Estrada, 1998; Mody, 1997, 2003; Servaes 2003, 2008; XXX, 1997).

3. The *participatory model*, on the other hand, incorporates the concepts in the framework of multiplicity. It stresses the importance of cultural identity of local communities and of *democratisation and participation at all levels*—international, national, local and individual. It points to a strategy, not merely inclusive of, but largely emanating from, the traditional 'receivers'. Paulo Freire (1983:76) refers to this as the right of all people to individually and collectively speak their word: "This is not the privilege of some few men, but the right of every (wo)man. Consequently, no one can say a true word alone—nor can he say it for another, in a prescriptive act which robs others of their words".

In order to share information, knowledge, trust, commitment, and a right attitude in development projects participation is very important in any decision-making process for development. Therefore, the International Commission for the Study of Communication Problems, chaired by the late Sean MacBride, argued that "this calls for a new attitude for overcoming stereotyped thinking and to promote more understanding of diversity and plurality, with full respect for the dignity and equality of peoples living in different conditions and acting in different ways" (MacBride, 1980:254). This model stresses reciprocal collaboration throughout all levels of participation (for more details, see Jacobson & Servaes, 1999; Papa, Singhal & Papa, 2006; Servaes, Jacobson & White, 1996).

Also, these newer approaches argue, the *point of departure must be the community* (see, for instance, Fuglesang, 1982; Geertz, 1973, 1983; Omoto, 2005; Servaes & Liu, 2007). It is at the community level that the problems of living conditions are discussed, and interactions with other communities are elicited. The most developed form of participation is self-management. This principle implies the right to participation in the planning and production of media content. However, not everyone wants to or must be involved in its practical implementation. More important is that participation is made possible in the decision-making regarding the subjects treated in the messages and regarding the selection procedures.

One of the fundamental hindrances to the decision to adopt the participation strategy is that it threatens existing hierarchies. However, participation does not imply that there is no longer a role for development specialists, planners, and institutional leaders. It only means that the viewpoint of the local groups of the public is considered before the resources for development projects are allocated and distributed, and that suggestions for changes in the policy are taken into consideration.

3. Research Priorities

1. Development communication in the *1958-1986 period* was generally greeted with enthusiasm and optimism. In her PhD-thesis Jo Ellen Fair

(1988; summarized in *Gazette*, 1989) examined 224 studies of communication and development published between 1958 and 1986, and found that models predicting either powerful effects or limited effects informed the research.

Communication has been a key element in the West's project of developing the Third World. In the one-and-a-half decades after Lerner's influential 1958 study of communication and development in the Middle East, communication researchers assumed that the introduction of media and certain types of educational, political, and economic information into a social system could transform individuals and societies from traditional to modern. Conceived as having fairly direct and powerful effects on Third World audiences, the media were seen as magic multipliers, able to accelerate and magnify the benefits of development. (Fair, 1989: 145)

Three directions for future research were suggested: (a) to examine the relevance of message content, (b) to conduct more comparative research, and (c) to conduct more policy research.

2. As a follow-up to this research, Jo Ellen Fair and Hemant Shah (1997) studied 140 journal articles, book chapters and books published in English between 1987 and 1996. Their findings are quite illuminating: "*In the 1987-1996 period, Lerner's modernization model completely disappears*. Instead, the most frequently used theoretical framework is participatory development, an optimist postmodern orientation, which is almost the polar opposite of Lerner who viewed mass communication as playing a top-down role in social change. Also vanishing from research in this latter period is the two-step flow model, which was drawn upon by modernization scholars" (Fair & Shah, 1997:10).

3. *Both periods* do make use of theories or approaches such as knowledge gap, indirect influence, and uses and gratifications. However, research appearing in the years from 1987-1996 can be characterized as much more theoretically diverse than that published between 1958-1986.

In the 1987-1996 study, the most frequent suggestion was "the need to conduct more policy research, including institutional analysis of development agency coordination. This was followed by the need to research and develop indigenous models of communication and development through participatory research" (Fair & Shah, 1997:19). Therefore, nobody was making the optimistic claims of the early years any longer.

4. Recently, Hemant Shah (2007) completed an analysis of 167 items (123 journal articles, 38 book chapters and 6 books) covering the *1997-2005 period*. Though the meta-research technique may still need some fine-tuning, some of the findings are noteworthy:

- *Most authors work at Western institutions* (70% compared to 48% fifteen years ago), rather than in the Non-Western world (29%) or at Inter-Governmental (IGO) or Non-Governmental (NGO) Organizations (1.5%). Of those working in the Non-Western World, 47% work in Asia, 33% in Africa, and only 7% in Latin America and 1.6% in the Middle East. This was clearly different in the past.
- *Funding for development communication has decreased over time*: While 36.2% of studies were funded in the previous periods, only 11% were funded in this period. "In the 1997-2005 time period, US university funding disappeared almost completely. In contrast, funding of development communication research from IGOs and NGOs and US government funding (exclusively from USAID, however) increased in 1997-2005" (Shah, 2007: 9).
- Surveys, secondary data analysis, content analysis and meta-research were the most popular quantitative methods used in 1997-2005; whereas on the qualitative side

these are interviews, case studies, observation, focus groups and ethnography.

- On the content side, *modernization theories remain dominant* (51%), followed by participatory development (38%), dependency (22%), feminist development (19%) and globalization (6%).
- A comparative analysis of media theories used to assess media impact on national development, leads to the following interesting observations: "First, *Lerner's model of media and development has reappeared in the 1997-2005 time period* after totally disappearing in the 1987-1996 period. Second, only two other theories from the traditional US-based behavioral science approach, social learning theory and knowledge gap, appear in the 1997-2005 period… The third trend to note is that the two most prominently mentioned theories in 1997-2005 – participatory communication and social learning – reflect two popular development communication project orientations that were mentioned as innovations in the 1987-1996 study: *participatory development and edu-tainment*" (Shah, 2007: 13). Shah explains the persistence of 'old' ideas, especially Lerner's model (1958, 1977), from a *technological deterministic* perspective: "Each new technological innovation in the postcolonial world since 1958 – television, satellites, microwave, computers, call centers, wireless technology – has been accompanied by determined hope that Lerner's modernization model will increase growth and productivity and produce modern cosmopolitan citizens" (Shah, 2007:24). Hence, a lot of the hype associated with the new media and ICTs seems to be "old wine in a new technology" (Cotton, Anderson & Tufekci, 2009).
- The *consequences of development communication* are very much associated with the more traditional views on modernization; that is, media activate modernity (45%), and media raise knowledge levels (42%). This more traditional perspective makes a strong return, as it was less pronounced during the 1987-1996 time frame. The three other consequences listed are more critical to modernization: media create participatory society (29%), media benefit certain classes (22%), and media create development problems (22%).

The *optimistic belief* that there are overall positive impacts of development communication on individuals, dominant in 1958-1986, has consistently dropped from 25% (in 1958-1986) to 6% (in 1997-2005). Increasingly, however, it is pointed out that *more attention needs to be paid to theory and research*. "Aside from the conclusions urging attention to development theory, the studies also urged more attention to development communication campaign planning by taking into account, as implied by other conclusions, local culture, gender issues, and multimedia delivery of information, and to improving research methods" (Shah, 2007: 20).

- *Suggestions for future research* prioritize the development of new development communication models and the examination of content relevance (both 27%), the need for indigenous models (24%), the study of new technologies (21%), more comparative research (18%), the need for more policy research (8%), and the development of a new normative framework (5%).

The findings by Jo Ellen Fair and Hemant Shah present us with a clear but at the same time complex picture of our field. The suggestions for future research may need to be complemented with other ones --new and old -- mentioned or discussed at other pages or meetings (see, for instance, Gumucio-Dragon & Tufte, 2006, Kim,

2005, Lie, 2003, Mody, 2003, Morris, 2001, Servaes, 2007b, 2008, UNESCO, 2003; or the recommendations by the Scientific Committee for the World Congress on Communication for Development, held in October 2006, in *Annex 1*).

However, *the implicit assumptions on which the so-called dominant modernization paradigm is built do still linger on and continue to influence the policy and planning-making discourse of major actors in the field of Communication for Development and Social Change (CDSC), especially when it comes to the role and impact of ICTs for Development and Social Change* (De Bruijn, Nyamnjoh & Brinkman, 2009).

INFORMATION AND COMMUNICATION TECHNOLOGIES FOR DEVELOPMENT AND SOCIAL CHANGE

To communication technology is attributed a direct impact on the economic development and the political organization of a society. Three points will be briefly explored: (1) the relationship between technology and culture; (2) the issue of technology transfer, which leads to three different views of technology transfer, and (3) its consequences for polymaking.

1. Technology and Culture

1. To technology is attributed a direct impact on the economic development, the political organization, and socio-cultural value system of a society. However, technology is called into existence by a particular set of historical circumstances that shape and define that technology (Holmes, 2005). One must understand that set of historical circumstances if one is to comprehend the effective *relationship between technology and society*: "Technology does not cause society. Nor does society cause technology. Rather, technology is developed in society, in the complex interplay of social forces that are at the same time both cause and effect" (Slack, 1984:XV). Therefore, contrary to popular belief, technology is *not* politically neutral and value free; technology definitely determines the socio-cultural structure and communication patterns of a given society (Hassan, 2008). In other words, Western technology cannot be borrowed without taking in aspects of Western culture at the same time. In my opinion, science and technology are much more than the mere instruments they were expected to be; they cannot be just borrowed or bought. Many Southern policymakers seem to assume that technical and economic progress is simply a *means to an end* and that it hardly affects the culture in which it occurs. It seems to me as if they believe that they can achieve Western-style progress and at the same time retain their culture and their morals or, at the least, most of the essential parts of them.

2. In the international field, accelerating technological developments suggest that opportunities for the expression of cultural differences can be ensured. In reality, however, these developments appear to be the instrument precisely for the destruction of these differences and for its replacement by a uniform, Western-dominated technological model. There are at least three implicit *basic values* in Western technology (Hofstede & Hofstede, 2005). First, Western technology shows little respect for myth, symbol or the power of the mysterious. Every phenomenon has to be broken down into component parts, tested and verified. Secondly, the technology is based on the cult of efficiency. The central considerations are productivity, cost-benefit ratio and the bottom line. Thirdly, the technology dominates and manipulates nature rather than being in harmony with it. Problem-solving is the goal; hence, reality is reduced to those dimensions which can be studied as problems needing solutions. *The values implicit in Western technology may therefore come into conflict with the pre-existing values of a non-Western environment.* In their interesting analysis of computerization in Southeast Asia, two

researchers at the East-West Center in Honolulu, Syed Rahim and Anthony Perkins (1987: 141), note that "the metaphorical concept structuring computer rhetoric seems to be coherent with some basic values of the mainstream Western industrial culture. We can easily associate the computer advertising concept with such value concepts as individualism, rationality, freedom of communication, sociability, salubrity, innovativeness and material progress. It is also interesting to note that the representation of the computer values is strikingly similar to the representation of certain values in the American popular media, for instance, in the Superman movies."

3. Furthermore, *science and technology penetrate other subsystems of society as well.* They are a whole system of explanation, that not only affects science and technology alone, but also seemingly unconnected areas like positive law, and a modern government and bureaucracy where traditional ascriptions, hierarchies and authorities are replaced by objective standards and decision making. Referring to its historically rooted adaptable culture, many Southern policymakers underestimate this overwhelming and comprehensive aspect of nowadays Western science and technology. In other words, technology is not politically neutral and value free. Technology definitely determines the socio-cultural structure and communication patterns of a given society. This also implies that technology may not be equated with mere technical equipment, but that it should refer to knowledge in all its varied applications. Technology produces, innovates and maintains industrial products as well as industrial production methods, basic as well as luxurious commodities, and creates material as well as immaterial results.

2. Three Different Views on Technology Transfer

1. Another fundamental point is the predominant *impact and role of transnational corporations in the production and distribution of technology* (Brynjolfsson & Saunders, 2009). In virtually all the so-called Third World countries, it are primarily the transnational suppliers and financiers that benefit from the advantages of technology transfer. Most of the Third World countries depend on transnationals for their own technical development capacity. These transnational corporations spend not only billions of dollars in advertising products, services and corporate images in a highly competitive market.

2. The problem of technology transfer can be looked at from the three above paradigmatic perspectives. *The first viewpoint builds on the modernization paradigm and integration model.* It maintains that technology produced in the West is appropriate to the needs of developing countries, and that the most efficient and adequate mechanism for transferring this technology is commercially, through transnational corporations. The rationale for this strategy is based on the proposition that, (a) commercial technology transfer through transnationals offer all necessary combinations of technology components within a complete system of technology development and transfer, and, (b) technology transfer needed to establish any new productive facility is a complicated process, requiring special knowledge and skills for each stage of its development. Moreover, this viewpoint claims that the alleged success of transnationals in transferring technology is due not only to their mastery of the whole 'technology package', but also to the fact that this package is closely integrated with management, marketing and financing skills. As a consequence one argues that any attempt to separate elements of technology transfer will either fail or result in considerably larger 'costs'.

3.The second point of view, growing out of the dependency school, can be said to be the anti-pole of the first position. It completely rejects the premises of this integration model. It states that, because the social and economic structures and problems of developing countries are different, technology available in the West, is not appropriate for Third World societies. All the developing countries that accepted technology imports with private capital investment are today more technologically dependent upon the outside world than ever, since they never have been developing their own indigenous scientific and technological capabilities. The only way to break out of the existing technological dependence, this viewpoint advocates, is to drastically limit imports of privately owned and controlled technology and to turn to inward-organizing broad technology import-substitution programs. These should not copy advanced technologies but rather involve the design of *'appropriate' technologies* and the fostering of technological cooperation and exchange among all developing countries, or, in other words, establish a so-called *South-to-South co-operation* (Khor, 2000).

4. The third position, which I advocate, tries to strike the right balance between the two above mentioned strategies. It states that developing countries need an *appropriate mix of technologies and not just appropriate 'home-made' technologies.* Such a mix may include modern foreign-produced and traditional indigenous technology, it may blend the 'latest and the best' with 'primitive technology'. Appropriateness is not measured in terms of the dimensions of that technology, but as a function of the relationship between technology and development (Verzola, 2004, Wilhelm, 2004). Transnational companies are invited to participate provided certain conditions are fulfilled. One such condition is that transnational technology transfer is directed and controlled by the host countries so as to contribute to the creation of domestic scientific and technological capability. Another condition should be that these transnational technology imports are not considered the exclusive or the most important channel of transfer. A real policy problem for Third World nations will be the need to strike a balance between self-reliance and strategic imports. Thus local industry in order to remain competitive may require the importation of advanced technology. The assessment process must be guided by the goal of autonomous development. Further, assessment must be based on a nuanced view of technology; that is, one must differentiate between the various kinds of technologies (i.e., technologies used in production, consumption and distribution), and the impact that these technologies are likely to have. One must ask: Who will benefit most from the importation of these technologies?

3. Consequences for Policymaking

1. The *convergence between telecommunications, broadcasting* and informatics occurred at three levels: at the levels of networks (infrastructure), service provision, and corporate organization (Wang, Servaes & Goonasekera, 2000).

Policy decisions or policy perspectives are *mainly technology and/or commercially driven*. A lot of attention focuses on the research concerning (and the implementation of) hardware. A lot of money is spent for the development of network infrastructure, broadcasting facilities etc. Second, regarding media ownership, we always seem to meet *the same players* in the different sectors on different global, regional and national levels. These are telecommunication operators, major publishing firms and media moguls. In most of the countries we observe that one or two of these actors (or a merger of them) control the ICT sector. Third, because of these concentration tendencies, national governments are afraid of *broadcasting monopolies.* Their legal reaction is the promulgation of anti-trust and anti-concentration laws. Last, the emphasis in public policy making is on hardware. Software/content development is heavily neglected.

Though a user-driven (and consequently more content oriented) and user-specific policy framework may be preferred, a more corporate driven economic rationale seems to become the norm. Both policy perspectives start from quite opposite assumptions, as shown in Table 1 (further developed in Burgelman & Verhoest, 1996):

Table 1.

Current corporate-driven policy	Preferred user-driven policy
• Agitated market/uncertain revenues	• 'Controlled' market/'guaranteed' revenues
• Competition	• 'New deal' type of policy
• Short term	• Long-term objectives
• Technology push/technology specific	• User-driven solutions
• The medium is the message	• Content oriented

The *regulation and control* of the process and structure of technology transfer is a very important issue that often does not receive proper consideration from policymakers. Such control mechanisms can be internal as well as external.

2. Observing the real process of technology transfer, *two related considerations* become obvious: on the one hand, the technological and other objectives of transnationals (and Western governments) differ considerably from the sometimes explicit but mostly implicit national objectives of developing countries; and on the other hand, many Third World nations continue to regard the transnationals as the major channel of international technological transfer. Vincent Lowe (1987), and Rahim & Perkins (1987), for instance, point out that none of the ASEAN countries had yet adopted a coherent and comprehensive policy in the field of information technology, and that "very little systematic research on the social and cultural problems of computerization has been carried out in the Asean region. (Moreover,) there is very little public discussion about it" (Rahim & Perkins, 1987: 162). These authors indicate that this may be due to (a) the rather recent explosive development of the field; (b) the unclarity as to where in public administration the responsibility for technological matters reside; and (c) the degree of technical complexity that demands a very specialized expertise. (d) Furthermore, in several countries the dominant climate of privatization and deregulation does not favor public policymaking. (e) Also, Third World countries are increasingly affected by the modernization recipe of economic growth by either donor countries or donor agencies (such as the World Bank, the IMF, or -- in the Southeast Asian context -- the Asian Development Bank) that implies the retreat of the state in favor of various forms of privatization. (f) Lastly, there appears to be a general concern to miss the so-called 'information revolution' and as a consequence the majority of policy decisions relate to the spending of public funds on the acquisition of the latest generation of computers and new technology in general. Most of the Third World governments fear to remain 'backward', and therefore often overlook the specific socio-economic and socio-cultural contexts in which technology transfer has to be considered.

3. UNESCO's former Assistant-Director-General, Alan Hancock (1984), argues that *successful technology transfer cannot occur unless a certain level of capability has already been built up in the recipient.* If this is not the case, the partnership will be unequal, and the transfer will be mechanical and imitative. He concludes that commercial transfers are of little use: no real transfer is involved, and the recipient is totally disadvantaged. Moreover, looking at domestic technologies, Hancock observes that indigenous forms do badly at first, when they are compared with imported models, and as a result they need a good deal of committed and resolute policy support if they are to have the chance to survive, to become entrenched and competitive.

ICTS AND THE MILLENNIUM DEVELOPMENT GOALS

1. The field of communication for social change is vast, and the models supporting it are as different as the ideologies that inspired them. However, generally speaking, we see two approaches: one aims to produce a common understanding among all the participants in a development initiative by

implementing a policy or a development project, that is, the *top–down model*; the other emphasizes engaging the grassroots in making decisions that enhance their own lives, or the *bottom–up model*. Despite the diversity of approaches, there is a consensus today on the need for grassroots participation in bringing about change at both social and individual levels. Therefore, as Sen (2004: 20) argues, "The deciding issue, ultimately, has to be one of democracy. An overarching value must be the need for participatory decision-making on the kind of society people want to live in, based on open discussion, with adequate opportunity for the expression of minority positions." Many believe that ICTs may contribute to this and the UN campaign to reach the Millennium Development Goals by 2015 is an appropriate objective to aim for (Chayko, 2008).

2. The Millennium Development Goals (MDG) initiative follows decades of debate over how nations might collaborate on long-term strategies for a global social agenda. Wealthy countries were asked to increase development aid, relieve the debt burden on poor countries, and give them fair access to markets and technology.

Each Millennium Development Goal (MDG) has its own set of targets and benchmarks that provide a measurable way to track its implementation (UNDP, 2006). However, questions can be raised about the feasibility and appropriateness of setting the same global targets for governments worldwide:

"The MDGs can justly claim to generate a bit of buzz about duties a government might otherwise neglect... Sadly, however, they cannot do what they purport to do, which is to provide credible benchmarks against which governments can be judged... Some goals cannot be met, others cannot be measured... The goals are supposed to be everyone's responsibility, which means they are no one's. Poor countries can blame rich ones for not stumping up enough cash; rich governments can accuse poor ones of failing to deserve more money" (The Economist, 2007: 13).

In essence, development is about the development of people and the transformation of society. The Millennium Development Goals (MDG) in general look at development as an 'engineering problem' to be solved from a top-down perspective. "Sixty years of countless reform schemes to aid agencies and dozens of different plans, and $2.3 trillion later, the aid industry is still failing to reach the beautiful goal (of making poverty history, JS). The evidence points to an unpopular conclusion: Big Plans will always fail to reach the beautiful goal" (Easterly, 2006: 11). *Table 2* presents a number of roles which Information and Communication Technologies (ICTs) could play to achieve the MDGs.

However, in the UN system conflicts seldom concern the MDGs as such, but rather the means of achieving them. In line with my above argument the MDGs should be assessed from a *people's perspective*. In other words: What can MDGs do for the poor and 'voiceless'? "Setting a prefixed (and grandiose) goal is irrational because there is no reason to assume that the goal is attainable at a reasonable cost with the available means" (Easterly, ibid.). Even UN Secretary-General, Ban Ki-Moon, expressed his disappointment and concern in the latest annual interim report on the MDGs: "The lack of any significant increase in official development assistance since 2004 makes it impossible, even for well-governed countries, to meet the MDGs" (UN, 2007:3).

It is therefore essential to start from the perspective of local communities and to cooperate with organizations (UN, governmental, NGOs, the public and the private sector, and civil society) that have developed a trust within a community in order to achieve sustainable change in society.

*Table 2. Role of ICTs in achieving MDGs**

MDGs Role of ICTs
1. Eradicate extreme poverty and hunger
• Increase access to market information and lower transaction costs for poor farmers and traders
• Increase efficiency, competitiveness and market access of developing country firms
• Enhance ability of developing countries to participate in global economy and to exploit comparative advantage in factor costs (particularly skilled labour)
2. Achieve universal primary education
3. Promote gender equality and empower women
• Increase supply of trained teachers through ICTs enhanced, distance training of teachers and networks that link teachers to their colleagues
• Improve efficiency and effectiveness of education ministries and related bodies through strategic application of technologies and ICTs enabled skill development
• Broaden availability of quality educational materials/resources through ICTs
• Deliver educational and literacy programmes specifically targeted to poor girls and women using appropriate technologies
• Influence public opinion on gender equality through information/ communication programmes using a range of ICTs
4. Reduce child mortality
5. Improve maternal health
6. Combat HIV/AIDS, malaria and other diseases
• Enhance delivery of basic and in-service training for health workers
• Increase monitoring and information sharing on disease and famine
• Increase access of rural care-givers to specialist support and remote diagnosis
• Increase access to reproductive health information, including information on AIDS prevention, through locally appropriate content in local languages
7. Ensure environmental sustainability
• Remote sensing technologies and communications networks permit more effective monitoring, resource management, mitigation of environmental risks
• Increase access to/awareness of sustainable development strategies, in areas such as agriculture, sanitation, water management, mining, etc.
• Greater transparency and monitoring of environmental abuses, enforcement of environmental regulations
• Facilitate knowledge exchange and networking among policy makers, practitioners and advocacy groups

continued on following page

Table 2. continued

MDGs Role of ICTs
8. Develop a global partnership for development
• Enable LDCs, landlocked countries and small islands to link up with the global market to accelerate their progression and full integration into the world economy
• Distance working facilitated by ICTs opens up opportunities to create service-sector jobs in developing countries in such industries as call centres, data entry and processing, and software development
• Telecentres provide access to telecommunications and create direct employment for men and women
• Improve youth learning skills, employability to meet the challenges of the knowledge-based global economy of the 21st century
• Provide online databases on drugs
• Combine low and high technology to achieve relative ubiquity of access to effective and affordable ICTs tools
• Promote digital literacy through e-learning
• Develop a critical mass of knowledge workers with the technical capabilities to provide and maintain ICTs infrastructure

* Adapted from Marker et al., 2002, and Siriginidi, 2009.

CONCLUSION

Today almost nobody would dare to make the optimistic claims of the early years any longer. The experience of the past fifty years has demonstrated that development is possible, but not inevitable.

As many others, we are in search of a new paradigm for development, one which looks at development as a *transformation of society*. In such a perspective "change is not an end in itself, but a means to other objectives. The changes that are associated with development provide individuals and societies more control over their own destiny. Development enriches the lives of individuals by widening their horizons and reducing their sense of isolation. It reduces the afflictions brought on by disease and poverty, not only increasing lifespans, but improving the vitality of life" (Stiglitz, 1998: 3). In the past, we have referred to this emerging paradigm as *multiplicity* (Servaes, 1999).

This perspective argues that considerations of communication needs to be explicitly built into development plans to ensure that a mutual sharing/learning process is facilitated. Such communicative sharing is deemed the best guarantee for creating successful transformative projects. Therefore, I define Communication for Development and Social Change as:

Communication for social change is the nurturing of knowledge aimed at creating a consensus for action that takes into account the interests, needs and capacities of all concerned. It is thus a social process, which has as its ultimate objective sustainable development at distinct levels of society.

Communication media and ICTs are important tools in achieving social change but their use is not an end in itself. Interpersonal communication and traditional group media must also play a fundamental role.

The new starting point is examining the processes of 'bottom-up' change, focusing on self-development of local communities. The basic assumption is that there are no countries or communities that function completely autonomously and that are completely self-sufficient, nor are there any nations whose development is exclusively determined by external factors. Every society is dependent in one way or another, both in form and in degree.

Attention is also needed to critically analyze the *content of development agendas*. An understanding of the way in which development projects both encounter and transform power relationships within (and between) the multiple stakeholders who are impacted by such projects; and an understanding of the way in which ICTs play a central part in building (or maintaining or changing) power relationships is needed.

REFERENCES

Brynjolfsson, E., & Saunders, A. (2009). *Wired for innovation. How information technology is reshaping the economy*. Cambridge, MA: MIT Press.

Burgelman, J.-C., & Verhoest, P. (1996). Trans-European information networks. *Telematics and Informatics*, *13*(2/3), 67–80. doi:10.1016/S0736-5853(96)00006-8

Chayko, M. (2008). *Portable communities. The social dynamics of online and mobile connectedness*. Albany, NY: State University of New York Press.

Cotton, S., Anderson, W., & Tufekci, Z. (2009). Old wine in a new technology, or a different type of digital divide? *New Media & Society*, *11*(7), 1163–1186. doi:10.1177/1461444809342056

De Bruijn, M., Nyamnjoh, F., & Brinkman, I. (Eds.). (2009). *Mobile phones: The new talking drums of everyday Africa*. Leiden, The Netherlands: African Studies Centre.

De Cuellar, J. P. (1995). *Our Creative Diversity. Report of the World Commission on Culture and Development*. Paris: UNESCO.

Easterly, W. (2006). *The White Man's Burden. Why the West's Efforts to Aid the Rest Have Done So Much Ill and So Little Good*. New York: Penguin Books.

Elliott, A., & Lemert, C. (2006). *The new individualism. The emotional costs of globalization*. London: Routlege.

Fair, J. E. (1988). *A meta-research of mass media effects on audiences in developing countries from 1958 through 1986*. Unpublished doctoral dissertation Indiana University, Bloomington, IN.

Fair, J. E. (1989). 29 years of theory and research on media and development: The dominant paradigm impact. *Gazette*, *44*, 129–150. doi:10.1177/001654928904400204

Fair, J. E., & Shah, H. (1997). Continuities and discontinuities in communication and development research since 1958. *Journal of International Communication*, *4*(2), 3–23.

Fraser, C., & Restrepo-Estrada, S. (1998). *Communicating for Development. Human Change for Survival*. New York: I.B. Tauris Publishers.

Freire, P. (1983). *Pedagogy of the Oppressed*. New York: Continuum.

Fuglesang, A. (1982). *About understanding. Ideas and observations on cross-cultural communication*. Uppsala, Sweden: Dag Hammarskjold Foundation.

Geertz, C. (1973). *The interpretation of cultures*. New York: Basic Books.

Geertz, C. (1983). *Local knowledge. Further essays in interpretative anthropology*. New York: Basic books.

Giddens, A. (1995). *Modernity and Self-Identity. Self and Society in the Late Modern Age*. Cambridge, UK: Polity.

Gumucio-Dragon, A., & Tufte, T. (Eds.). (2006). *Communication for Social Change Anthology: Historical and contemporary readings*. South Orange, NJ: CFSC Consortium.

Halavais, A. (2009). *Search engine society*. Cambridge, UK: Polity.

Hancock, A. (1984). *Technology transfer and communication*. Paris: UNESCO.

Harindranath, R. (2006). *Perspectives on Global Cultures*. Berkshire, UK: Open University Press.

Harvey, M. (Ed.). (2005). *Media matters. Perspectives on Advancing Governance & Development from the Global Forum for Media Development*. London: Internews Europe.

Hassan, R. (2008). *The Information Society*. Cambridge, UK: Polity Press.

Hofstede G. & Hofstede. (2005). *Cultures and organizations. Software of the mind*. London: McGraw Hill.

Holmes, D. (2005). *Communication Theory. Media, Technology and Society*. London: Sage.

Jacobson, T., & Servaes, J. (Eds.). (1999). *Theoretical Approaches to Participatory Communication*. Cresskill NJ: Hampton.

Khor, M. (2000). *Globalization and the South. Some critical issues*. Penang, Malayasia: Third World Network.

Kim, Y. Y. (2005). Inquiry in intercultural and development communication . *The Journal of Communication*, *55*(3), 554–577. doi:10.1111/j.1460-2466.2005.tb02685.x

Lerner, D. (1958). *The passing of traditional society. Modernizing the Middle East*. New York: Free Press.

Lerner, D., & Schramm, W. (Eds.). (1967). *Communication and change in the developing countries*. Honolulu, HI: University Press of Hawaii.

Lie, R. (2003). *Spaces of Intercultural Communication. An interdisciplinary Introduction to Communication, Culture, and Globalizing/Localizing Identities*. Cresskill, NJ: Hampton Press.

Lowe, V. (1987). *Dependency within Bounds: Media and Information Technology policies within the ASEAN region*. Bangkok: Institute of Asian Studies, Chulalongkorn University.

MacBride, S. (Ed.). (1980). *Many Voices, One World: Communication and Society. Today and Tomorrow*. Paris: UNESCO.

Marker, P., McNamara, K., & Wallace, L. (2002). *The significance of information and communication technologies for reducing poverty.* London: DFID. Retrieved December 18, 2008 from http://www.dfid.gov.uk/pubs/files/ictpoverty.pdf

Mody, B. (Ed.). (1997). Communication and Development: Beyond Panaceas. *The Journal of International Communication*, *4*(2), 138.

Mody, B. (Ed.). (2003). *International and Development Communication. A 21st Century Perspective*. Thousand Oaks, CA: Sage.

Morris, N. (2001). *Bridging the Gap: An examination of Diffusion and Participatory Approaches in Development Communication*. Washington, DC: The Change Project/USAID.

Omoto, A. (Ed.). (2005). *Processes of Community Change and Social Action*. Mahwah, NJ: LEA.

Papa, M., Singhal, A., & Papa, W. (2006). *Organizing for Social Change. A dialectic journey of theory and praxis*. New Delhi, India: Sage.

Rahim, S., & Pennings, A. (1987). *Computerization and Development in Southeast Asia*. Singapore: Asian Mass Communication and Information Center (AMIC).

Robertson, R. (1992). *Globalization: Social theory and global culture*. London: Sage.

Rogers, E. M. (1983). *The diffusion of innovations* (3rd ed.). New York: The Free Press.

Rogers, E. M. (1986). *Communication Technology: The new media in society*. New York: The Free Press.

Scholte, J. A. (2005). *Globalization. A critical introduction*. New York: Palgrave.

Schramm, W. (1954). *The process and effects of mass communication*. Urbana, IL: University of Illinois Press.

Schramm, W. (1964). *Mass media and national development. The role of information in the developing countries*. Stanford, CA: Stanford University Press.

Schramm, W., & Lerner, D. (Eds.). (1976). *Communication and change. The last ten years - and the next*. Honolulu, HI: University Press of Hawaii.

Sen, A. (2000). *Development as freedom*. New York: Anchor Books.

Sen, A. (2004). Cultural liberty and human development . In Fukuda-Parr, S. (Ed.), *Human development report: Cultural liberty in today's diverse world*. New York: United Nations Development Programme.

Servaes, J. (1997). Development Communication: What the 'Masters' Say [special issue]. *The Journal of Development Communication*, *8*(2), 179.

Servaes, J. (1999). *Communication for Development. One World, Multiple Cultures*. Cresskill, NJ: Hampton Press.

Servaes, J. (Ed.). (2003). *Approaches to Development. Studies on Communication for Development*. Paris: UNESCO Publishing House.

Servaes, J. (2007a).Harnessing the UN system into a common approach on Communication for Development. *The International Communication Gazette, 69*(6), 483-507. Retrieved from http://gaz.sagepub.com

Servaes, J. (Ed.). (2007b). *Communication for Development. Making a Difference*. Background paper for the World Congress on Communication for Development, Rome, 25-27 October 2006.

Servaes, J. (Ed.). (2008). *Communication for Development and Social Change*. Los Angeles: Sage.

Servaes, J., Jacobson, T., & White, S. (Eds.). (1996). *Participatory Communication for Social Change*. New Delhi: Sage.

Servaes, J., & Liu, S. (Eds.). (2007). *Moving targets. Mapping the paths between communication, technology and social change in communities*. Penang, Malayasia: Southbound.

Shah, H. (2007). *Meta-Research of Development Communication Studies, 1997-2005: Patterns and Trends Since 1958*. Paper presented at ICA Conference, San Francisco, May.

Siriginidi, S. R. (2009). Achieving millennium development goals: Role of ICTS innovations in India. *Telematics and Informatics*, *26*(2), 127–143. doi:10.1016/j.tele.2008.02.001

Slack, J. (1984). *Communication Technologies and Society*. Norwood, MA: Ablex.

Stiglitz, J. (1998). *Towards a New Paradigm for Development: Strategies, Policies, and Processes*. Prebisch Lecture at UNCTAD, Geneva, 19 October.

The Economist. (2007, 7 July). The eight commandments. *The Economist,* (pp. 26-29).

Tomlinsion, J. (1999). *Globalization and Culture*. Cambridge, MA: Polity Press.

UNESCO. (2003). *International Experts Brainstorming Meeting on Development Communication* (New Delhi, 1-3 September 2003), Paris: UNESCO.

United Nations. (2007). *The Millennium Development Goals Report 2007*. New York: UN.

United Nations Development Programme. (2006). *Communication for Empowerment: developing media strategies in support of vulnerable groups*, Oslo, Norway. Retrieved from www.worldbank.org/wbi/news/docs/Severinoonfreepress.htm

Verzola, R. (2004). *Towards a political economy of information*. Quezon City, Philippines: Foundation for Nationalist Studies.

Wang, G., Servaes, J., & Goonasekera, A. (Eds.). (2000). *The New Communications Landscape. Demystifying media globalization*. London: Routledge.

Wilhelm, A. (2004). *Digital Nation. Toward an Inclusive Information Society*. Boston: The MIT Press.

ENDNOTE

1 Jan Servaes, Nicholas Carah, Martin Hadlow, Eric Louw, Pradip Thomas (University of Queensland), Silvia Balit (independent consultant), Maria Celeste Cadiz (University of the Philippines Los banos), Nabil Dajani (American University of Beirut), Cees Hamelink (University of Amsterdam), Tom Jacobson (Temple University), Ullamaija Kivikuru (University of Helsinki), John Mayo (Florida State University), Rafael Obregon (Ohio University), Doug Storey (John Hopkins University), Thomas Tufte (Roskilde University), and Karin Gwinn Wilkins (University of Texas at Austin).

APPENDIX

RECOMMENDATIONS BY THE SCIENTIFIC COMMITTEE OF THE WORLD CONGRESS ON COMMUNICATION FOR DEVELOPMENT

Members[1] of the Scientific Committee of the World Congress on Communication for Development (Rome, 25-27 October 2006), identified the following main challenges for communication for development and social change (Servaes, 2007b: 112-114) to be recognized as a field in its own right and to be adopted systematically in development initiatives:

1. *Good governance, transparency, accountability and development communication go hand in hand.* Good governance and a good government are not the same. Good governance is based on the participation of all people concerned. Decentralization of governmental institutions does not necessarily imply people's participation. Decentralization does not always mean democratization. In reality the motives for decentralizing may hide a wish of central powers to get rid of certain responsibilities while tightening their control. This blurs the lines of accountability. For this reason, local media have a crucial role to play in facilitating a mutual understanding between those in power and the communities.
2. Participatory concepts in the context of communication for development can *be* complex and challenging. Communities consist of fluid interests and shifting relationships.
3. *Participation can take place at different levels*: (a) decision making; (b) benefits; (c) evaluation; and (d) implementation. Participation is about changing power relations. While empowering one group, it may do the opposite to another. Meaningful participation requires organization around common interests and awareness on how to handle power relations.
4. It is important to reinforce *independent and pluralistic media* to foster good governance and transparency. Print media can play a special role in society as they are sometimes more independent and pluralistic than radio and television. However, all media need to be sensitised and become more participatory. Currently there is often a gap between what media report and the realities of a country. Pure commercialism avoids tackling the crucial issues of a country because such issues do not sell. It undermines the role of media as watchdogs. *Press freedom is never guaranteed, not even in a democracy.*
5. Communication for development has not made full use of the *potential of radio*, which in some regions could be the most effective participatory tool. Radio has the highest penetration in many rural areas in developing countries. It is not too late to rediscover radio. In particular community radio (often linked to the global world through the Internet) has proven its ability to make participation effective and sustainable. Therefore, also ICTs are an important tool to facilitate good governance provided that application and operation systems are made available in local languages.
6. *Policies and resources* - communication for development initiatives need to be properly enabled by concerted actions, and adequate policies and resources. These should consider longer timescales. It is essential to bridge the digital divide by supporting community access to relevant information in their own language and at an affordable cost, for example through community telecentres/multimedia centers. This should also involve support for the production of content by the local communities.

It is crucial to encourage the production of diverse local content in local languages for the media and ICTs, bearing in mind the potential of interactive technologies to carry multimedia content.

7. National governments should implement a *legal and supportive framework* favoring the right to free expression and the emergence of free and pluralistic information systems, including the recognition of the specific and crucial role of community media in providing access to communication for isolated and marginalized groups. There is a need to influence policy on communication for development through advocacy, not only with governments and international agencies but also within development agencies, private corporations and civil society partners, for communication for development to be successful.
8. *Building alliances*. There is a need for effective linkages which give voices to the poorest and have the ability to engage with policy and influence decision-making on sustainable development. To this end, special attention should be given to fostering local, national and regional communication for development processes.
9. *New global partnerships* are necessary with the media, development agencies, universities and governments. It is important to identify possibilities for convergence and for complementing existing work and to coordinate and document such work via a truly independent scientific body.

Chapter 16
Health Hazards of Mobile Information Communication Technologies

Sohayla M. Attalla
Mansoura University, Egypt

INTRODUCTION

During recent years, the prevalence of mobile phone users has increased, and concerns have been raised that the use has increased dramatically without sufficient understanding of the potential adverse health effects. Mobile phones are used by almost everybody today. Kids, adults, old people, all are using mobile phones. One of the most common sights we see these days is that of people with their mobile phones next to their ears. A boon for better communication! But are we aware about the dangers these mobile phones cause to our health? Are we carrying a hazard with us everyday? The erection of mobile telephone base stations in inhabited areas has raised concerns about possible health effects caused by emitted microwaves. During the last decade, use of radio frequency (RF) applications like mobile phones and other wireless devices, has increased remarkably. This has triggered numerous studies related to possible health risks due to the exposure of RF electromagnetic (EM) fields.

What is RF Energy and How Can It Affect the Body?

RF energy, also called radio waves, is a form of electromagnetic radiation. Electromagnetic radiation can be ionizing (high-frequency) or non-ionizing (low-frequency). RF energy belongs to the non-ionizing type of electromagnetic radiation. It is known that ionizing radiation, such as that produced by x-ray machines, can present a health

DOI: 10.4018/978-1-61692-818-6.ch016

risk at high levels of exposure. However, it is not yet known whether non-ionizing radiation poses a cancer risk (FDA, 2003).

The most important use of RF energy is for telecommunications (FDA, 2003). In the United States, cellular telephones operate in a frequency ranging from about 1,800 to 2,200 megahertz (MHz) (Ahlbom et al., 2004). In that range, the radiation produced is in the form of non-ionizing RF energy. AM/FM radios, VHF/UHF televisions, and cordless telephones (telephones that have a base unit connected to the telephone wiring in a house) operate at somewhat lower radio frequencies than cellular telephones; microwave ovens, radar, and satellite stations operate at somewhat higher radio frequencies (FDA, 2003). RF energy produces heat, which can increase body temperature and damage those parts exposed to it (FDA, 2003 and Ahlbom et al., 2004). It is generally agreed that the amount of RF energy encountered by the general public is too low to produce significant tissue heating or an increase in body temperature. However, it is also agreed that further research is needed to determine what effects, if any, low-level non-ionizing RF energy has on the body and whether it is dangerous to people (FDA, 2003).

Various studies indicate that the emissions from a cell phone can be extremely harmful, causing genetic damage, tumors, memory loss, and increased blood pressure and weakening of the immune system. On the other hand, the manufacturers of this technology confirm that they are strict to the recommended low levels of healthy exposure and no health hazards can result from exposure. But the fact that this technology radiation is invisible, intangible, and enters and leaves our bodies without our knowledge makes it even more intimidating.

So, the following pages will try to expose the scientific facts with or against the health risk of this technology without any conflict of interests.

HEALTH HAZARDS OF CELL PHONE BASE STATIONS

The erection of mobile telephone base stations in inhabited areas has raised concerns about possible health effects caused by emitted microwaves. Hutter et al. (2006) studied the Subjective symptoms, sleeping problems, and cognitive performance in subjects living near mobile phone base stations. Results showed that the Total high-frequency electromagnetic fields (HF-EMF) and exposure related to mobile telecommunication were far below recommended levels (max. 4.1 mW/m2). Distance from antennae was 24-600 m in the rural area and 20-250 m in the urban area. Average power density was slightly higher in the rural area (0.05 mW/m2) than in the urban area (0.02 mW/m2). Despite the influence of confounding variables, including fear of adverse effects from exposure to HF-EMF from the base station, there was a significant relation of some symptoms to measured power density; this was highest for headaches. Perceptual speed increased, while accuracy decreased insignificantly with increasing exposure levels. There was no significant effect on sleep quality. It was concluded that despite very low exposure to HF-EMF, effects on wellbeing and performance cannot be ruled out.

Several surveys have found increases of symptoms depending upon proximity to electromagnetic sources such as mobile phone base stations. A 2002 survey study by Santini et al. in France found a variety of self-reported symptoms for people who reported that they were living within 300 metres (984 ft) of GSM cell towers in rural areas, or within 100 m (328 ft) of base stations in urban areas. Fatigue, headache, sleep disruption and loss of memory were among the symptoms reported (Santini et al., 2003). Similar results have been obtained with GSM cell towers in Spain (Navarro et al., 2003), Egypt (Abdel-Rassoul et al., 2007), Poland (Bortkiewicz et al., 2004) and Austria (Hutter et al., 2006). It is, however, important to note that these surveys do not show statistically significant clustering or causality and those complaining of adverse symptoms may be displaying the nocebo effect, unless this is controlled in the study. There are significant challenges in conducting studies of populations near base stations, especially in assessment of individual exposure (Neubauer et al., 2007). However, a study conducted at the University of Essex and another in Switzerland concluded that mobile phone masts were unlikely to be causing these short term effects in a group of volunteers who complained of such symptoms (Regel et al., 2006 and Eltiti et al., 2007).

Schreier et al. (2006) assessed the prevalence of self-reported symptoms attributed to electromagnetic fields (EMF) and other environmental exposures in the general population of Switzerland. RESULTS revealed a prevalence of 5% for electromagnetic hypersensitivity (EHS) in the study sample. The most common health complaints among EHS individuals were sleep disorders (43%) and headaches (34%), which were mostly attributed to power lines and mobile phone handsets. In addition, 53% were worried about adverse health effects from EMF, without attributing their own health symptoms to them. These large proportion of the population who is concerned or attributed own symptoms to EMF may cause societal conflicts given the ubiquity of EMF in our everyday life.

Some users of mobile handsets have reported feeling several unspecific symptoms during and after its use; ranging from burning and tingling sensations in the skin of the head and extremities, fatigue, sleep disturbances, dizziness, loss of mental attention, reaction times and memory retentiveness, headaches, malaise, tachycardia (heart palpitations), to disturbances of the digestive system. Reports have noted that all of these symptoms can also be attributed to stress and that current research cannot separate the symptoms from nocebo effects.

However, the effects of electromagnetic fields may be related not only to intensity or duration of exposure but also to other parameters, such as frequency or modulation (Belyav, 2005). To classify a clinical symptom as psychological, first we must exclude biochemical changes that could be triggered by the electromagnetic fields and cause neurobehavioral responses. This is supported by studies that show changes in neurotransmitters [e.g., acetylcholine (Ahmed et al., 2004), γ-aminobutyric acid (Kolomytkin et al., 1994), glutamate (Wieraszko et al., 2004), histamine and somatostatin (Johansson, 2006) as well as their correlation with the clinical symptoms.

To identify the possible neurobehavioral deficits among inhabitants living nearby mobile phone base stations, a cross-sectional study was conducted on (85) inhabitants living nearby the first mobile phone station antenna in Menoufiya governorate, Egypt, a control group (80) participants were matched with the exposed for age, sex, occupation and educational level. All participants completed a structured questionnaire and neurobehavioral test battery (NBTB) [involving tests for visuomotor speed, problem solving, attention and memory]. Results revealed prevalence of neuropsychiatric complaints as headache (23.5%), memory changes (28.2%), dizziness (18.8%), tremors (9.4%), depressive symptoms (21.7%), and sleep disturbance (23.5%) were significantly higher among exposed inhabitants than controls: (10%), (5%), (5%), (0%), (8.8%) and (10%) respectively. The NBTB indicated that the exposed inhabitants exhibited a significantly lower performance than controls in one of the tests of attention and short-term auditory memory. Also, the inhabitants opposite the station exhibited a lower performance in the problem solving test (block design) than those under the station. All inhabitants exhibited a better performance in the tests of visuomotor speed and test of attention than controls. However, the last available measures of RFR emitted from the first mobile phone base station antennae in Menoufiya governorate were less than the allowable standard level (Abdel-Rassoul et al., 2007).

On the other hand, Eltiti et al. (2007) concluded that short-term rf-emf [radio frequency electromagnetic fields] exposure from mobile phone technology is not related to the levels of well-being or physical symptoms in IEI-EMF [idiopathic environmental intolerance with attribution to electromagnetic fields] individuals. But Zinelis 2008 criticized this conclusion because in this study by Eltiti et al. (2007), the intensity of the radiation emitted by the mobile phone base station was 1 μW/cm2 (5 mW/m2 for 900 MHz and 5 mW/m2 for 1,800 MHz). The authors assumed that the participants would not react to higher intensities such as 10 or 20 μW/cm2, or even to intensities up to 900 μW/cm2, which are used in mobile phone technology.

HEALTH HAZARDS OF CELL PHONE BASE STATIONS AND HUMAN HEAD

Part of the radio waves emitted by a mobile telephone handset is absorbed by the human head. The rate at which radiation is absorbed by the human body is measured by the Specific Absorption Rate (SAR), and its maximum levels for modern handsets have been set by governmental regulating agencies in many countries. In the USA, the SAR limit has been set as 1.6 W/kg. In Europe, the limit is 2 W/kg, averaged over a volume of 10 grams of tissue. SAR data for specific mobile phones, along with other useful information, can be found directly on manufacturers' websites, as well as on third party web sites (Mobile Phones UK, 2008).

Thermal Effects

One well-understood effect of microwave radiation is dielectric heating, in which any dielectric material (such as living tissue) is heated by rotations of polar molecules induced by the electromagnetic field. In the case of a person using a cell phone, most of the heating effect will occur at the surface of the head, causing its temperature to increase by a fraction of a degree. In this case, the level of temperature increase is an order of magnitude less than that obtained during the exposure of the head to direct sunlight. The brain's blood circulation is capable of disposing of excess heat by increasing local blood flow. However, the cornea of the eye does not have this temperature regulation mechanism and exposure of 2-3 hours' duration has been reported to produce cataracts in rabbits' eyes at SAR values from 100-140W/kg, which produced lenticular temperatures of 41°C. Premature cataracts have not been linked with cell phone use, possibly because of the lower power output of mobile phones (ICNIRP, 1998).

Cell Phone and Cancer

There is a general concern on the possible hazardous health effects of exposure to radiofrequency electromagnetic radiations (RFR) emitted from mobile phone base station antennas on the human nervous system.

Though there is no evidence of cell phones causing cancer or any such illness, but the suspicion, or fear of the same is not baseless either. The electromagnetic radiation from cell phones does have a potential link to cancer. Studies have been conducted suggesting that rats that have been exposed to microwaves similar to the sort generated by mobile phones but more powerful, showed breaks in their DNA which could indicate an adverse effect. Also, mice exposed to radiation for 18 months developed brain tumours. Though of course, these studies are not concrete proof.

Originally, the focus of epidemiologic research was the possible association between mobile phone use and risk of brain tumours, because of elevated exposure to the radiofrequency electromagnetic fields by the temporal region of the brain close to the antenna of the mobile phone. So far, the majority of published studies have not indicated an increased risk of brain tumors (NRPB, 2003, Lonn et al., 2003 & Christensen et al., 2005).

Because widespread cellular phone use is little more than a decade old, there has been limited opportunity to examine its long-term health effects. So, there are few published large scale epidemiologic studies. Hardell et al., 1999 made a large case-control study comparing 233 brain cancer patients diagnosed between 1994 and 1996 in the Stockholm and Uppsala regions of Sweden, and 466 controls. Muscat et al., 2000 compared 469 brain cancer patients diagnosed between 1994 and 1998 in New York, Providence, and Boston, with 422 controls. Inskip at al. (2001) made a larger study that compared 782 brain cancer patients diagnosed in Phoenix, Boston, and Pittsburgh between 1994 and 1998 with 799 controls. All three case-control studies had similar results.

- The brain cancer patients did not report more cellular phone use overall than the controls.
- When individual types of brain cancer were considered, none was consistently associated with cell phone use.
- When specific locations of tumours within the brain were considered separately, no associations with cell phone use were observed.
- None of the studies showed a "dose response relationship" between cell phone use and brain tumor risk.
- None of the studies showed a clear link between the side of the head on which the brain cancer occurred and the side on which the cellular phone was used. Hardell, et al. found an insignificant but intriguing increase in ipsilateral temporal lobe tumours (the site of highest RF exposure).

A large cohort of studies has compared cell phone use among brain cancer patients and individuals free of brain cancer. This study linked data on all of the 420,095 cellular telephone users in Denmark between 1982 and 1995 to the Danish Cancer Registry. Results stated that cellular phone use was not associated with an increased risk of developing brain tumours overall; nor was there an association with any brain tumour subtypes or with tumours in any anatomical location within the brain. As in the case-control studies, there was no correlation between brain tumour risk and RF dose, as assessed by duration of cellular phone use, date since first subscription, age at first subscription, or type of cellular phone used (Johansen et al., 2001).

In summary, there is now considerable epidemiologic evidence that shows no consistent association between cellular phone use and brain cancer

Some animal studies have reported a lack of RF-induced DNA damage (Repacholi, 1997 and Malyapa et al., 1998), while others proved presence of this effect (Lai and Singh, 1995). The incidence of lymphoma was increased following whole-body RF irradiation in a transgenic animal model (Repacholi, 1997).

In 2006 a large Danish study about the connection between mobile phone use and cancer incidence was published. It followed over 420,000 Danish citizens for 20 years and showed no increased risk of cancer (Schüz, et al., 2006_a).

The ongoing INTERPHONE collaboration is a multicenter, comprehensive study on mobile phones and cancer. It is coordinated by the International Agency on Research on Cancer (IARC), a specialized cancer agency of the WHO, and researchers in 13 countries are taking part using a common protocol. The INTERPHONE protocol is a population-based, case–control study correlating head and neck tumors with mobile-phone use by persons 30–59 years of age who reside in the study regions. Exposure assessment is based on individual records of cell phone use. Because of pooling of data from all participating centers, the study is statistically powerful (Valberg et al., 2007). The following studies of long time exposure have been published:

- A Danish study (2004) that took place over 10 years and found no evidence to support a link (Schüz et al., 2006_b).
- A Swedish study (2005) that draws the conclusion that "the data do not support the hypothesis that mobile phone use is related to an increased risk of glioma or meningioma" (Lönn et al., 2005).
- The risk of acoustic neuroma in relation to mobile phone use has been assessed via six population-based, shared-protocol, case–control studies in four Nordic countries and the United Kingdom. The authors concluded that there was no association of risk with duration of use, lifetime cumulative hours of use, or number of calls, for phone use overall or for analogue or digital phones separately (Schoemaker et al., 2005).
- A German study (2006) that states "In conclusion, no overall increased risk of glioma or meningioma was observed among these cellular phone users; however, for long-term cellular phone users, results need to be confirmed before firm conclusions can be drawn" (Schüz et al., 2006_a and Berg et al., 2006).
- A joint study conducted in northern Europe that draws the conclusion that "Although our results overall do not indicate an increased risk of glioma in relation to mobile phone use, the possible risk in the most heavily exposed part of the brain with long-term use needs to be explored further before firm conclusions can be drawn" (Lahkola et al., 2007).
- A Swedish scientific team at the Karolinska Institute conducted an epidemiological study (2004) that suggested that regular use of a mobile phone over a decade or more was associated with an increased risk of acoustic neuroma, a type of benign brain tumor. The increase was not noted in those who had used phones for fewer than 10 years (Lönn et al., 2004).
- The INTERPHONE study group from Japan published the results of a study of brain tumour risk and mobile phone use. They used a new approach: determining the SAR inside a tumour by calculating the radiofrequency field absorption in the exact tumour location. Cases examined included glioma, meninigioma, and pituitary adenoma. They reported that the overall odds ratio (OR) was not increased and that there was no significant trend towards an increasing OR in relation to SAR-derived exposure Takebayashi et al., 2008).

There have also been other studies of mobile telephone users, particularly on brain tumors (and less often on other cancers and on symptoms). Results of these studies to date give no consistent or convincing evidence of a causal relation between RF exposure and any adverse health effect (Hepworth et al., 2006 & Lai and Singh, 1995). A 4-year British survey released in 2006 showed no link between regular, long-term use of cell phones and the most common type of brain tumor, glioma (Ahlbom et al., 2004).

There have been reports in the media and claims in the courts that radiofrequency (RF) emissions from mobile phones are a cause of cancer, and there have been numerous public objections to the sitting of mobile phone base antennas because of a fear of cancer. Moulder,et al. (2005) reviewed the current state of evidence concerning whether the RF energy used for wireless communication might be carcinogenic. Relevant studies were identified by searching MedLine with a combination of exposure and endpoint terms. Where there were multiple studies, preference was given to recent reports, to positive reports of effects and to attempts to confirm such positive reports. Biophysical considerations indicate that there is little theoretical basis for anticipating that RF energy would have significant biological effects at the power levels used by modern mobile phones and their base station antennas. The epidemiological evidence for a causal association between cancer and RF energy is weak and limited. Animal studies have provided no consistent evidence that exposure to RF energy at non-thermal intensities causes or promotes cancer. Extensive *in vitro* studies have found no consistent evidence of genotoxic potential, but *in vitro* studies assessing the epigenetic potential of RF energy are limited. Overall, a weight-of-evidence evaluation shows that the current evidence for a causal association between cancer and exposure to RF energy is weak and unconvincing. However, the existing epidemiology is limited and the possibility of epigenetic effects has not been thoroughly evaluated, so that additional research in those areas will be required for a more thorough assessment of the possibility of a causal connection between cancer and the RF energy from mobile telecommunications (Moulder et al., 2005).

Audiovestibular Functions

A double blind study tested subjects who complained of various symptoms regarding the audiovestibular functions after prolonged mobile telephone use. There were no significant transient evoked otoacoustic emissions (TEOAE) changes from baseline to post-exposure recording for any of the exposures and no significant differences in the TEOAEs' change from baseline to post exposure between cases and controls. The video-oculography (VOG) did not identify any effect of the exposure on the vestibular end organ in either cases or controls. In conclusion, 30 min exposure to mobile phone RF did not show any immediate effects on vestibulocochlear function (Bamiou et al., 2008).

Effect of Mobile Phone on Metallic Implants in the Body

Virtanen et al. (2006) concluded that a conductive object in tissues e.g., metallic implants may cause notable local enhancement of the EM field and thus enhanced power absorption. The degree of enhancement depends, for example, on the orientation, the dimensions, the shape, and the location of the implant. However, in most of the cases, the field enhancement has not been strong enough to cause remarkable excess heating (more than 1 °C) of tissues.

Effect of Mobile Phone on Parotid Gland

Another organ of interest is the parotid gland, which is located over the jaw bone in front of the ear. This gland is likely to be exposed to radiofrequency radiation during mobile phone use if the handset is held close to the ear. Hence, if radiofrequency radiation has a carcinogenic effect, the exposure might be associated with parotid gland tumor. The few epidemiologic studies that have investigated the association between the use of mobile phones and the risk of malignant parotid gland tumours did not identify any association (Lonn et al., 2005; Auvinen et al., 2002 and Hardell et al., 2004). However, these studies have limitations due to the small number of exposed cases (Hardell et al., 1999 and Virtanen et al., 2006) and the short duration of phone use among study subjects (Lonn et al., 2005; Auvinen et al., 2002 and Hardell et al., 2004).

The etiology of tumours in the parotid gland is largely unknown. Many etiologic factors have been suggested (e.g., smoking, ultraviolet radiation, and occupation exposures), but the only established risk factors are ionizing radiation and history of prior cancer (Belsky et al., 1975; Spitz et al., 1984; Preston-Martin et al., 1988; Horn-Ross et al., 1997 and Saku et al., 1997). After adjustment of these factors, Lönn et al. (2006) applied their study to explore the risk of parotid gland tumors in mobile phone users. Findings do not indicate any association between mobile phone use and the risk of malignant or benign parotid gland tumours. On the other hand, results of other study suggest an association between cellular phone use and parotid gland tumours (Sadetzki et al., 2008).

Chemical and Hormonal Effects of Mobile Phone

Ahmed et al. (2004) and Lai et al. (1992, 1994) concluded that the response depends on the duration of the radiation exposure. After 1 hr of exposure, alterations of certain biochemicals, which could be producing the symptoms, may or may not occur. For example, an increase in acetylcholinesterase activity is responsible for the levels of acetylcholine and with other neurotransmitters responsible for cognitive functions; with further exposure, this activity increases in two areas of the brain, the hippocampus and the striatum. Also, Johansson (2006) reported that electromagnetic fields may stimulate mast cells, which produce histamine, and then symptoms are produced in the skin and other organs.

Exposure to 900 MHz and 1800 MHz electromagnetic field (EMF) showed no effect on serum nocturnal melatonin levels of adult male Sprague-Dawley rats when exposures were performed 30 min/day, for five days/week for four weeks (Koyu, 2005).

Pregnancy and Mobile Phone

The effects of continuous exposure of chick embryos and young chickens to the electromagnetic fields (EMFs) emitted by video display units (VDUs) and GSM cell phone radiation were investigated. Permanent exposure to the EMFs radiated by a VDU was associated with significantly increased fetal loss (47-68%) and markedly depressed levels of circulating specific antibodies (lgG), corticosterone and melatonin. It has been shown that under chronic exposure conditions, GSM cell phone radiation was harmful to chick embryos, stressful for healthy mice and, in this species, synergistic with cancer insofar as it depleted stress hormones (Bastide et al., 2001).

Research published in 2004 by a team at the University of Athens had a reduction in reproductive capacity in fruit flies exposed to 6 minutes of 900 MHz pulsed radiation for five days (Panagopoulos et al., 2004). Subsequent research, again conducted on fruit flies, was published in 2007, with the same exposure pattern but conducted at both 900 MHz and 1800 MHz, and had similar changes in reproductive capacity with no significant difference between the two frequencies (Panagopoulos et al., 2007).

Mobile Phones and Children

Because of their smaller heads, thinner skulls and higher tissue conductivity, children may absorb more energy from a given phone than adults. A groundbreaking scientific study published this week in the peer-reviewed Australasian Journal of Clinical Environmental Medicine warns that wireless communication technology may be responsible for accelerating the rise in autism among the world's children (Aust. J. Clin. Env. Med.,2007).

Exposure to cell phones prenatally-and, to a lesser degree, postnatally-was associated with behavioral difficulties such as emotional and hyperactivity problems around the age of school entry. These associations may be noncausal and may be due to unmeasured confounding. If real, they would be of public health concern given the widespread use of this technology (Divan et al., 2008).

Blood Brain Barrier Effects

Swedish researchers from Lund University (Salford, Brun, Perrson, Eberhardt, and Malmgren) have studied the effects of microwave radiation on the rat brain. They found a leakage of albumin into the brain via a permeated blood-brain barrier (Salford et al., 2003). Other groups have not confirmed these findings in cell (Franke et al., 2007) or animal studies (Kuribayashi et al., 2007).

Genotoxic Effects

In 1995, Lai and Singh reported damaged DNA after two hours of microwave radiation at levels deemed safe according to government standards in the journal *Bioelectromagnetics* (Harrill, 2005). Reviews of in vitro genotoxicity studies have generally concluded that RF is not genotoxic and that studies reporting positive effects had experimental deficiences (Vijayalaxmi et al., 2008).

Sleep and EEG Effects

Some studies have claimed to show that mobile phone signals affect sleep patterns and possibly delay sleep onset during exposure (Hung et al., 2007). In another clinical study, carried out by Sweden's Karolinska Institute and Wayne State University in the US, the authors concluded an association between RF exposure and adverse effects on sleep quality within certain sleep stages, though participants were unable to determine if they had been exposed to actual radiation or sham exposure (Arnetz et al., 2007). The UK National Health Service (NHS) (2008) criticized this research because of the small sample size used, and because of the 53% of participants who reported sensitivity to mobile use, a proportion unlikely to be representative of the general population. The NHS also criticized the press for inaccurate reporting of the study.

Other Effects

Cardiovascular System

It is possible that electromagnetic field generated by mobile phones may have an influence on the autonomic nervous system and modulates the function of circulatory system. It was shown that the call with a mobile phone may change the autonomic balance in healthy subjects leading to changes in heart rate during the call with a mobile phone (Andrzejak et al., 2008).

Headaches, Heating Effects, Fatigue

Studies indicate that the longer the people used mobile phones, the more likely they were to report symptoms such as hot ears, burning skin, headaches and fatigue. But these studies usually did not include a control group (that is people who do not use mobile phones, to make a comparison); therefore the symptoms reported could have been caused by any number of other factors in the mobile phones users' environment, such as working with computers, stress, driving or reading.

Cinel et al. (2008) did not find consistent evidence suggesting that exposure to mobile phone REFs affect subjective symptoms e.g., headache, fatigue. Even though they acknowledge that more research is needed.

Memory

Numerous studies have investigated the potential effects of electromagnetic fields (EMFs) emitted by GSM mobile phones (900 MHz to 1800 MHz) on cognitive functioning, but results have been equivocal. Results of the meta-analysis suggest that EMFs may have a small impact on human attention and working memory (Barth et al., 2008).

Posture (Holding Phone Between Raised Shoulder and Ear)

Some researchers claim that holding a mobile phone between the raised shoulder and the ear could have a damaging effect on muscles, bones, tendons and discs. These problems would apply equally to a cordless phone or a landline phone as to a mobile phone and are the effect of bad posture.

OCCUPATIONAL HEALTH HAZARDS

Telecommunication workers who spend time at a short distance from the active equipment, for the purposes of testing, maintenance, installation, etc. may be at risk of much greater exposure than the general population. Many times base stations are not turned off during maintenance, because that would affect the network, so people work near "live" antennas.

A variety of studies over the past 50 years have been done on workers exposed to high RF radiation levels; studies including radar laboratory workers, military radar workers, electrical workers, and amateur radio operators. Most of these studies found no increase in cancer rates over the general population or a control group. Many positive results could have been attributed to other work environment conditions, and many negative results of reduced cancer rates also occurred (Moulder et al., 1999).

CONCLUSION

Globalisation is the new mantra. In this age, it is very difficult not to have technology. But with technology, come certain hazards. The only way to beat these is again, better technology. Electromagnetic radiation is everywhere. More and more wireless communication services (cellular phones, paging, wireless Internet) are expected so is the artificial electromagnetic radiation. It seems that there is no way to reverse this trend. Scientists and engineers are developing better and safer wireless systems and devices. Smaller cell size, better base station antennas and other more advanced technologies will allow future cell phones to radiate much lower power. So one can only hope that cell phone hazards will be reduced.

Inhabitants living nearby mobile phone base stations are at risk for developing neuropsychiatric problems and some changes in the performance of neurobehavioral functions either by facilitation or inhibition. So, revision of standard guidelines for public exposure to RER from mobile phone base station antennas and using of NBTB for regular assessment and early detection of biological effects among inhabitants around the stations are recommended.

SUGGESTIONS FOR PROPHYLAXIS

- Cell phones should be used for emergencies, and not for long conversations.
- A small chip-like cell phone microwave radiation protection device is available, which has the ability to absorb electromagnetic energy waves from your mobile phone. It helps in reducing the potential harmful effects of these emissions to the human body.
- Using a mobile headset is a good idea, you don't have to hold phones next to your ears all the time
- Use a hands free mobile car kit while driving, without taking your hands off the steering wheel.
- Mobile phone users should limit their exposure to harmful radio frequencies by cutting the length of calls.
- Mobile phones should not be used in Intensive Care Units of hospitals as they can pose a danger to patients by interfering with the working of pacemakers and defibrillators.
- People with hearing aids should not use mobile phones.
- Base stations, which have low powered antennae on their terrace to communicate with cell phones, should not be located near children's schools and playgrounds.
- Always make sure the phones are kept in not your front pockets but in the pockets of your trousers or pants, away from the heart. specially when using pace-makers
- Avoid using vibration mode of the mobile phone.
- Avoid overuse of mobile phones.
- Use landlines more than mobile phones if possible.

- Buy mobile phones from reputated companies.
- Faulty mobile phones should be repaired soon.
- Avoid using mobile phones during pregnancy

REFERENCES

Abdel-Rassoul, G., Abou El-Fateh, O., Abou Salem, M., Michael, A., Farahat, F., El-Batanouny, M., & Salem, E. (2007). Neurobehavioral effects among inhabitants around mobile phone base stations. *Neurotoxicology*, *28*(2), 434–340. doi:10.1016/j.neuro.2006.07.012

Ahlbom, A., Green, A., & Kheifets, L. (2004). Epidemiology of health effects on radiofrequency exposure. *Environmental Health Perspectives*, *112*(14), 1741–1754.

Ahmed, N., Asaad, A., Aboul-Ezz, H., & Radwan, N. (2004). Effect of exposure to electromagnetic radiation from mobile phone on acetylcholinesterase activity in the hippocampus and striatum of young and adult male rats. In P. Kostarakis, (ed.), *Biological Effects of TMFs: Third International Workshop*, Kos, Greece, 4–8 October 2004, Ioannina, Greece: University of Ioannina, (pp. 924–930).

Andrzejak, R., Poreba, R., Poreba, M., Derkacz, A., Skalik, R., & Gac, P. (2008). The influence of the call with a mobile phone on heart rate variability parameters in healthy volunteers. *Industrial Health*, *46*(4), 409–417. doi:10.2486/indhealth.46.409

Arnetz, B. B., Akerstedt, T., Hillert, L., Lowden, A., Kuster, N., & Wiholm, C. (2007). The Effects of 884 MHz GSM Wireless Communication Signals on Self-reported Symptom and Sleep (EEG)- An Experimental Provocation Study. *PIERS Online*, *3*(7), 1148–1150. doi:10.2529/PIERS060907172142

Aust. J. Clin. Env. Med. (2007). Australasian Journal of Clinical Environmental Medicine warns that wireless communication technology may be responsible for accelerating the rise in autism among the world's children. *J.Aust.Coll.Nutr.& Env.Med, 26* (2), 3 – 7.

Auvinen, A., Hietanen, M., & Luukkonen, R. (2002). Brain tumors and salivary gland cancers among cellular telephone users. *Epidemiology (Cambridge, Mass.)*, *13*, 356–359. doi:10.1097/00001648-200205000-00018

Bamiou, D., Ceranic, B., Cox, R., Watt, H., Chadwick, P., & Luxon, L. M. (2008). Mobile telephone use effects on peripheral audiovestibular function: A case-control study. *Bioelectromagnetics*, *29*(2), 108–117. doi:10.1002/bem.20369

Barth, A., Winker, R., Ponocny-Seliger, E., Mayrhofer, W., Ponocny, I., Sauter, C., & Vana, S. (2008). A meta-analysis for neurobehavioural effects due to electromagnetic field exposure emitted by GSM mobile phones. *Occupational and Environmental Medicine*, *65*, 342–346. doi:10.1136/oem.2006.031450

Bastide, M., Youbicier-Simoa, B. J., Lebecq, J. C., & Giaimis, J. (2001). Toxicologic Study of Electromagnetic Radiation Emitted by Television and Video Display Screens and Cellular Telephones on Chickens and Mice. *Indoor and Built Environment*, *10*(5), 291–298.

Belsky, J. L., Takeichi, N., & Yamamoto, T. (1975). Salivary gland neoplasms following atomic radiation: additional cases and reanalysis of combined data in a fixed population, 1957–1970. *Cancer*, *35*, 555–559. doi:10.1002/1097-0142(197502)35:2<555::AID-CNCR2820350240>3.0.CO;2-G

Belyav, I. (2005). Non-thermal biological effects of microwaves. *Microwave Review,* 13–29.

Berg, G., Spallek, J., Schuz, J., Schlehofer, B., Bohler, E., & Schlaefer, K. (2006). Occupational exposure to radio frequency/microwave radiation and the risk of brain tumors: interphone study group, Germany. *American Journal of Epidemiology*, *164*(6), 538–548. doi:10.1093/aje/kwj247

Bortkiewicz, A., Zmyslony, M., Szyjkowska, A., & Gadzicka, E. (2004). Subjective symptoms reported by people living in the vicinity of cellular phone base stations [review]. *Medycyna Pracy*, *55*(4), 345–352.

Christensen, H. C., Schu¨z, J., & Kosteljanetz, M. (2005). Cellular telephones and risk for brain tumors: a population-based, incident case-control study. *Neurology*, *64*, 1189–1195.

Cinel, C., Russo, R., Boldini, A., & Fox, E. (2008). Exposure to Mobile Phone Electromagnetic Fields and Subjective Symptoms: A Double-Blind Study. *Psychosomatic Medicine*, *70*, 345–348. doi:10.1097/PSY.0b013e31816521f1

Divan, H.A., Kheifets, L., Obel, C., Olsen, J. (2008). Prenatal and postnatal exposure to cell phone use and behavioral problems in children. *Epidemiology. Jul, 19*(4), 523-9.

Eltiti, S., Wallace, D., Ridgewell, A., Zougkou, K., Russo, R., & Sepulveda, F. (2007). Does short-term exposure to mobile phone base station signals increase symptoms in individuals who report sensitivity to electromagnetic fields? A double-blind randomized provocation study. *Environmental Health Perspectives*, *115*(11), 1603–1608. doi:10.1289/ehp.10286

FDA. (2003). *Cell Phone Facts: Consumer Information on Wireless Phones*. Retrieved May 10, 2005, from http://www.fda.gov/cellphones/qa.html

Franke, (2007). Electromagnetic fields (GSM 1800) do not alter blood-brain barrier permeability to sucrose in models in vitro with high barrier tightness. *Bioelectromagnetics*, *26*(7), 529–535. doi:10.1002/bem.20123

Hardell, L., Hallquist, A., & Hansson Mild, K. (2004). No association between the use of cellular or cordless telephones and salivary gland tumours. *Occupational and Environmental Medicine*, *61*, 675–679. doi:10.1136/oem.2003.011262

Hardell, L., Nasman, A., & Pahlson, A. (1999). Use of cellular telephones and the risk for brain tumors: A case-control study. *International Journal of Oncology*, *15*, 113–116.

Harrill, R. (2005). Wake-up Call. *The University of Washington Alumni Magazine,* (March). Retrieved on 2008-05-31.

Hepworth, S. J., Schoemaker, M. J., Muir, K. R., Swerdlow, A. J., van Tongeren, M. J., & McKinney, P. A. (2006). Mobile phone use and risk of glioma in adults: case-control study. *BMJ (Clinical Research Ed.)*, *332*, 883–887. doi:10.1136/bmj.38720.687975.55

Horn-Ross, P. L., Ljung, B., & Morrow, M. (1997). Environmental factors and the risk of salivary gland cancer. *Epidemiology (Cambridge, Mass.)*, *8*, 414–419. doi:10.1097/00001648-199707000-00011

HPA. (2008). Exposures from Base Stations. *Mobile Telephony and Health.* Health Protection Agency. Retrieved on 2008-01-31.

Hung, C. S., Anderson, C., Horne, J. A., & McEvoy, P. (2007). Mobile phone 'talk-mode' signal delays EEG-determined sleep onset. *Neuroscience Letters*, *421*(1), 82–86. doi:10.1016/j.neulet.2007.05.027

Hutter, H. P., Moshammer, H., Wallner, P., & Kundi, M. (2006). Subjective symptoms, sleeping problems, and cognitive performance in subjects living near mobile phone base stations. *Occupational and Environmental Medicine*, *63*(5), 307–313. doi:10.1136/oem.2005.020784

ICNIRP International Commission on Non-Ionizing Radiation Protection. (1998). Guidelines For Limiting Exposure To Time-Varying Electric, Magnetic, And Electromagnetic Fields (up to 300 GHz), International Commission on Non-Ionizing Radiation Protection. *Health Physics*, *74*(4), 494–522.

Inskip, P. D., Tarone, R. E., & Hatch, E. E. (2001). Cellular telephone use and brain tumors. *The New England Journal of Medicine*, *344*, 79–86. doi:10.1056/NEJM200101113440201

Johansen, C., Boice, J. D. Jr, McLaughlin, J. K., & Olsen, J. H. (2001). Cellular telephones and cancer – a nationwide cohort study in Denmark. *Journal of the National Cancer Institute*, *93*, 203–207. doi:10.1093/jnci/93.3.203

Johansson, O. (2006). Electrohypersensitivity: state-of-the-art of a functional impairment. *Electromagnetic Biology and Medicine*, *25*, 245–255. doi:10.1080/15368370601044150

Kolomytkin, O., Kuznetsov, V., Yurinska, M., Zharikova, A., & Zharikov, S. (1994). Response of brain receptor systems to microwave energy exposure . In Frey, A. (Ed.), *On the Nature of Electromagnetic Field Interactions with Biological Systems* (pp. 195–206). Austin, TX: R.G. Landes Co.

Koyu, A., Fehmi Ozguner, F., Cesur, G., Gokalp, O., Mollaoglu, H., Caliskan, S., & Delibas, N. (2005). No effects of 900 MHz and 1800 MHz electromagnetic field emitted from cellular phone on nocturnal serum melatonin levels in rats. *Toxicology and Industrial Health*, *21*(10), 27–31. doi:10.1191/0748233705th212oa

Kuribayashi, (2007). Lack of effects of 1439 MHz electromagnetic near field exposure on the blood-brain barrier in immature and young rats. *Bioelectromagnetics*, *26*(7), 578–588. doi:10.1002/bem.20138

Lahkola, A., Auvinen, A., Raitanen, J., Schoemaker, M. J., Christensen, H. C., & Feychting, M. (2007). Mobile phone use and risk of glioma in 5 North European countries. *International Journal of Cancer*, *120*(8), 1769–1775. doi:10.1002/ijc.22503

Lai, H., Carino, M., Horita, A., & Guy, A. (1992). Opioid receptor subtypes that mediate a microwave induced decrease in central cholinergic activity in the rat. *Bioelectromagnetics*, *13*, 237–246. doi:10.1002/bem.2250130308

Lai, H., Horita, A., & Guy, A. (1994). Microwave irradiation affects radial-arm maze performance in the rat. *Bioelectromagnetics*, *15*, 95–104. doi:10.1002/bem.2250150202

Lai, H., & Singh, N. P. (1995). Acute low-intensity microwave exposure increases DNA single-strand breaks in rat brain cells. *Bioelectromagnetics*, *16*, 204–210. doi:10.1002/bem.2250160309

Lönn, S., Ahlbom, A., Christensen, H. C., Johansen, C., Schüz, J., & Edström, S. (2006). Mobile Phone Use and Risk of Parotid Gland Tumor. *American Journal of Epidemiology*, *164*, 637–643. doi:10.1093/aje/kwj242

Lönn, S., Ahlbom, A., & Hall, P. (2005). Long-term mobile phone use and brain tumor risk. *American Journal of Epidemiology*, *161*(6), 526–535. doi:10.1093/aje/kwi091

Lönn, S., Ahlbom, A., Hall, P., & Feychting, M. (2004). Mobile phone use and the risk of acoustic neuroma [in English]. *Epidemiology (Cambridge, Mass.)*, *15*(6), 653–659. doi:10.1097/01.ede.0000142519.00772.bf

Malyapa, R. S., Ahern, E. W., & Straube, W. L. (1998). DNA damage in rat brain cells after in vivo exposure to 2450 MHz electromagnetic radiation and various methods of euthanasia. *Radiation Research*, *149*, 637–645. doi:10.2307/3579911

Mobile Phones UK. (2008). *Mobile Phones UK web site*. Landmark Internet Ltd. Retrieved on 2008-01-19., of phones from 2005 and earlier at "The Complete SAR List For All Phones (Europe)". On-Line-Net - Web Design & Internet Services (as SARValues.com). Retrieved on 2008-01-19. (a listing of US phones from 2005 and earlier, using the US 1 g standard, is also available at the SARValues site)

Moulder, J.E., Erdreich, L.S., Malyapa, R.S., Merritt, J., & Pickard, W.F., Vijayalaxmi. (1999). Cell phones and cancer: what is the evidence for a connection? *Radiation Research*, *151*(5), 513–531. doi:10.2307/3580028

Moulder, J. E., Foster, K. R., Erdreich, L. S., & McNamee, J. P. (2005). Mobile phones, mobile phone base stations and cancer. *International Journal of Radiation Biology*, *81*(3), 189–203. doi:10.1080/09553000500091097

National Health Service. (2008). Mobile effect on sleep. *Behind the Headlines*. National Health Service (2008-01-21). Retrieved on 2008-01-22.

Navarro, E. A., Segura, J., Portolés, M., & Gómez-Perretta de Mateo, C. (2003). The Microwave Syndrome: A Preliminary Study in Spain. *Electromagnetic Biology and Medicine*, *22*(2), 161–169. doi:10.1081/JBC-120024625

Neubauer, (2007). Feasibility of future epidemiological studies on possible health effects of mobile phone base stations. *Bioelectromagnetics*, *28*(3), 224–230. doi:10.1002/bem.20298

NRPB (National Radiological Protection Board). (2003). *Health effects from radiofrequency electromagnetic fields: report of an independent advisory group on non-ionising radiation*. Chilton, UK: National Radiological Protection Board (Health Protection Agency), 2003. Retrieved from http:// www. hpa.org.uk /radiation /publications /documents_ of_ nrpb /pdfs /doc_14_2.pdf

Panagopoulos, D. J., Chavdoula, E. D., Karabarbounis, A., & Margaritis, L. H. (2007). Comparison of bioactivity between GSM 900 MHz and DCS 1800 MHz Mobile Telephony Radiation. *Electromagnetic Biology and Medicine*, *26*(1), 33–44. doi:10.1080/15368370701205644

Panagopoulos, D. J., Karabarbounis, A., & Margaritis, L. H. (2004). Effect of GSM 900 MHz mobile phone radiation on the reproductive capacity of Drosophila melanogaster. *Electromagnetic Biology and Medicine*, *23*(1), 29–43. doi:10.1081/JBC-120039350

Preston-Martin, S., Thomas, D. C., & White, S. C. (1988). Prior exposure to medical and dental x-rays related to tumors of the parotid gland. *Journal of the National Cancer Institute*, *80*, 943–949. doi:10.1093/jnci/80.12.943

Radiofrequency electromagnetic field exposure and non-specific symptoms of ill health: A systematic review. (n.d.). Röösli, Environmental Research. Available online 21 March 2008 at http://dx.doi.org/10.1016/j.envres.2008.02.003

Regel, (2006). UMTS Base Station-Like Exposure, Well Being and Cognitive Performance. *Environmental Health Perspectives*, *114*(8). doi:10.1289/ehp.8934

Repacholi, M. H. (1997). Radiofrequency field exposure and cancer: What do the laboratory studies suggest? *Environmental Health Perspectives*, *105*(suppl 6), 7565–1568. doi:10.2307/3433671

Sadetzki, S., Chetrit, A., Jarus-Hakak, A., Cardis, E., Deutch, Y., & Duvdevani, S. (2008). Cellular Phone Use and Risk of Benign and Malignant Parotid Gland Tumors—A Nationwide Case-Control Study. *American Journal of Epidemiology*, *167*(4), 457–467. doi:10.1093/aje/kwm325

Saku, T., Hayashi, Y., & Takahara, O. (1997). Salivary gland tumors among atomic bomb survivors, 1950–1987. *Cancer*, *79*, 1465–1475. doi:10.1002/(SICI)1097-0142(19970415)79:8<1465::AID-CNCR4>3.0.CO;2-A

Salford, L. G., Brun, A. E., Eberhardt, J. L., Malmgren, L., & Persson, B. R. R. (2003). Nerve Cell Damage in Mammalian Brain after Exposure to Microwaves from GSM Mobile Phones. *Environmental Health Perspectives*, *111*(7), 881–883. doi:10.1289/ehp.6039

Santini, R., Santini, P., Danze, J. M., LeRuz, P., & Seigne, M. (2003). Survey Study of People Living in the Vicinity of Cellular Phone Base Stations. *Electromagnetic Biology and Medicine*, *22*(1), 41–49. doi:10.1081/JBC-120020353

Schoemaker, M. J., Swerdlow, A. J., Ahlbom, A., Auvinen, A., Blaasaas, K. G., & Cardis, E. (2005). Mobile phone use and risk of acoustic neuroma: results of the Interphone case-control study in five North European countries. *British Journal of Cancer*, *93*(7), 842–848. doi:10.1038/sj.bjc.6602764

Schreier, N., Huss, A., & Röösli, M. (2006). The prevalence of symptoms attributed to electromagnetic field exposure: a cross-sectional representative survey in Switzerland. *Sozial- und Praventivmedizin*, *51*(4), 202–209. doi:10.1007/s00038-006-5061-2

Schüz, J., Böhler, E., Berg, G., Schlehofer, B., Hettinger, I., & Schlaefer, K. (2006). Cellular phones, cordless phones, and the risks of glioma and meningioma (Interphone Study Group, Germany). *American Journal of Epidemiology*, *163*(6), 512–520. doi:10.1093/aje/kwj068

Schüz, J., Jacobsen, R., Olsen, J. H., Boice, J. D., McLaughlin, J. K., & Johansen, C. (2006). Cellular Telephone Use and Cancer Risk: Update of a Nationwide Danish Cohort. *Journal of the National Cancer Institute*, *98*(23), 1707–1713. doi:10.1093/jnci/djj464

Spitz, M. R., Tilley, B. C., & Batsakis, J. G. (1984). Risk factors for major salivary gland carcinoma. A case-comparison study. *Cancer*, *54*, 1854–1859. doi:10.1002/1097-0142(19841101)54:9<1854::AID-CNCR2820540915>3.0.CO;2-1

Takebayashi, T., Varsier, N., Kikuchi, Y., Wake, K., Taki, M., & Watanabe, S. (2008). Mobile phone use, exposure to radiofrequency electromagnetic field, and brain tumour: a case-control study. *British Journal of Cancer*, *98*(3), 652–659. doi:10.1038/sj.bjc.6604214

Valberg, P. A., van Deventer, T. E., & Repacholi, M. H. (2007). Workgroup Report: Base Stations and Wireless Networks—Radiofrequency (RF) Exposures and Health Consequences. *Environmental Health Perspectives*, *115*(3), 416–424. doi:10.1289/ehp.9633

Vijayalaxmi, (2008). Genetic Damage in Mammalian Somatic Cells Exposed to Radiofrequency Radiation: A Meta-analysis of Data from 63 Publications (1990–2005). *Radiation Research*, *169*(5), 561–574. doi:10.1667/RR0987.1

Virtanen, H., Keshvari, J., & Lappalainen, R. (2006). Interaction of radio frequency electromagnetic fields and passive metallic implants - A brief review. *Bioelectromagnetics*, *27*(6), 431–439. doi:10.1002/bem.20224

Wieraszko, A., Armani, J., Hanna, A., Maqsood, N., Raja, H., & Hogan, M. (2004). Changes in neurotransmitter turnover and second messenger levels in brain tissue exposed to magnetic fields. In P. Kostarakis, (Ed.), *Biological Effects of EMFs: Third International Workshop*, Kos Greece, October 4–8, (pp. 614–632). Ioannina, Greece: University of Ioannina.

Zinelis, S. A. (2008). Short-term exposure to mobile phone base station signals. *Environmental Health Perspectives*, *116*(2), A62. doi:10.1289/ehp.10733

Compilation of References

Abdel-Rassoul, G., Abou El-Fateh, O., Abou Salem, M., Michael, A., Farahat, F., El-Batanouny, M., & Salem, E. (2007). Neurobehavioral effects among inhabitants around mobile phone base stations. *Neurotoxicology*, *28*(2), 434–340. doi:10.1016/j.neuro.2006.07.012

Aberg, L. (2001). Attitudes . In Barjonet, P. E. (Ed.), *Traffic psychology today* (pp. 119–137). USA: Kluwer Academic Publisher.

Ackroyd, S., & Thompson, P. (1999). *Organisational misbehaviour*. Thousand Oaks, CA: Sage.

Aconex (2009, June 17). Aconex selected for world's second largest terminal. Retrieved July 2009 from http://www.aconex.com/Corporate/News/Latest-news/Aconex-selected-for-DIAL.html

Adams, J. S. (1965). Inequity in social exchange . In Berkowitz, L. (Ed.), *Advances in experimental social psychology* (*Vol. 2*, pp. 267–299). New York: Academic Press. doi:10.1016/S0065-2601(08)60108-2

Aderinoye, R. A., Ojokheta, K. O., & Olojede, A. A. (2007). Integrating Mobile Learning into Nomadic Education Programmes in Nigeria: Issues and perspectives. *International Review of Research in Open and Distance Learning*, *8*(2), 1–17.

Agarwal, R., & Bayus, B. L. (2002). The market evolution and sales takeoff of product innovations. *Management Science*, *48*(4), 1024–1041. doi:10.1287/mnsc.48.8.1024.167

Ahlbom, A., Green, A., & Kheifets, L. (2004). Epidemiology of health effects on radiofrequency exposure. *Environmental Health Perspectives*, *112*(14), 1741–1754.

Ahmed, N., Asaad, A., Aboul-Ezz, H., & Radwan, N. (2004). Effect of exposure to electromagnetic radiation from mobile phone on acetylcholinesterase activity in the hippocampus and striatum of young and adult male rats. In P. Kostarakis, (ed.), *Biological Effects of TMFs: Third International Workshop*, Kos, Greece, 4–8 October 2004, Ioannina, Greece: University of Ioannina, (pp. 924–930).

Ahuja, G. (2000). Collaboration networks, structural holes and innovation: A longitudinal study. *Administrative Science Quarterly*, *45*, 425–455. doi:10.2307/2667105

Ahuja, V., Yang, J., & Shankar, R. (2006, November). Strategic use of ICT for construction organisations - Requirements and implementation issues. In *Proceedings of INCITE / ITCSED 2006 Conference* (Vol. 1, pp. 235-250). New Delhi, India.

Ahuja, V., Yang, J., & Shankar, R. (2009). Study of ICT Adoption for Building Project Management in the Indian Construction Industry. *Automation in Construction*, *18*(4), 415–423. doi:10.1016/j.autcon.2008.10.009

Airport-technology.com (n.d.). *Indira Gandhi International Airport Delhi, India*. Retrieved July 2009 from airport-technology.com/projects/delhi/

Ajzen, I. (1985). From intentions to actions: A theory of planned behavior . In Kuhl, J., & Beckmann, J. (Eds.), *Action Control: From Cognition to Behavior* (pp. 11–39). Heidelberg: Springer.

Ajzen, I. (1991). The theory of planned behavior. *Organizational Behavior and Human Decision Processes*, *50*, 179–211. doi:10.1016/0749-5978(91)90020-T

Ajzen, I. (1991). The theory of planned behaviour. *Organizational Behavior and Human Decision Processes*, *50*, 179–211. doi:10.1016/0749-5978(91)90020-T

Ajzen, I., & Fishbein, M. (1980). *Understanding attitudes and predicting social behaviour*. Englewood Cliffs, NJ: Prentice-Hall.

Alexander, J., Coble, R., & Elliott, B. (1997). Hand-held communication for construction supervision. In [Minneapolis, MN.]. *Proceedings of ASCE Construction Congress*, *V*, 972–979.

Aleyidieno, S. (1985). Education and occupational diversification among young learners: The problem of harmonising tradition practices with the lessons of our colonial heritage. In *Issues on Development: Proceedings of a Seminar held in Zaria*. January 12-13. Zaria, Nigeria: Ahmadu Bello University Press.

Ally, M. (2004). Using learning theories to design instruction for mobile learning devices . In Attewell, J., & Savill-Smith, C. (Eds.), *Mobile learning anytime everywhere: A book of papers from MLEARN* (pp. 5–9). London: Learning and Skills Development Agency.

Ally, M. (2007). Guest Editorial: Mobile Learning. *International Review of Research in Open and Distance Learning*, *8*(2), 1–4.

Ally, M. (2007). Mobile learning. *International Review of Research in Open and Distance Learning*, *8*(2), 1492–3831.

Alm, H., & Nilsson, L. (1995). The effects of a mobile telephone task on driver behaviour in a car following situation. *Accident; Analysis and Prevention*, *27*(5), 707–715. doi:10.1016/0001-4575(95)00026-V

Almeida, P., & Kogut, B. (1999). Localization of knowledge and the mobility of engineers in regional networks. *Management Science*, *45*(7), 905–917. doi:10.1287/mnsc.45.7.905

AmeInfo. (2007). Saudi Arabia's Internet users spend over $3.28bn in B2C e-commerce during 2007. Retrieved November 2008 from http://www.ameinfo.com/143272.html

Andrews, T., Smyth, R., Tynan, B., Vale, D., & Caladine, R. (2008). Rich Media Technologies and uncertain futures- developing sustainable, scalable models. In *Proceedings ASCILITE 2008 Hello! Where are you in the landscape of Educational Technology*, Melbourne November 30-December 3.

Andrzejak, R., Poreba, R., Poreba, M., Derkacz, A., Skalik, R., & Gac, P. (2008). The influence of the call with a mobile phone on heart rate variability parameters in healthy volunteers. *Industrial Health*, *46*(4), 409–417. doi:10.2486/indhealth.46.409

Aoki, K., & Downes, E. J. (2003). An analysis of young people's use of and attitudes toward cell phones. *Telematics and Informatics, 20,* 349-364. In M. Campbell (2005). The impact of the mobile phone on young people's social life. Paper presented to the Social Change in the 21st Century Conference. Retrieved April 2009 from http://eprints.qut.edu.au/3492/1/3492.pdf

APS. (2004). *Psychological Aspects of Mobile Phone Use among Adolescents*. Australian Psychological Society.

Arato, M. (2003). A famous nonlinear stochastic equation: Lotka-Volterra model with diffusion. *Mathematical and Computer Modelling*, *38*, 709–726. doi:10.1016/S0895-7177(03)90056-2

Arellano, M., & Bond, S. (1991). Some Tests of Specification for Panel Data: Monte Carlo Evidence and an Application to Employment Equations. *The Review of Economic Studies*, *58*(2), 277–297. doi:10.2307/2297968

Argote, L., & Ingram, P. (2000). Knowledge transfer: A basis for competitive advantage in firms. *Organizational Behavior and Human Decision Processes*, *82*(1), 150–169. doi:10.1006/obhd.2000.2893

Arnetz, B. B., Akerstedt, T., Hillert, L., Lowden, A., Kuster, N., & Wiholm, C. (2007). The Effects of 884 MHz GSM Wireless Communication Signals on Self-reported Symptom and Sleep (EEG)- An Experimental Provocation Study. *PIERS Online*, *3*(7), 1148–1150. doi:10.2529/PIERS060907172142

Ashford, B. E., & Gibbs, B. W. (1990). The Double-Edge of Organizational Legitimation. *Organization Science*, *1*(2), 177–194. doi:10.1287/orsc.1.2.177

Ashforth, B. E., & Anand, V. (2003). The Normalization Of Corruption In Organizations. *Research in Organizational Behavior*, *25*, 1–52. doi:10.1016/S0191-3085(03)25001-2

Aust. J. Clin. Env. Med. (2007). Australasian Journal of Clinical Environmental Medicine warns that wireless communication technology may be responsible for accelerating the rise in autism among the world's children. *J.Aust.Coll.Nutr.& Env.Med, 26* (2), 3 – 7.

Auvinen, A., Hietanen, M., & Luukkonen, R. (2002). Brain tumors and salivary gland cancers among cellular telephone users. *Epidemiology (Cambridge, Mass.)*, *13*, 356–359. doi:10.1097/00001648-200205000-00018

Avgerou, C., Ciborra, C., & Land, F. (Eds.). (2004). *The social study of information and communication technology*. Oxford: Oxford University Press.

Bacheldor B. (2002, March 18). Handheld system assesses damage to see how buildings survived, *Information Week*.

Baecker, R. (2003). *A principled design for scalable internet visual communications with rich media, interactivity, and structured archives*. Paper presented at the Proceedings of the 2003 conference of the Centre for Advanced Studies on Collaborative research Toronto, Ontario, Canada.

Baharoglu, D., & Leitmann, J. (1998). Coping Strategies for Infrastructure: How Turkey's SpontaneousSettlements Operate in the Absence of Formal Rules. *Habitat International*, *22*(2), 115–135. doi:10.1016/S0197-3975(97)00313-5

Bamiou, D., Ceranic, B., Cox, R., Watt, H., Chadwick, P., & Luxon, L. M. (2008). Mobile telephone use effects on peripheral audiovestibular function: A case-control study. *Bioelectromagnetics*, *29*(2), 108–117. doi:10.1002/bem.20369

Bandura, A. (1997). *Self-Efficacy: The Exercise of Control*. New York: Freeman.

Bandusiri, A. S. W. (2009). Statistical overview of the telecommunications sector in Sri Lanka. Retrieved from http://www.itu.int/

Banister, E. N., & Hogg, M. K. (2004). Negative symbolic consumption and consumers' dive for self-esteem: The case of the fashion industry. *European Journal of Marketing*, *38*(7), 850–868. doi:10.1108/03090560410539285

Barney, J. B., & Hesterly, W. (1996). Organizational economics: Understanding the relationship betweenorganizations and economic analysis . In Clegg, S., Hardy, C., & Nord, W. R. (Eds.), *Handbook of Organization Studies*. Thousand Oaks, CA: Sage Publications.

Barro, R. J. (1991). Economic Growth in a Cross Section of Countries. *The Quarterly Journal of Economics*, *106*(2), 407–443. doi:10.2307/2937943

Barro, R. J., & Lee, J. (2001). International Data on Educational Attainment: Updates and Implications. *Oxford Economic Papers*, *53*(3), 541–563. doi:10.1093/oep/53.3.541

Barth, A., Winker, R., Ponocny-Seliger, E., Mayrhofer, W., Ponocny, I., Sauter, C., & Vana, S. (2008). A meta-analysis for neurobehavioural effects due to electromagnetic field exposure emitted by GSM mobile phones. *Occupational and Environmental Medicine*, *65*, 342–346. doi:10.1136/oem.2006.031450

Bass, F. M. (1969). A new product growth model for consumer durables. *Management Science*, *15*, 215–227. doi:10.1287/mnsc.15.5.215

Bastide, M., Youbicier-Simoa, B. J., Lebecq, J. C., & Giaimis, J. (2001). Toxicologic Study of Electromagnetic Radiation Emitted by Television and Video Display Screens and Cellular Telephones on Chickens and Mice. *Indoor and Built Environment*, *10*(5), 291–298.

Baumeister, R. F., & Heatherton, T. F. (1996). Self-Regulation Failure: An Overview. *Psychological Inquiry*, *7*(1), 1–15. doi:10.1207/s15327965pli0701_1

Baumeister, R. F., Heatherton, T. F., & Tice, D. M. (1993). When Ego Threats Lead to Self-Regulation Failure: Negative Consequences of High Self-Esteem. *Journal of Personality and Social Psychology*, *64*(1), 141–156. doi:10.1037/0022-3514.64.1.141

Bazykin, A. (1998). Nonlinear Dynamics of Interacting Populations . In Khibnik, A. I., & Krauskopf, B. (Eds.), *World Scientific series on nonlinear science (Series A, 11)*. River Edge, NJ: World Scientific.

Becker, G. S. (1968). Crime and punishment: an economic approach. *The Journal of Political Economy*, *76*(2), 169–217. doi:10.1086/259394

Becta (2004). *Becta Home Page*. Retrieved April 27, 2009 from http://www.becta.org.uk/subsections/foi/documents/technology_and_education_research/ handheld_computers.doc

Bell, G. (2005). The age of the thumb: A cultural reading of mobile technologies from Asia. In P. Glotz, S. Bertschi, & C. Locke (Eds.), *Thumb Culture: The Meaning of Mobile Phones for Society* (pp. 67-88). Transcript, Verlag, Bielefeld.

Belsky, J. L., Takeichi, N., & Yamamoto, T. (1975). Salivary gland neoplasms following atomic radiation: additional cases and reanalysis of combined data in a fixed population, 1957–1970. *Cancer*, *35*, 555–559. doi:10.1002/1097-0142(197502)35:2<555::AID-CNCR2820350240>3.0.CO;2-G

Belt, H. (2001). Wireless Devices in the Classroom. *Learning and Training Innovations*. Retrieved April 3, 2009 from http://www.ltimagazine.com/ltimagazine/article

Belyav, I. (2005). Non-thermal biological effects of microwaves. *Microwave Review,* 13–29.

Berg, G., Spallek, J., Schuz, J., Schlehofer, B., Bohler, E., & Schlaefer, K. (2006). Occupational exposure to radio frequency/microwave radiation and the risk of brain tumors: interphone study group, Germany. *American Journal of Epidemiology*, *164*(6), 538–548. doi:10.1093/aje/kwj247

Berger, G. (2004). Putting perspective on cellphone euphoria. *Mail & Guardian*. Retrieved April 27, 2009 from http://archive.mg.co.za

Bhabha, H. (1985). Signs taken for wonders: questions of ambivalence and authority under a tree outside Delhi, May 1917. *Critical Inquiry*, *12*, 144–165. doi:10.1086/448325

Bhargava, S. C. (1989). Generalized Lotka-Volterra equations and the mechanism of technological substitution. *Technological Forecasting and Social Change*, *35*, 319–326. doi:10.1016/0040-1625(89)90068-1

Bianchi, A., & Phillips, J. G. (2005). Psychological predictors of problem mobile phone use. *Cyberpsychology & Behavior*, *8*(1), 39–51. doi:10.1089/cpb.2005.8.39

Bijker, W. E. (1995). *Of bicycles, bakelites, and bulbs toward a theory of sociotechnical change*. Massachusetts: MIT Press.

Blundell, R. W., & Bond, S. (1998). Initial Conditions and Moment Restrictions in Dynamic Panel Data Models. *Journal of Econometrics*, *87*(1), 115–143. doi:10.1016/S0304-4076(98)00009-8

Boguraev, B., & Kennedy, C. (1997). Salience-based content characterization of text documents. In *Proceedings of a Workshop on Intelligent Scalable Text Summarization Sponsored by the Association for Computational Linguistics*, Madrid, Spain.

Bond, S. R. (2002). Dynamic Panel Data Models: A Guide to Micro Data Methods and Practice. *Portuguese Economic Journal*, *1*(2), 141–162. doi:10.1007/s10258-002-0009-9

Bortkiewicz, A., Zmyslony, M., Szyjkowska, A., & Gadzicka, E. (2004). Subjective symptoms reported by people living in the vicinity of cellular phone base stations [review]. *Medycyna Pracy*, *55*(4), 345–352.

Botelho, A., & Pinto, L. C. (2004). The diffusion of cellular phones in Portugal. *Telecommunications Policy*, *28*, 427–437. doi:10.1016/j.telpol.2003.11.006

Brown, B., Green, N., & Harper, R. (2002). *Wireless World: Social, Cultural and Interactional Issues in Mobile Communications and Computing*. London: Springer.

Brown, I., Cajee, Z., Davies, D., & Stroebel, S. (2003). Cell phone banking: predictors of adoption in South Africa – an exploratory study. *International Journal of Information Management*, *23*, 381–394. doi:10.1016/S0268-4012(03)00065-3

Browne, L. (2003). *Technology explosion, early 20th century*. Retrieved from http://www.economicadventure.org/gazette/ch3.pdf

Broyd, T. (1999). What Impact can IT Realistically have on Construction? In *Proceedings of Berkeley-Stanford CE&M Workshop – Defining a Research Agenda*. Stanford, Berkeley, USA: University of Stanford and University of California.

Bruner, G. C., & Kumar, A. (2005). Explaining consumer acceptance of handheld Internet devices. *Journal of Business Research*, *58*(5), 553–558. doi:10.1016/j.jbusres.2003.08.002

Brusque, C., & Alauzet, A. (2008). Analysis of the individual factors affecting mobile phone use while driving in France: Socio-demographic characteristics, car and phone use in professional and private contexts. *Accident; Analysis and Prevention*, *40*(1), 35–44. doi:10.1016/j.aap.2007.04.004

Brynjolfsson, E., & Saunders, A. (2009). *Wired for innovation. How information technology is reshaping the economy*. Cambridge, MA: MIT Press.

Burgelman, J.-C., & Verhoest, P. (1996). Trans-European information networks. *Telematics and Informatics*, *13*(2/3), 67–80. doi:10.1016/S0736-5853(96)00006-8

Burkink, T. (2002). Cooperative and voluntary wholesale groups: channel coordination and interfirm knowledge transfer. *Supply Chain Management: An International Journal*, *7*(2), 60–70. doi:10.1108/13598540210425812

Caladine, R., Andrews, T., Tynan, B., Smyth, R., & Vale, D. (in press). New communications Options: a renaissance in IP use . In Veletsianos, G. (Ed.), *Using Emerging Technologies in Distance Education*. Athabasca University Press.

Calisir, F., & Lehto, M. R. (2002). Young drivers' decision making and safety belt use. *Accident; Analysis and Prevention*, *34*, 793–805. doi:10.1016/S0001-4575(01)00079-3

Campbell, J. L. (2004). *Institutional Change and Globalization*. Princeton: Princeton University Press.

Campbell, M. (2005). *The impact of the mobile phone on young people's social life*. Paper presented to the Social Change in the 21st Century Conference. Retrieved April 2009 from http://eprints.qut.edu.au/3492/1/3492.pdf

Canel, E. (1992). New social movement theory and resource mobilization: the need for integration . In Carroll, W. (Ed.), *Organizing Dissent: Social Movements in Theory and Practice* (pp. 22–51). Toronto: Garamond.

Carroll, J. (2006). 'What's in it for me?': Taking m-government to the people. Paper presented at the 19th Bled eConference, Bled, Slovakia.

Carver, C. S., & Scheier, M. F. (1998), *On the Self-Regulation of Behavior*. Cambridge, UK: Cambridge.

Castells, M. (2000). *The Rise of the Network Society* (2nd ed.). Malden: Blackwell Publishing.

Castells, M. (2001). *The internet galaxy: Reflections on the Internet business and society*. Oxford: Oxford University Press.

Caudill, J. G. (2007). The growth of m-learning and the growth of mobile computing: Parallel developments. *International Review of Research in Open and Distance Learning*, *8*(2), 1–12.

Cavus, N., & Ibrahim, D. (2009). M-Learning: An experiment in using SMS to support learning new English language words. *British Journal of Educational Technology*, *40*(1), 78–91. doi:10.1111/j.1467-8535.2007.00801.x

Central Government (The Central Government of the People's Republic of China). (2006). The Eleventh Five-year Plan. Retrieved June 12, 2009, from http://www.gov.cn/ztzl/2006-03/16/content_228841.htm

Chambers, R. (1983). *Rural Development: Putting the Last First*. Prentice Hall.

Chapman, S., & Schofield, W. N. (1998). Lifesavers and Samaritans: emergency use of cellular (mobile) phones in Australia. *Accident; Analysis and Prevention*, *30*(6), 815–819. doi:10.1016/S0001-4575(98)00034-7

Chayko, M. (2008). *Portable communities. The social dynamics of online and mobile connectedness*. Albany, NY: State University of New York Press.

Chen, G., & Kotz, D. (2000). *A Survey of context-aware mobile computing research* (TR2000-381). Hanover, NH, USA, Dartmouth College: 16.

Chen, Y. S., Kao, T. C., Sheu, J. P., & Chiang, C. Y. (2002). A mobile scaffolding-aid-based bird-watching learning system. In M. Milrad, H.U. Hoppe, & Y. Kinshuk (Eds.), *IEEE International Workshop on Wireless and Mobile Technologies in Education* (pp. 15-22). Los Alamitos, CA: IEEE Computer Society.

Cheng, B. H. C., & Atlee, J. M. (2007). Research Directions in Requirements Engineering. In *Proceedings of Workshop on the Future of Software Engineering held at ICSE 2007* (pp. 285-303).

Choi, S., Lee, M., & Chung, G. (2001). Competition in Korean mobile telecommunications market: Business strategy and regulatory environment. *Telecommunications Policy*, *25*, 125–138. doi:10.1016/S0308-5961(00)00075-6

Christensen, H. C., Schu¨z, J., & Kosteljanetz, M. (2005). Cellular telephones and risk for brain tumors: a population-based, incident case-control study. *Neurology*, *64*, 1189–1195.

Chung, S., Singh, H., & Lee, K. (2000). Complementarity, status similarity, and social capital and drivers of alliance formation. *Strategic Management Journal*, *21*, 1–22. doi:10.1002/(SICI)1097-0266(200001)21:1<1::AID-SMJ63>3.0.CO;2-P

Church, J., & Gandal, N. (1993). Complementary network effects and technological adoption. *International Journal of Industrial Organization*, *11*(2), 239–260. doi:10.1016/0167-7187(93)90006-X

Cieslik, A., & Kaniewsk, M. (2004). Telecommunications infrastructure and regional economic development: The case of Poland. *Regional Studies*, *38*(6), 713–725. doi:10.1080/003434042000240996

Cilingir, D., & Kushchu, I. (2004). E-government and m-government: Concurrent leaps by Turkey. In D. Remenyi (Ed.) *Proceedings of European Conference on E-Government (ECEG 2004)* (pp. 813-821). Trinity College, Dublin, June 17-18. Department of Taoiseach, Dublin, Ireland. Reading, UK: Academic Conferences International.

Cinel, C., Russo, R., Boldini, A., & Fox, E. (2008). Exposure to Mobile Phone Electromagnetic Fields and Subjective Symptoms: A Double-Blind Study. *Psychosomatic Medicine*, *70*, 345–348. doi:10.1097/PSY.0b013e31816521f1

Clarke, I., Hallsworth, A., Jackson, P., de Kervenoael, R., Del Aguila, R. P., & Kirkup, M. (2006). Retail restructuring and consumer choice 1. Long-term local changes in consumer behavior: Portsmouth, 1980-2002. *Environment & Planning A*, *38*(1), 25–46. doi:10.1068/a37207

Clegg, S. (1975). *Power, Rule and Domination: A Critical and Empirical Understanding of Power in Sociological Theory and Organization Life*. London: Routledge and Kegan Paul.

Clegg, S. (1989). Radical revisions: Power, discipline and organizations. *Organization Studies*, *10*(1), 97–115. doi:10.1177/017084068901000106

CNNIC. (2007). *Survey report on the Internet development in China*. Retrieved October 5, 2009 from http://www.cnnic.net.cn/uploadfiles/pdf/2008/1/17/104156.pdf

CNNIC. (2009). *Survey report on the Internet development in China*. Retrieved October 5, 2009 from http://www.cnnic.net.cn/uploadfiles/pdf/2009/1/13/92458.pdf

Colazzo, L., Molinari, A., Ronchetti, M., & Trifonova, A. (2003). *Towards a multi-vendor mobile learning management system*. Retrieved April 24, 2009 from: http://trifonova.net/docs/Towards%20a%20multi-vendor%20Mobile%20LMS%20(long).pdf

Coleman, J. W., & Ramos, L. L. (1998). Subcultures and deviant behavior in the organisational context. *Research in the Sociology of Organizations*, *15*, 3–34.

Collinson, D. (1994). Strategies of resistance . In Jermier, J., Knoghts, D., & Nord, W. (Eds.), *Resistance and Power in Organizations*. Thousand Oaks, CA: Sage.

ConcordECC, *31*(1), 3.

Corlett, D., Sharples, M., Bull, S., & Chan, T. (2005). Evaluation of a mobile learning organiser for university students. *Journal of Computer Assisted Learning*, *21*, 162–170. doi:10.1111/j.1365-2729.2005.00124.x

COST269. (2003, 3-5 September 2003). *The good, the bad and the irrelevant: the user and the future of ICTs*. Helsinki (Finland).

Cotton, S., Anderson, W., & Tufekci, Z. (2009). Old wine in a new technology, or a different type of digital divide? *New Media & Society*, *11*(7), 1163–1186. doi:10.1177/1461444809342056

Courcoubetis, C., & Weber, R. R. (2003). *Pricing communication networks: Economics, technology and modeling*. Wiley. doi:10.1002/0470867175

Courpasson, D. (2000). Managerial Strategies of Domination: Power in SoftBureaucracies. *Organization Studies*, *21*(1), 141–162. doi:10.1177/0170840600211001

Craig-Lees, M., & Hill, C. (2002). Understanding voluntary simplifiers. *Psychology and Marketing*, *19*(2), 187–210. doi:10.1002/mar.10009

Crocker, J., Major, B., & Steele, C. (1998). Social stigma . In Gilbert, D., Fiske, S. T., & Lindzey, G. (Eds.), *Handbook of social psychology* (4th ed., pp. 504–553). Boston: McGraw Hill.

Cronin, F. J., Parker, E. B., Colleran, E. K., & Gold, M. A. (1991). Telecommunications infrastructure and economic growth: An analysis of causality. *Telecommunications Policy*, *15*, 529–535. doi:10.1016/0308-5961(91)90007-X

Cui, G. Z. (2001). M-learning-a new direction of modern educational technology. *Journal of Modern Educational Technology*, *6*, 15–18.

Dacin, M. T., Oliver, C., & Roy, J.-P. (2007). The legitimacy of strategic alliances: An in-stitutional perspective. *Strategic Management Journal*, *28*, 169–187. doi:10.1002/smj.577

Dalgaard, C.-J., Hansen, H., & Tarp, F. (2004). On the Empirics of Foreign Aid and Growth. *The Economic Journal*, *114*(496), 191–216. doi:10.1111/j.1468-0297.2004.00219.x

Danziger, J. N., Dutton, W. H., Kling, R., & Kraemer, K. L. (1982). *Computers and politics: High technology in American local governments*. New York: Columbia University Press.

Datta, A., & Agarwal, S. (2004). Telecommunications and economic growth: a panel data approach. *Applied Economics*, *36*(15), 1649–1654. doi:10.1080/0003684042000218552

Davison, R. M., Vogel, D. R., & Harris, R. W. (2005). The e-transformation of Western China. *Communications of the ACM*, *48*(4), 62–66. doi:10.1145/1053291.1053320

Dawes, S. S. (2009). Governance in the digital age: A research and action framework for an uncertain future. *Government Information Quarterly*, *26*, 257–264. doi:10.1016/j.giq.2008.12.003

De Bruijn, M., Nyamnjoh, F., & Brinkman, I. (Eds.). (2009). *Mobile phones: The new talking drums of everyday Africa*. Leiden, The Netherlands: African Studies Centre.

De Cuellar, J. P. (1995). *Our Creative Diversity. Report of the World Commission on Culture and Development*. Paris: UNESCO.

Dedeoglu, A. O. (2004). The symbolic use of mobile telephone among Turkish consumers. *Journal of Euromarketing, 13*(2/3), 143–163. doi:10.1300/J037v13n02_08

Dedeoğlu, A. O. (2004). The Symbolic Use of Mobile Telephone Among Turkish Consumers. *Journal of Euromarketing, 13*(2/3), 143–163. doi:10.1300/J037v13n02_08

Deegan, C., Rankin, M., & Voght, P. (2000). Firms' Disclosure Reactions to Major Social Incidents: Australian Evidence. *Accounting Forum, 24*(1), 101–130. doi:10.1111/1467-6303.00031

Dekimpe, M. G., Parker, P. M., & Sarvary, M. (1998). Staged estimation of international diffusion models: An application to global cellular telephone adoption. *Technological Forecasting and Social Change, 57*(1-2), 105–132. doi:10.1016/S0040-1625(97)00085-1

Delhi Indira Gandhi International Airport. (n.d.). Retrieved July 2009 from www.newdelhiairport.in/about-dial.asp

Delhi Indira Gandhi International Airport. (n.d.). Retrieved July 2009 from www.newdelhiairport.in/milestones.asp

Designing useful mobile services for Africa (2009). Retrieved from http://googleblog.blogspot.com/2009/07/designing-useful-mobile-services-for.html

Deutsch, M. (1985). *Distributive Justice*. New Haven, CT: Yale University Press.

Devi, C. (2005). New hotlink services for youths. *New Straits Times . Computimes, 15*, 3.

Dewey, J. (1938). *Experience and Education*. New York: Macmillan.

Dholakia, R. R., & Harlam, B. (1994). Telecommunications and economic development: Econometric analysis of the US experience. *Telecommunications Policy, 18*(6), 470–477. doi:10.1016/0308-5961(94)90015-9

Dieckmann, U., Marrow, P., & Law, R. (1995). Evolutionary cycling in predator-prey interactions: Population dynamics and the red queen. *Journal of Theoretical Biology, 176*, 91–102. doi:10.1006/jtbi.1995.0179

DiMaggio, P. J., & Powell, W. W. (1983). The iron cage revisited: Institutional isomorphism and collective rationality in organizational fields. *American Sociological Review, 48*, 147–160. doi:10.2307/2095101

Divan, H.A., Kheifets, L., Obel, C., Olsen, J. (2008). Prenatal and postnatal exposure to cell phone use and behavioral problems in children. *Epidemiology. Jul, 19*(4), 523-9.

Donner, J. (2007). The use of mobile phones by microentrepreneurs in Kigali, Rwanda: Changes to social and business networks. *Information Technologies and International Development, 3*(2), 3–19. doi:10.1162/itid.2007.3.2.3

Donner, J. (2008). Research Approaches to Mobile Use in the Developing World: A Review of the Literature. *The Information Society, 24*(3), 140–159. doi:10.1080/01972240802019970

Doring, N., & Gundolf, A. (2005). Your life in snapshots: Mobile weblogs (moblogs) . In Glotz, P., Bertschi, S., & Locke, C. (Eds.), *Thumb Culture: The Meaning of Mobile Phones for Society* (pp. 211–224). New Brunswick, NJ: Transaction Publishers.

Dourish, P. (2006). Implications for design. In *Proceedings of SIGCHI Conference on Human Factors in Computing Systems* (pp. 541-550).

Dowling, J., & Pfeffer, J. (1975). Organizational Legitimacy: Social Values and Organizational Behavior. *Pacific Sociological Review, 18*(1), 122–1366.

Downing, A. J. (1985). Road Accidents in Pakistan and the need for improvements in driver training and traffic law enforcement. In: PTRC. Summer Annual Meeting, University of Sussex, July 15-18, 1985, Proceedings of Seminar H. London: PTRC Education and Research Services.

Downing, A. J., Baguley, C. J., & Hills, B. L. (1991). *Road safety in developingcountries: an overview*. Crowthome, UK: Transport and Road Research Laboratory.

Duncombe, R. A. (2006) Analysing ICT applications for poverty reduction via Micro-enterprise using the livelihoods framework. Development Informatics Working Paper Series, No.27, Institute for Development Policy and Management, University of Manchester. Retrieved June 2, 2009 from http://www.sed.manchester.ac.uk/idpm/publications/wp/di/index.htm

Dunleavy, P., Margetts, H., Bastow, S., & Tinkler, J. (2004). Government IT performance and the power of the IT industry: A cross-national analysis. Paper presented at the Annual meeting of American Political Science Association, Chicago.

Dutta, A. (2001). Telecommunications and economic activity: An analysis of Granger causality. *Journal of Management Information Systems*, *17*(4), 71–95.

Easterly, W. (2006). *The White Man's Burden. Why the West's Efforts to Aid the Rest Have Done So Much Ill and So Little Good.* New York: Penguin Books.

Economides, N. (1991). Compatibility and the creation of shared networks . In Guerin-Calvert, M. E., & Wildman, S. S. (Eds.), *Electronic services networks: A business and public policy challenge*. New York.

Eisenhardt, K. M., & Schoonhoven, C. B. (1996). Resource-based view of strategic alliance formation: Strategic and social effects in entrepreneurial firms. *Organization Science*, *7*, 136–150. doi:10.1287/orsc.7.2.136

Elliott, A., & Lemert, C. (2006). *The new individualism. The emotional costs of globalization*. London: Routlege.

Elliott, R. (1994). Addictive Consumption: Function and Fragmentation in Postmodernity. *Journal of Consumer Policy*, *17*(2), 150–180. doi:10.1007/BF01016361

Elsbach, K. D. (2003). Organizational perception management. *Research in Organizational Behavior*, *25*, 297–332. doi:10.1016/S0191-3085(03)25007-3

Elsbach, K. D., & Sutton, R. I. (1992). Acquiring organizational legitimacy through illegitimate actions: A marriage of institutional and impression management theories. *Academy of Management Journal*, *35*(4), 699–738. doi:10.2307/256313

Eltiti, S., Wallace, D., Ridgewell, A., Zougkou, K., Russo, R., & Sepulveda, F. (2007). Does short-term exposure to mobile phone base station signals increase symptoms in individuals who report sensitivity to electromagnetic fields? A double-blind randomized provocation study. *Environmental Health Perspectives*, *115*(11), 1603–1608. doi:10.1289/ehp.10286

Elzarka, H., & Bell, L. (1997). Development of pen-based computer field applications . *Journal of Computing in Civil Engineering*, *11*(2), 140–143. doi:10.1061/(ASCE)0887-3801(1997)11:2(140)

Esbjornsson, M., Juhlin, O., & Weilenmann, A. (2007). Drivers Using Mobile Phones in Traffic: An Ethnographic Study of Interactional Adaptation. *International Journal of Human-Computer Interaction*, *22*(1-2), 37. doi:10.1207/s15327590ijhc2201-02_3

Escobar, M., & Fernandez, A. L. (1994). *Paulo Freire on Higher Education: A Dialogue at the National University of Mexico*. New York: State University of New York Press.

Esselaar, S., & Stork, C. (2005). Mobile cellular telephone: Fixed-line substitution in sub-Saharan Africa. *South African Journal of Information and Communication*, *6*, 64–73.

Ettlinger, N. (2003). Cultural economic geography and a relational and microspace approach to trusts, rationalities, networks, and change in collaborative workplaces. *Journal of Economic Geography*, *3*, 145–171. doi:10.1093/jeg/3.2.145

Euromonitor (2009). *Global Market Information Database*. Euromonitor International.

Evci, C., Ciliz, K., Anarim, E., & Sankur, B. (2004). Wireless networks in Turkey: A jewel in the crowd. *Alcatel Telecommunications Review 3rd Quarter*.

Ezeomah, C. (1982). Movements and demography of Fulani Nomads and their implications for education development. In *Proceedings of the 1st Annual Conference on the Education of Nomads in Nigeria, Jos, Nigeria*.

Fair, J. E. (1988). *A meta-research of mass media effects on audiences in developing countries from 1958 through 1986.* Unpublished doctoral dissertation Indiana University, Bloomington, IN.

Fair, J. E. (1989). 29 years of theory and research on media and development: The dominant paradigm impact. *Gazette, 44*, 129–150. doi:10.1177/001654928904400204

Fair, J. E., & Shah, H. (1997). Continuities and discontinuities in communication and development research since 1958. *Journal of International Communication, 4*(2), 3–23.

Fan, Y., & Li, D. (2006). Mobile government development research (yidong dianzi zhengwu fazhan yanjiu). E-government, 2006(9), 48-52.

Farrell, C. J. (1993). A theory of technological progress. *Technological Forecasting and Social Change, 44*, 161–178. doi:10.1016/0040-1625(93)90025-3

FDA. (2003). *Cell Phone Facts: Consumer Information on Wireless Phones*. Retrieved May 10, 2005, from http://www.fda.gov/cellphones/qa.html

Fellbaum, C. (1998). *WordNet: An electronic lexical database*. Cambridge, MA: MIT Press.

Filos, E. (2002). European collaborative R&D projects related to the "Smart organization". A first evaluation of activities and implications for construction. In Turk & Scherer (Eds.), *Proceedings of the Conference on eWork and eBusiness in AEC* (pp. 27-32). Swets and Zietilinger, Lisse.

Fishbein, M., & Ajzen, I. (1975). *Belief, Attitude, Intention, and Behavior: An Introduction to Theory and Research*. Reading, MA: Addison-Wesley.

Fleming, P., & Spicer, A. (2007). *Contesting the corporation*. Cambridge, UK: Cambridge University Press. doi:10.1017/CBO9780511628047

Foley, P. (2004). Does the internet help to overcome social exclusion? *Electronic . Journal of E-Government, 2*(2), 139–146.

Fombrun, C. J. (1996). *Reputation: Realizing Value from the Corporate Image*. Harvard Business School.

Fortunati, L. (1998). The ambigious image of the mobile phone . In Haddon, T. L. (Ed.), *Communications on the Move: The Experience of Mobile Telephony in the 1990s*.

Fortunati, L., Katz, J. E., & Riccini, R. (2003). *Mediating the Human Body. NJ*. Mahwah: LEA.

Fourt, L. A., & Woodlock, J. W. (1960). Early prediction of market success for new grocery products. *Journal of Marketing, 25*(October), 31–38. doi:10.2307/1248608

Franke, (2007). Electromagnetic fields (GSM 1800) do not alter blood-brain barrier permeability to sucrose in models in vitro with high barrier tightness. *Bioelectromagnetics, 26*(7), 529–535. doi:10.1002/bem.20123

Fraser, C., & Restrepo-Estrada, S. (1998). *Communicating for Development. Human Change for Survival*. New York: I.B. Tauris Publishers.

Fraunholz, B., & Unnithan, C. (2004). Critical Success factors in mobile communications - a comparative roadmap for Germany and India. *International Journal of Mobile Communications, 2*(1), 87–101. doi:10.1504/IJMC.2004.004489

Freire, P. (1983). *Pedagogy of the Oppressed.* New York: Continuum.

Frembong, G. (2009). Mobile telephone opportunities: the case of micro- and small enterprises in Ghana. *Info - The journal of policy, regulation and strategy for telecommunications, 11*(2), 79-94.

Frohlich, D. M., Rachovides, D., Riga, K., Bhat, R., Frank, M., Edirisinghe, E., et al. (2009). StoryBank: mobile digital storytelling in a development context. In *Proceedings of the 27th international conference on Human factors in computing systems* (pp. 1761-1770). Boston: ACM.

Fuglesang, A. (1982). *About understanding. Ideas and observations on cross-cultural communication*. Uppsala, Sweden: Dag Hammarskjold Foundation.

Fuller, S., & Sattineni, A. (2002). Investigating the use of wearable computers in construction. In *Proceedings of the First International Conference on Construction in the 21st Century Challenges and Opportunities inManagement and Technology* (pp. 583-588), Miami, Florida.

Gabriel, Y. (1999). Beyond happy families: A critical reevaluation of the control resistance-identity triangle. *Human Relations*, *52*(2), 179–203. doi:10.1177/001872679905200201

Ganesh, J., Kumar, V., & Subramaniam, V. (1997). Learning effect in multinational diffusion of consumer durables: An exploratory investigation. *Journal of the Academy of Marketing Science*, *25*(3), 214–228. doi:10.1177/0092070397253003

Garai, A., & Shadrach, B. (2006). Opportunities and Challenges . In *One World South Asia*. Taking ICT to Every Indian Village.

Gatignon, H., & Robertson, T. S. (1991). Innovative decision processes . In Robertson, T. S., & Kassarjian, H. H. (Eds.), *Handbook of Consumer Behavior* (pp. 316–348). Englewood Cliffs, NJ: Prentice-Hall.

Gausepohl, K. A. (2008). *Investigation of Storytelling as a Requirements Elicitation Method for Medical Devices*. Masters Thesis in Industrial and Systems Engineering, Virginia Polytechnic Institute.

Geddes, S. (2004). Mobile learning in the 21st century: benefit for learners. *The Knowledge Tree: An e-Journal of Learning Innovation*, 6.

Geertz, C. (1973). *The interpretation of cultures*. New York: Basic Books.

Geertz, C. (1983). *Local knowledge. Further essays in interpretative anthropology*. New York: Basic books.

Georgiev, T., Georgieva, E., & Trajovski, G. (2006). *Transitioning from e-Learning to m-Learning: Present issues and future challenges*. Retrieved April 28, 2009 from http://ieeexplore.ieee.org/iel5/10923/34372/01640716.pdf?tp=&isnumber=34372&arnumber=1640716.

Geser, H. (2005). Is the Cell Phone Undermining the Social Order? Understanding Mobile Technology from a Sociological Perspective. In P. Glotz, S. Bertschi, & C. Locke (Eds.), *Thumb Culture: The Meaning of Mobile Phones for Society* (pp. 23-37). Transcript, Verlag, Bielefeld.

Ghyasi, A. F., & Kushchu, I. (2004). Uses of mobile government in developing countries, unpublished, *Mobile Government Lab.*

Gibbs, G. R. (2002). *Qualitative Data Analysis: Explorations with NVivo*. Buckingham: Open University Press.

Giddens, A. (1991). *Modernity and self-identity*. Cambridge: Polity Press.

Giddens, A. (1992). *The transformation of intimacy*. Cambridge: Polity Press.

Giddens, A. (1995). *Modernity and Self-Identity. Self and Society in the Late Modern Age*. Cambridge, UK: Polity.

Gintis, H., Bowles, S., Boyd, R., & Fehr, E. (2005). *Moral sentiments and material interests*. Cambridge, MA: MIT Press.

Glaser, B. (1992). *Basics of grounded theory analysis: Emergence vs. forcing*. Mill Valley, CA: Sociology Press.

Glaser, B. G. (2005). *The Grounded Theory Perspective III: Theoretical coding*. Sociology Press.

Glaser, B. G., & Strauss, A. (1967). *The Discovery of Grounded Theory: Strategies for Qualitative Research*. Chicago, IL: Aldine Publishing Co.

GMID. (n.d.). Retrieved from http://portal.euromonitor.com/PORTAL/tandc.aspx

Goffman, E. (1963). *Stigma: Notes on the Management of Spoiled Identity*. Prentice-Hall.

Goggin, G. (2008). *Mobile phone cultures*. London: Routledge.

Golder, P. N., & Tellis, G. J. (1997). Will it ever fly? Modeling the takeoff of really new consumer durables. *Marketing Science*, *16*(3), 253–270. doi:10.1287/mksc.16.3.256

Gollwitzer, P. M., & Bayer, U. (1999). Deliberative Versus Implemental Mindsets in the Control of Action . In Chaiken, S., & Trope, Y. (Eds.), *Dual-Process Theories in Social* (pp. 403–422). New York: Guilford.

Gomez, S. (2007). Scroll to "e" for Education. *Times Higher Education Supplement*, *1780*, 13.

Goodman, M. J., Tijerina, L., Bents, F. D., & Wierwille, W. W. (1999). Using cellular phones in vehicles. *Transportation Human Factors*, *1*, 3–42. doi:10.1207/sthf0101_2

Goodman, P. S., & Friedman, A. (1971). An Examination of Adams' Theory of Inequity. *Administrative Science Quarterly*, *16*(3), 271–288. doi:10.2307/2391900

Grajek, M., & Kretschmer, T. (2009). Usage and diffusion of cellular telephony, 1998-2004. *International Journal of Industrial Organization*, *27*, 238–249. doi:10.1016/j.ijindorg.2008.08.003

Granovetter, M. S. (1985). Economic action and social structure: The problem of embeddedness. *American Journal of Sociology*, *91*, 481–510. doi:10.1086/228311

Greenberg, J. (1990). Employee theft as a reaction to underpayment inequity: The hidden cost of pay cuts. *The Journal of Applied Psychology*, *75*(5), 561–568. doi:10.1037/0021-9010.75.5.561

Griliches, Z. (1957). Hybrid corn: An exploration in the economics of technological change. *Econometrica*, *25*, 501–522. doi:10.2307/1905380

Grinter, R., & Eldridge, M. (2001). y do tngrs luv 2 txt msg. In *Proceedings of the Seventh European Conference on Computer-Supported Cooperative Work ECSCW* (vol. 1, pp. 219-238).

Gruber, H. (2001). Competition and innovation: The diffusion of mobile telecommunications in central and Eastern Europe. *Information Economics and Policy*, *13*, 19–34. doi:10.1016/S0167-6245(00)00028-7

Gruber, H., & Verboven, F. (2001). The diffusion of mobile telecommunications services in the European Union. *European Economic Review*, *45*, 577–588. doi:10.1016/S0014-2921(00)00068-4

Gubbins, M. (2004). *Global IT spending by sector*. Computing.

Gudykunst, W. B. (1993). Toward a Theory of Effective Interpersonal and Intergroup Communication. In Wiseman & Koester (Eds.), *Intercultural Communication Competence*. Thousand Oaks, CA: Sage.

Gulati, R. (1999). Network location and learning: The influence of network resources and firm capabilities on alliance formation. *Strategic Management Journal*, *20*(5), 397–420. doi:10.1002/(SICI)1097-0266(199905)20:5<397::AID-SMJ35>3.0.CO;2-K

Gulf News (2009, February 22). iPhone launch could spur mobile commerce growth in Gulf.

Gulf News (2009b, April 24). Apple and eBay beat analysts' expectations.

Gulf News (2010, August 2). BlackBerry to go offline.

Gulf News. (2009c, January 17). UAE at top of Arab ICT Index, by Naushad K. Cherrayil, January 17. Retrieved April 2009 from http://www.gulfnews.com/business/Technology/10276038.html

Gulf Times (2008, August 21). Internet and Mobile phones spur economic development.

Gumucio-Dragon, A., & Tufte, T. (Eds.). (2006). *Communication for Social Change Anthology: Historical and contemporary readings*. South Orange, NJ: CFSC Consortium.

Gunavardhana, D. (1991). *The Gajaman story: Gajaman puvatha*. Sridevi Printers.

Gunawardena, C. (2008). Gender equity in higher education in Sri Lanka. *International Studies in Education*. Retrieved from http://iise.pitt.edu/publications/ncontents/v9/n1/ise200904.pdf

Gyimah-Brempong, K. (2002). Corruption, economic growth, and income inequality in Africa. *Economics of Governance*, *3*(3), 183–209. doi:10.1007/s101010200045

Ha, I., Yoon, Y., & Choi, M. (2007). Determinants of adoption of mobile games under mobile broadband wireless access environment. *Information & Management, 44*, 276–286. doi:10.1016/j.im.2007.01.001

Haapaniemi, T. P., & Mäkinen, S. J. (2007). Determining the takeoff point in adoptions of innovations: A comparison of content and discrimination analysis. *International Journal of Technology Marketing, 2*(1), 65–80. doi:10.1504/IJTMKT.2007.011586

Hahn, H. P., & Kibora, L. (2008). The domestication of the mobile phone. Oral society and new ICT in Burkina Faso. *The Journal of Modern African Studies, 46*, 87–109. doi:10.1017/S0022278X07003084

Halavais, A. (2009). *Search engine society*. Cambridge, UK: Polity.

Hall, S. (1990). Culture identity and diaspora . In Rutherford, J. (Ed.), *Identity: Community, Culture, Difference* (pp. 222–237). London: Lawrence and Wishart.

Hallin, A., & Lundevall, K. (2007). mCity: User focused development of mobile services within the city of Stockholm. In I. Kushchu (Ed.), *Mobile Government: An Emerging Direction in E-Government.* New York: IGI Publishing.

Hancock, A. (1984). *Technology transfer and communication*. Paris: UNESCO.

Hardell, L., Hallquist, A., & Hansson Mild, K. (2004). No association between the use of cellular or cordless telephones and salivary gland tumours. *Occupational and Environmental Medicine, 61*, 675–679. doi:10.1136/oem.2003.011262

Hardell, L., Nasman, A., & Pahlson, A. (1999). Use of cellular telephones and the risk for brain tumors: A case-control study. *International Journal of Oncology, 15*, 113–116.

Harindranath, R. (2006). *Perspectives on Global Cultures*. Berkshire, UK: Open University Press.

Harmer, J. A. (2003). Mobile Multimedia Services. *BT Technology Journal, 21*(3), 169–180. doi:10.1023/A:1025175518841

Harrill, R. (2005). Wake-up Call. *The University of Washington Alumni Magazine,* (March). Retrieved on 2008-05-31.

Harvey, M. (Ed.). (2005). *Media matters. Perspectives on Advancing Governance & Development from the Global Forum for Media Development*. London: Internews Europe.

Hassan, R. (2008). *The Information Society*. Cambridge, UK: Polity Press.

Haynes, D., & Prakash, G. (1991). Introduction: The entanglement of power and resistance . In Haynes, D., & Prakash, G. (Eds.), *Contesting power: Resistance and everyday social relations in South Asia*.

Head, J. G. (1990). On merit wants: reflections on thr evolution, normative status and policy relevance of a controversial public finance concept. In G. Brennan & C. Walsh (Eds.), *Rationality individuals and public policy* (pp. 211-244). Centre for research on Federal Financial Relations, Australian National University.

Hearit, K. M. (1995). 'Mistakes Were Made': Organizations, Apologia, and Crises of Social Legitimacy. *Communication Studies, 46*(1-2), 1–17.

Heckhausen, H., & Gollwitzer, P. M. (1987). Thought Contents and Cognitive Functioning in Motivational Versus Volitional States of Mind. *Motivation and Emotion, 11*, 101–120. doi:10.1007/BF00992338

Heeks, R. (2001). Understanding e-governance for development. I-government working paper series, 11. Retrieved December 200 from http://idpm.man.ac.uk/wp/igov/igov_wp11.htm

Heeks, R. (2008). CT4D 2.0: The Next Phase of Applying ICT for International Development. *IEEE Computer, 41*(6), 26–33.

Heeks, R. B., & Bailur, S. (2006) Analysing eGovernment research: perspectives, philosophies, theories, methods and practice. *iGovernment Working Paper Series*, Paper No.16, Institute for Development Policy and Management (IDPM), the University of Manchester. Retrieved May 2, 2009 from http://www.sed.manchester.ac.uk/idpm/publications/wp/igov/educigov.htm

Heeks, R., & Stanforth, C. (2007). Understanding e-government project trajectories from an actor-network perspective. *European Journal of Information Systems*, *16*(2), 165–177. doi:10.1057/palgrave.ejis.3000676

Henwood, K., & Pidgeon, N. (2003). Grounded theory in psychological research in qualitative research . In Camic, P. M., Rhodes, J. E., & Yardley, L. (Eds.), *Psychology: Expanding Perspectives in Methodology and Design* (pp. 131–155). Washington, DC: American Psychological Association. doi:10.1037/10595-008

Hepworth, S. J., Schoemaker, M. J., Muir, K. R., Swerdlow, A. J., van Tongeren, M. J., & McKinney, P. A. (2006). Mobile phone use and risk of glioma in adults: case-control study. *BMJ (Clinical Research Ed.)*, *332*, 883–887. doi:10.1136/bmj.38720.687975.55

Hirschman, E. C. (1992). The consciousness of addiction: Toward a general theory of compulsive consumption. *The Journal of Consumer Research*, *19*, 155–179. doi:10.1086/209294

Hjorth, L. (2009). "It's complicated": A case study of women and mobile intimacy. gbc2009 proceedings/papers/P204

Hoch, S. J., & Loewenstein, G. (1991). Time-Inconsistent Preferences and Consumer Self-Control. *The Journal of Consumer Research*, *17*(March), 492–507. doi:10.1086/208573

Hodge, J. (2005). Tariff structures and access substitution of mobile cellular for fixed-line in South Africa. *Telecommunications Policy*, *29*(7), 493–505. doi:10.1016/j.telpol.2005.05.001

Hoff, J. (2000). Technology and social change: The path between technological determinism, social constructivism and new institutionalism . In Hoff, J., Horrocks, I., & Tops, P. (Eds.), *Democratic governance and new technology: Technologically mediated innovations in political practice in Western Europe*. London: Routledge.

Hofstede G. & Hofstede. (2005). *Cultures and organizations. Software of the mind*. London: McGraw Hill.

Hogg, M. K., & Michell, P. C. N. (1996). Identity, self and consumption: A conceptual framework. *Journal of Marketing Management*, *12*(7), 629–644. doi:10.1080/0267257X.1996.9964441

Holak, S. L., & Lehman, D. R. (1990). Intentions and the dimensions of innovation: An exploratory model. *Journal of Product Innovation Management*, *7*(1), 59–73. doi:10.1016/0737-6782(90)90032-A

Holden, S. H., Norris, D. F., & Fletcher, P. D. (2003). Electronic government at the local level: Progress to date and future issues. *Public Performance & Management Review*, *26*(4), 325–344. doi:10.1177/1530957603026004002

Hollander, E. P. (1993). Legitimacy, power, and influence: a perspective on relational features of leadership . In Chemers, M. M., & Ayman, R. (Eds.), *Leadership Theory and Research: Perspectives and Directions* (pp. 29–47). Orlando, FL: Academic Pres.

Holliday, I., & Yep, R. (2005). E-government in China. *Public Administration and Development*, *25*(3), 239–249. doi:10.1002/pad.361

Holmes, D. (2005). *Communication Theory. Media, Technology and Society*. London: Sage.

Honold, P. (1999). Learning how to use a cellular phone: Comparison between German and Chinese users. *Technical Communication*, *46*(2), 196–205.

Hood, C. (1991). A Public Management for All Seasons? *Public Administration*, *69*(1), 3–19. doi:10.1111/j.1467-9299.1991.tb00779.x

Horn-Ross, P. L., Ljung, B., & Morrow, M. (1997). Environmental factors and the risk of salivary gland cancer. *Epidemiology (Cambridge, Mass.)*, *8*, 414–419. doi:10.1097/00001648-199707000-00011

Hoyer, W. D., & MacInnis, D. J. (2001). *Consumer Behaviour*. Boston: Houghton Mifflin Company.

HPA. (2008). Exposures from Base Stations. *Mobile Telephony and Health.* Health Protection Agency. Retrieved on 2008-01-31.

Huang, C., Fisher, N., & Broyd, T. (2002, September). Development of a triangular TPC model to support adoption of construction integrated system. In D. Greenwood (Ed.), *Proceedings of ARCOM 18th Annual Conference* (Vol. 2, pp. 427-438). University of Northumbria, UK

Hughes, N., & Lonie, S. (2007). M-PESA: mobile learning for the "uneducated" in Kenya. *Innovations: Technology, Governance, Globalization*, *2*(1-2), 63–81. doi:10.1162/itgg.2007.2.1-2.63

Hulme, M., & Peters, S. (2003). *Me, My phone and I: The Role of The Mobile Phone*. 3GSM World Congress Cannes: Teleconomy Research House.

Hung, C. S., Anderson, C., Horne, J. A., & McEvoy, P. (2007). Mobile phone 'talk-mode' signal delays EEG-determined sleep onset. *Neuroscience Letters*, *421*(1), 82–86. doi:10.1016/j.neulet.2007.05.027

Hunt, J., & Manning, P. K. (1991). The Social Context of Police Lying. *Symbolic Interaction*, *14*, 51–70. doi:10.1525/si.1991.14.1.51

Hussain, K., Al-Shakarchi, J., Mahmoudi, A., Al-Mawlawi, A., & Marshall, T. (2006). Mobile phones and driving: a follow-up. *Journal of Public Health*, *28*(4), 395–396. doi:10.1093/pubmed/fdl057

Hutter, H. P., Moshammer, H., Wallner, P., & Kundi, M. (2006). Subjective symptoms, sleeping problems, and cognitive performance in subjects living near mobile phone base stations. *Occupational and Environmental Medicine*, *63*(5), 307–313. doi:10.1136/oem.2005.020784

Ibarguen, G. (2003). Liberating the radio spectrum in Guatemala. *Telecommunications Policy*, *27*, 543–554. doi:10.1016/S0308-5961(03)00050-8

ICNIRP International Commission on Non-Ionizing Radiation Protection. (1998). Guidelines For Limiting Exposure To Time-Varying Electric, Magnetic, And Electromagnetic Fields (up to 300 GHz), International Commission on Non-Ionizing Radiation Protection. *Health Physics*, *74*(4), 494–522.

IGW. (2007, December 7) InterGovWorld. The top 10 developments in the world of e-commerce. Retrieved April 2009 from http://www.intergovworld.com/article/badbf45b0a0104080180bfe1b7c9af9b/pg1.htm

Indiapressagency.com. (2006, July 18). *Indian SMEs shining bright: Survey*. Retrieved July 2009 from http://indiabusinessweek.com/News/Economy/sme.html

Inskip, P. D., Tarone, R. E., & Hatch, E. E. (2001). Cellular telephone use and brain tumors. *The New England Journal of Medicine*, *344*, 79–86. doi:10.1056/NEJM200101113440201

International Country Risk Guide. (2006). *The ICRG Researchers Dataset*. The PRS Group, Inc.

International Telecommunications Union. (2003). *Mobile Overtakes Fixed: Implications for Policy and Regulation*. Retrieved April 15, 2009, from http://www.itu.int/osg/spu/ni/mobileovertakes/Resources/Mobileovertakes_Paper.pdf

Ireland, R. D., Hitt, M. A., & Vaidyanath, D. (2002). Alliance management as a source of competitive advantage. *Journal of Management*, *28*(3), 413–446. doi:10.1177/014920630202800308

ITP.NET. (2008). Mashreq Bank today launched what it says is the most rich mobile banking service in the region, November 10. Retrieved September 2009 from http://www.itp.net/537746-mashreq-launches-mobile-banking

Ivatury, G., & Pickens, M. (2006) Mobile phone learning: evidence from South Africa, Consultative Group to Assist the Poor (CGAP). The World Bank and United Nations Foundation, Washington, D.C. Retrieved May 14, 2009 from http://www.cgap.org

Iyer, R., & Muncy, J. A. (2009). Purpose and object of anti consumption. *Journal of Business Research*, *62*(2), 160–168. doi:10.1016/j.jbusres.2008.01.023

Jacobs, H. H. (1976). Effects of alcohol on sustained attention in the presence of competing stimuli. *Dissertation Abstracts International*, 37, 3113.

Jacobson, T., & Servaes, J. (Eds.). (1999). *Theoretical Approaches to Participatory Communication*. Cresskill NJ: Hampton.

Jacoby, J., Berning, C. K., & Dietvorst, T. F. (1977, April). What About Disposition? *Journal of Marketing*, 22–28. doi:10.2307/1250630

Jain, D., & Rao, R. C. (1990). Effect of price on the demand for durables. *Journal of Business & Economic Statistics*, *8*, 163–170. doi:10.2307/1391978

Jalava, J., & Pohjola, M. (2007). ICT as a source of output and productivity growth in Finland. *Telecommunications Policy*, *31*, 463–472. doi:10.1016/j.telpol.2007.05.011

James, J., & Versteeg, M. (2007). Mobile phones in Africa: How much do we really know? *Social Indicators Research*, *84*, 117–126. doi:10.1007/s11205-006-9079-x

Jang, S.-L., Dai, S.-C., & Sung, S. (2005). The pattern and externality effect of diffusion of mobile telecommunications: The case of the OECD and Taiwan. *Information Economics and Policy*, *17*, 133–148. doi:10.1016/j.infoecopol.2004.05.001

Janssen, W., & Tenkink, E. (1988). Risk Homeostatis theory and its critics: time for an agreement. *Ergonomics*, *31*(4), 429–433. doi:10.1080/00140138808966689

Jarzabkowski, P., Balogun, J., & Seidl, D. (2007). Startegizing: the challenges of practice perspective. *Human Relations*, *60*(1), 5–27. doi:10.1177/0018726707075703

Jaselskis, E. J., & El-Misalami, T. (2003). Implementing radio frequency identification in the construction process. *Journal of Construction Engineering and Management*, *129*(6), 680–688. doi:10.1061/(ASCE)0733-9364(2003)129:6(680)

Jensen, M. (2002). The African Internet: A Status Report. Retrieved June 1, 2009 from http://www3.sn.apc.org/Africa/afstat.htm

Jentoft, S. (2000). Legitimacy and disappointment in fisheries management. *Marine Policy*, *24*(2), 141–148. doi:10.1016/S0308-597X(99)00025-1

Jessop, G. (2008). Who's on the line? Policing and enforcing laws relating to mobile phone use while driving. *International Journal of Law, Crime and Justice*, *36*(3), 135–152. doi:10.1016/j.ijlcj.2008.03.001

Jha, R., & Majumdar, S. K. (1999). A matter of connections: OECD telecommunications sector productivity and the role of cellular technology diffusion. *Information Economics and Policy*, *11*, 243–269. doi:10.1016/S0167-6245(99)00017-7

Jiang, Y. C. (2008). *Design and development of mobile game in learning model*. Unpublished master's thesis, East China Normal University, Shanghai, China.

Johansen, C., Boice, J. D. Jr, McLaughlin, J. K., & Olsen, J. H. (2001). Cellular telephones and cancer – a nationwide cohort study in Denmark. *Journal of the National Cancer Institute*, *93*, 203–207. doi:10.1093/jnci/93.3.203

Johansson, O. (2006). Electrohypersensitivity: state-of-the-art of a functional impairment. *Electromagnetic Biology and Medicine*, *25*, 245–255. doi:10.1080/15368370601044150

Jorgenson, D., & Nomura, K. (2005). The industry origins of Japanese economic growth. *Journal of International Economics*, *19*, 482–542.

Jost, J. T., Banaji, M. R., & Nosek, B. A. (2004). A decade of system justification theory: Accumulated evidence of conscious and unconscious bolstering of the status quo. *Political Psychology*, *25*, 881–919. doi:10.1111/j.1467-9221.2004.00402.x

Juniu, S. (2003). Implementing wireless technology in the classroom: the iPAQ project. *College Planning and Management*, *6*(10), 38–40.

Kale, P., Singh, H., & Perlmutter, H. (2000). Learning and protection of proprietary assets in strategic alliances: Building relational capital. *Strategic Management Journal*, *21*(3), 217–237. doi:10.1002/(SICI)1097-0266(200003)21:3<217::AID-SMJ95>3.0.CO;2-Y

Kalish, S. (1985). A new product adoption model with price, advertising, and uncertainty. *Management Science*, *31*, 1569–1585. doi:10.1287/mnsc.31.12.1569

Katranuschkov, P., Scherer, R. J., & Turk, Z. (2001). Intelligent services and tools for concurrent engineering – An approach towards the next generation of collaboration platforms. *ITcon*, *6*, 111–128.

Katz, J. (2005). Mobile Communication and the Transformation of Daily Life. In P. Glotz, S. Bertschi, & C. Locke (Eds.), *Thumb Culture: The Meaning of Mobile Phones for Society* (pp. 171-184). Transcript, Verlag, Bielefeld.

Katz, J. E. (2003). *Machines that Become Us: The Social Context of Personal Communication Technology*. NJ: Transaction, New Brunswick.

Katz, J. E., & Sugiyama, S. (2006). Mobile phones as fashion statements: evidence from student surveys in the US and Japan. *New Media & Society*, *8*(2), 321–337. doi:10.1177/1461444806061950

Katz, J., & Aakhus, M. (2002). *Perpetual contact: Mobile communication, private talk, public performance*. Cambridge: Cambridge University Press. doi:10.1017/CBO9780511489471

Katz, M. L., & Shapiro, C. (1985). Network effects, competition, and compatibility. *The American Economic Review*, *75*(3), 424–440.

Keegan, D. (2003). *The future of learning: From eLearning to mLearning.* Hagen: Femstudienforchung, Germany. Retrieved May 11, 2009 from http://learning.ericsson.net/mlearning2/old_sites/book.html

Kemper, T. D. (1966). Representative Roles and the Legitimation of Deviance. *Social Problems*, *13*(3), 288–298. doi:10.1525/sp.1966.13.3.03a00040

Kenis, P., & Knoke, D. (2002). How organizational field networks shape interorganizationaltie-formation rates. *Academy of Management Review*, *27*(2), 275–293. doi:10.2307/4134355

Kerr, D. (2003). "We Know What the Problem Is": Using Oral History to Develop a Collaborative Analysis of Homelessness from the Bottom Up. *The Oral History Review*, *30*(1), 27–45. doi:10.1525/ohr.2003.30.1.27

Khanna, T., Gulati, R., & Nohria, N. (1998). The dynamics of learning alliances: Competition, cooperation, and relative scope. *Strategic Management Journal*, *19*(3), 193–210. doi:10.1002/(SICI)1097-0266(199803)19:3<193::AID-SMJ949>3.0.CO;2-C

Khor, M. (2000). *Globalization and the South. Some critical issues*. Penang, Malayasia: Third World Network.

Kim, J., Lee, D., & Ahn, J. (2006). A dynamic competition analysis on the Korean mobile phone market using competitive diffusion model. *Computers & Industrial Engineering*, *51*(1), 174–182. doi:10.1016/j.cie.2006.07.009

Kim, M.-S., & Kim, H. (2004). Innovation diffusion of telecommunications: General patterns, diffusion clusters and differences by technological attribute. *International Journal of Innovation Management*, *8*(2), 223–241. doi:10.1142/S136391960400099X

Kim, M.-S., & Kim, H. (2007). Is there early take-off phenomenon in diffusion of IP-based telecommunications services? *The International Journal of Management Science*, *35*, 727–739.

Kim, N., Bridges, E., & Srivastava, R. K. (1999). A simultaneous model for innovative product category sales diffusion and competitive dynamics. *International Journal of Research in Marketing*, *16*, 95–111. doi:10.1016/S0167-8116(98)00026-3

Kim, S. H., Mims, C., & Holmes, K. P. (2006). An introduction to current trends and benefits of mobile wireless technology use in higher education. *AACE Journal, 14*(1), 77–100.

Kim, Y. Y. (2005). Inquiry in intercultural and development communication . *The Journal of Communication, 55*(3), 554–577. doi:10.1111/j.1460-2466.2005.tb02685.x

Kimoto, K., Endo, K., Iwashita, S., & Fujiwara, M. (2005). The application of PDA as mobile computing system on construction management. *Automation in Construction, 14*, 500–511. doi:10.1016/j.autcon.2004.09.003

Klein, H. (1999). Tocqueville in cyberspace: Using the internet for citizen associations. *The Information Society, 15*(4), 213–220. doi:10.1080/019722499128376

Knights, D., & McCabe, D. (2000). Ain't misbehavin? Opportunities for resistance under new forms of "quality" management. *Sociology, 34*(3), 421–436.

Kolomytkin, O., Kuznetsov, V., Yurinska, M., Zharikova, A., & Zharikov, S. (1994). Response of brain receptor systems to microwave energy exposure . In Frey, A. (Ed.), *On the Nature of Electromagnetic Field Interactions with Biological Systems* (pp. 195–206). Austin, TX: R.G. Landes Co.

Kondratova, I. (2004). Voice and multimodal technology for the mobile worker. *ITcon, 9*, 345–353.

Kopomaa, T. (2000). *City in your pocket: Birth of the mobile information society*. Helsinki: Gaudeamus.

Kosasih, D., Robinson, R. & Snell, J. (1987). *A review of some recent geometric road standards and their application to developing countries*. TRRL, Research Report 114. Crowthorne: Transport and Road Research Laboratory.

Koseoglu, O. O., & Bouchlaghem, D. (2006, November). Design and construction collaboration using mobile visualization. Vol. II, *Proceedings of INCITE / ITCSED 2006 Conference* (Vol. 2, pp. 201-218) New Delhi, India.

Koski, H., & Kretschmer, T. (2005). Entry, standards and competition: Firm strategies and the diffusion of mobile telephony. *Review of Industrial Organization, 26*, 89–113. doi:10.1007/s11151-004-4085-0

Kostopoulos, G. (2003). E-government in the Arabian Gulf: a vision toward reality. In *ACM International Conference Proceeding Series; Vol. 130, Proceedings of the 2003 annual national conference on Digital government research* (pp. 1-7). Boston, MA.

Koyu, A., Fehmi Ozguner, F., Cesur, G., Gokalp, O., Mollaoglu, H., Caliskan, S., & Delibas, N. (2005). No effects of 900 MHz and 1800 MHz electromagnetic field emitted from cellular phone on nocturnal serum melatonin levels in rats. *Toxicology and Industrial Health, 21*(10), 27–31. doi:10.1191/0748233705th212oa

Kozinets, R. V., & Handelman, J. M. (2004). Adversaries of consumption: Consumer movements, activism, and ideology. *The Journal of Consumer Research, 31*(3), 691–704. doi:10.1086/425104

Kreutzer, T. (2008). Assessing Cell Phone Usage in a South African Township School. Paper presented at e/merge 2008, July 7 - 18, 2008 in Cape Town, South Africa Retrieved April 2009 from http://emerge2008.net/access/metaobj/content/group/emerge2008/Presentations/opening%20worlds/Mobile%20Technology%20Usage

Kreutzer, T. (2009). *Generation mobile: online and digital media usage on mobile phones among low-income urban youth in South Africa*. Cape Town: University of Cape Town.

Kubey, R. W., Lavin, M. J., & Barrows, J. R. (2001). Internet use and collegiate academic performance decrements: early findings. *The Journal of Communication, 51*(2), 366. doi:10.1111/j.1460-2466.2001.tb02885.x

Kukulska-Hulme, A., & Traxler, J. (Eds.). (2005). *Mobile Learning: A Handbook for Educators and Trainers* (F. Lockwood, Ed.). London: Routledge.

Kukulska-Hulme, A., Evans, D., & Traxler, J. (2005). *Landscape study in wireless and mobile learning in the post-16 sector*. Bristol: Joint Information Systems Committee.

Kuladinithi, K., Timm-Giel, A., & Gorg, C. (2004). Mobile ad-hoc communications in AEC industry. *ITcon*, *9*, 313–323.

Kumar, R. R., De, J. B., & Radhakrishna, K. (2006, November). Human resource development in construction industry through information technology (IT). In *Proceedings of INCITE / ITCSED 2006 Conference* (Vol. 3, pp. 303-326), New Delhi, India.

Kumas, E. (2007). *E-devlet kapısı ve risk değerlendirme metodolojisi. 12.* Ankara: Turkiye'de Internet Konferansi.

Kuran, N. H. (2005). *Türkiye icin e-devlet modeli*. Istanbul, Turkey: Bilgi Universitesi Yayınları.

Kuribayashi, (2007). Lack of effects of 1439 MHz electromagnetic near field exposure on the blood-brain barrier in immature and young rats. *Bioelectromagnetics*, *26*(7), 578–588. doi:10.1002/bem.20138

Kushchu, I. (2007). *Mobile government: An emerging direction in e-government*. Hershey: IGI Publishing.

Kushchu, I., & Kuscu, H. (2003). From e-government to m-government: Facing the inevitable. In *Proceedings of European Conference on E-Government (ECEG 2003), Trinity College, Dublin, July 3-4* (pp. 253-260). Reading, UK: Academic Conferences International.

Kushla-Hulme, A., & Traxler, J. (2007). Designing for mobile and wireless learning . In Beetham, H., & Sharpe, R. (Eds.), *Rethinking Pedagogy for a digital age: Designing and delivering e-learning*. London: Routledge.

Kvale, S. (1996). *Inter Views: An introduction to qualitative research interviewing*. Thousand Oaks, CA: Sage.

Kvavik, R. B. (2005). *Convenience, communications, and control: how students use technology*. Retrieved July 18, 2009 from www.educause.edu/ConvenienceCommunicationsandControl%3AHowStudentsUseTechnology/6070

LaBrasca, S. (2009, July). E-Learning without boarders. *Learning Circuits*. Retrieved July 20, 2009 from http://www.astd.org/LC/2009/0709_labrasca.htm

Laclau, E., & Mouffe, C. (1985). *Hegemony and socialist strategy*. London: Verso.

Laclau, E., & Mouffe, C. (2001). *Hegemony and socialist strategy: Towards a radical democratic politics* (2nd ed.). London: Verso.

Lacoh'ee, H., & Wakeford, N., I., P. (2003). A social history of the mobile telephone with a view of its future. *BT Technology Journal*, *21*(3), 203–211. doi:10.1023/A:1025187821567

Lahkola, A., Auvinen, A., Raitanen, J., Schoemaker, M. J., Christensen, H. C., & Feychting, M. (2007). Mobile phone use and risk of glioma in 5 North European countries. *International Journal of Cancer*, *120*(8), 1769–1775. doi:10.1002/ijc.22503

Lai, C.-H., Yan, J. C., Ho, C. W., & Chan, T. W. (2007). Affordances of mobile technologies for experiential learning: the interplay of technology and pedagogical practices. *Journal of Computer Assisted Learning*, *23*, 326–337. doi:10.1111/j.1365-2729.2007.00237.x

Lai, H., & Singh, N. P. (1995). Acute low-intensity microwave exposure increases DNA single-strand breaks in rat brain cells. *Bioelectromagnetics*, *16*, 204–210. doi:10.1002/bem.2250160309

Lai, H., Carino, M., Horita, A., & Guy, A. (1992). Opioid receptor subtypes that mediate a microwave induced decrease in central cholinergic activity in the rat. *Bioelectromagnetics*, *13*, 237–246. doi:10.1002/bem.2250130308

Lai, H., Horita, A., & Guy, A. (1994). Microwave irradiation affects radial-arm maze performance in the rat. *Bioelectromagnetics*, *15*, 95–104. doi:10.1002/bem.2250150202

Lajunen, T., & Räsänen, M. (2004). Can social psychological models be used to promote bicycle helmet use among teenagers? A comparison of the Health Belief Model, Theory of Planned Behaviour and the Locus of Control. *Journal of Safety Research, 35*, 115–123. doi:10.1016/j.jsr.2003.09.020

Laouris, Y., & Eteokleous, N. (2005). *We need an educationally relevant definition of mobile learning*. Paper presented at mLearn 2005, the 4th World Conference on Mobile Learning. Cape Town, South Africa. Retrieved June 6, 2009, from http://www.mlearn.org.za/CD/papers/Laouris%20&%20Eteokleous.pdf

Lapointe, L., & Rivard, S. (2003). A multilevel model of resistance to information technology implementation. Retrieved from http://cat.inist.fr/?aModele=afficheN&cpsidt=15877245

Larsen & Toubro Limited (n.d.). *Information Systems*. Retrieved October 2009 from lntecc.com/html/information_systems.htm

Larsen, E., & Rainie, L. (2002). The rise of the e-citizen: How people use e-government agencies' web sites. Washington, DC: Pew Internet & American Life Project, 3.

Laudon, K., & Laudon, J. (2007). *Management Information Systems: Managing the digital firm*. NJ: Prentice Hall.

Law, L. (1997). A matter of choice: discourses on prostitution in the Philippines . In Manderson, L., & Jolly, M. (Eds.), *Sites of Desire/Economies of Pleasure: Sexualities in Asia and the Pacific* (pp. 233–261). Chicago: University of Chicago Press.

Lazarus, R. S., & Folkman, S. (1984). *Stress, appraisal, and coping*. New York: Springer.

Lee, K. S. (2006). More youngsters using handphones. *New Straits Times, 25*, 6.

Lee, M. S. W., Fernadez, K. V., & Hyman, M. R. (2009a). Anti-consumption: An overview and research agenda. *Journal of Business Research, 62*, 145–147. doi:10.1016/j.jbusres.2008.01.021

Lee, M. S. W., Motion, J., & Conroy, D. (2009b). Anti-consumption and brand avoidance. *Journal of Business Research, 62*, 169–180. doi:10.1016/j.jbusres.2008.01.024

Lee, M., & Chan, A. (2007). Pervasive, lifestyle-integrated mobile learning for distance learners: an analysis and unexpected results from a podcasting study. *Open Learning, 22*(5), 201–218.

Lee, N., & Cadogan, J. W. (2009). Sales force social exchange in problem resolution situations. *Industrial Marketing Management, 32*(3), 355–372. doi:10.1016/j.indmarman.2008.02.002

Lee, S., Kim, M., & Park, Y. (2009). ICT co-evolution and Korean ICT strategy: An analysis based on patent data. *Telecommunications Policy, 33*(5-6), 253–271. doi:10.1016/j.telpol.2009.02.004

Lee, T., & Jun, J. (2005). Contextual perceived usefulness? Toward an understanding of mobile commerce acceptance. In *Proceedings of the International Conference on Mobile Business* (ICMB'05)

Leffingwell, D., & Widrig, D. (2003). *Managing software requirements: a use case approach*. Boston: Addison-Wesley.

Lehner, F., & Nosekebel, H. (2002). The role of mobile devices in e-learning – first experience with a e-learning environment. In M. Milrad, H.U. Hoppe & Y. Kinshuk (Eds.), *IEEE International Workshop on Wireless and Mobile Technologies in Education* (pp. 103-106). Los Alamitos, CA: IEEE Computer Society.

Lehr, D. (2007). *Going Wireless for Development: How Mobile Devices are transforming development at the base of the pyramid*. Working paper. Acumen Fund. Retreived July 5, 2009 from Dialling for Development_Working paper_jG2w63Q2.pdf.

Lehto, M. R. (1991). A proposed conceptual model of human behaviour and its implications for design of warning. *Perceptual and Motor Skills, 73*, 595–611. doi:10.2466/PMS.73.5.595-611

Leite, J. C. S. P. (2003). *Perspectives on Software Requirements*. Springer.

Leopoldina, F., & Manganelli, A. M. (2002). *Young People and the Mobile Phone*. Retrieved December 4, 2006, from http://www.itu.int/osg/spu/ni/ubiquitous/Papers/Youth_and_mobile_2002.pdf

Leppaniemi, M., & Karjaluoto, H. (2008). Mobile marketing: from marketing strategy to mobile marketing campaign implementation. *International Journal of Mobile Marketing*, *3*(1), 50–61.

Lerner, D. (1958). *The passing of traditional society. Modernizing the Middle East*. New York: Free Press.

Lerner, D., & Schramm, W. (Eds.). (1967). *Communication and change in the developing countries*. Honolulu, HI: University Press of Hawaii.

Lerner, M. J. (1980). *The belief in a just world: A fundamental delusion*. New York: Plenum Press.

Leung, L., & Wei, R. (2000). More than just talk on the move: Uses and gratifications of the cellular phone. *Journalism & Mass Communication Quarterly*, *77*(2), 308–320.

Li, M. (2005). Mobile government, extending government services (Yidong zhengwu, kuozhan zhengfu fuwu). China Computer User (Zhongguo jisuanji yonghu), 2005(31), 45-46.

Liao, C. (2006). *A comprehensive m-learning platform*. Retrieved April 28, 2009 from http://all.zcom.com/mag2/shehuikexue/wenkejiaoti/30982/200608X/15388980

Lie, R. (2003). *Spaces of Intercultural Communication. An interdisciplinary Introduction to Communication, Culture, and Globalizing/Localizing Identities*. Cresskill, NJ: Hampton Press.

Lim, B.-L., Choi, M., & Park, M.-C. (2003). The late take-off phenomenon in the diffusion of telecommunication services: Network effect and the critical mass. *Information Economics and Policy*, *15*, 537–557. doi:10.1016/S0167-6245(03)00037-4

Lin, A., & Tong, A. (2008). Text-messaging culture of college girls in Hong Kong: SMS as resources for achieving intimacy and gift-exchange with multiple functions . In Goggin, G. (Ed.), *Mobile Phone Cultures* (pp. 158–170). London: Routledge.

Lindlof, T. R., & Taylor, B. C. (2002). *Qualitative communication research methods*. Thousand Oaks, CA: Sage.

Ling, R. (1999). We release them little by little: maturation and gender identity as seen in the use of mobile telephone. *1999 International Symposium on Technology and Society (ISTAS`99*, July 29-31, 1999 Rutgers University, New Brunswick, New Jersey. Retrieved September 2009 from http://www.richardling.com/papers/1999_we_release_them.pdf

Ling, R. (2008). *New tech, new ties: How mobile communication is reshaping social cohesion*. Cambridge, MA: MIT Press.

Ling, R. S. (2004). *Mobile connection: The cell phone's impact on society*. San Francisco: Morgan Kaufmann.

Littlefield, L. (2004). *Psychosocial Aspects of Mobile Phone Use Among Adolescents*. Melbourne: The Australian Psychological Society.

Liu, H. Y. (2007). *The design and development of web-based SMS system for course answering*. Unpublished master's thesis, South China Normal University, Guangzhou, Guangdong, China.

Liu, J. S., Li, Q., & Liu, J. M. (2007). Review of mobile learning. *Journal of E-education Research*, *7*, 21–25.

Liu, T., Wang, H., Liang, J., Chan, T., Ko, H., & Yang, J. (2003). Wireless and mobile technologies to enhance teaching and learning. *Journal of Computer Assisted Learning*, *19*(3), 371–382. doi:10.1046/j.0266-4909.2003.00038.x

Lönn, S., Ahlbom, A., & Hall, P. (2005). Long-term mobile phone use and brain tumor risk. *American Journal of Epidemiology*, *161*(6), 526–535. doi:10.1093/aje/kwi091

Lönn, S., Ahlbom, A., Christensen, H. C., Johansen, C., Schüz, J., & Edström, S. (2006). Mobile Phone Use and Risk of Parotid Gland Tumor. *American Journal of Epidemiology*, *164*, 637–643. doi:10.1093/aje/kwj242

Lönn, S., Ahlbom, A., Hall, P., & Feychting, M. (2004). Mobile phone use and the risk of acoustic neuroma [in English]. *Epidemiology (Cambridge, Mass.)*, *15*(6), 653–659. doi:10.1097/01.ede.0000142519.00772.bf

Lorencova, V. (2009). YouTube dilemmas: The appropriation of user-generated online videos in teaching learning. *Current in Teaching and Learning*, *2*, 62–71.

Lotka, A. J. (1925). *Elements of Physical Biology*. Baltimore: Williams and Wilkins.

Lotka, A. J. (1956). *Elements of Physical Biology. 1924, republished as*. Elements of Mathematical.

Lowe, V. (1987). *Dependency within Bounds: Media and Information Technology policies within the ASEAN region*. Bangkok: Institute of Asian Studies, Chulalongkorn University.

Lu, J., & Weber, I. (2007). State, power and mobile communication: A case study of China. *New Media & Society*, *9*(6), 925–944. doi:10.1177/1461444807082640

Lun, M. H. (2007). *Theory study and application design of mobile learning content*. Unpublished master's thesis, Beijing Jiaotong University, Beijing, China.

MacBride, S. (Ed.). (1980). *Many Voices, One World: Communication and Society. Today and Tomorrow*. Paris: UNESCO.

MacDonald, B. (2009). Appropriate Distance Education Media in the Philippines and Mongolia. *International Review of Research in Open and Distance Learning*, *10*(2).

Madden, G. & Savage, S.J. (2000). Telecommunications and economic growth. *International Journal of Social Economics, 21*(7/8/9/10), 893-906.

Madden, G., & Savage, S. J. (1998). CEE telecommunications investment and economic growth. *Information Economics and Policy*, *10*, 173–195. doi:10.1016/S0167-6245(97)00020-6

Madden, G., Coble-Neal, G., & Dalzell, B. (2004). A dynamic model of mobile telephony subscription incorporating a network effect. *Telecommunications Policy*, *28*, 133–144. doi:10.1016/j.telpol.2003.12.002

Maddrell, J. (2008). *The effective of backchannel interactions on cognitive load*. Research paper. Retrieved July 2, 2009, from http://www.scribd.com/doc/3727549/Backchannel-Research-Paper-Jennifer-Maddrell.

Magdic, A., Rebolj, D., & Suman, N. (2004). Effective control of unanticipated on-site events: A pragmatic, human-oriented problem solving approach. *ITcon*, *9*, 409–418.

Mahajan, V., & Peterson, R. (1978). Innovation diffusion in a dynamic potential adopter population . *Management Science*, *24*, 1589–1597. doi:10.1287/mnsc.24.15.1589

Mahajan, V., & Peterson, R. A. (1979). First-purchase diffusion models of new-product acceptance. *Technological Forecasting and Social Change*, *15*, 127–146. doi:10.1016/0040-1625(79)90073-8

Mahajan, V., Muller, E., & Bass, F. M. (1996). Timing diffusion and substitution of successive generations of technological innovations: The IBM Mainframe Case. *Technological Forecasting and Social Change*, *51*, 109–132. doi:10.1016/0040-1625(95)00225-1

Mahajan, V., Muller, E., & Wind, Y. (Eds.). (2000). *New-Product Diffusion Models*. Boston: Kluwer Academic Press.

Mahajan, V., Sharma, S., & Buzzel, R. D. (1993). Assessing the impact of competitive entry on market expansion and incumbent sales. *Journal of Marketing*, *57*, 39–52. doi:10.2307/1251853

Mahler, A., & Rogers, E. M. (1999). The diffusion of interactive communication innovations and the critical mass: The adoption of telecommunications services by German Banks. *Telecommunications Policy, 23*(10-11), 719–740. doi:10.1016/S0308-5961(99)00052-X

Malayapa, R. S., Ahern, E. W., Bi, C., Struabe, W. L., LaRegina, M., Pickard, W. F., & Roti Roti, J. L. (1998). DNA damage in rat brain cells after in vivo exposure to 2450 MHz electromagnetic radiation and various methods of euthanasia. *Radiation Research, 149*, 637. doi:10.2307/3579911

Malaysian Communications and Multimedia Commission. (2009). Facts and figures. Retrieved July 14, 2009 from http://www.skmm.gov.my/facts_figures/stats/ViewStatistic.asp?cc=52036251&srid=50919742

Maliranta, M., & Rouvinen, P. (2004). *Informational mobility and productivity - Finnish evidence. Discussion Papers 919*. The Research Institute of the Finnish Economy.

Malyapa, R. S., Ahern, E. W., & Straube, W. L. (1998). DNA damage in rat brain cells after in vivo exposure to 2450 MHz electromagnetic radiation and various methods of euthanasia. *Radiation Research, 149*, 637–645. doi:10.2307/3579911

Manfield, E. (1961). Technical change and the rate of imitation. *Econometrica, 29*, 741–766. doi:10.2307/1911817

Mankiw, N. G., Romer, D., & Weil, D. N. (1992). A Contribution to the Empirics of Economic Growth. *The Quarterly Journal of Economics, 107*(2), 407–437. doi:10.2307/2118477

Marcelle, G. M. (2004). *Technological learning*. UK: Edward Elgar.

March, J. G., & Olsen, J. P. (1989). *Rediscovering Institutions: The Organizational Foundation of Politics*. New York: Free Press.

March, J., & Olsen, J. P. (2006). Elaborating the "New institutionalism" . In Rhodes, R. A. W., Binder, S. A., & Rockman, B. A. (Eds.), *The Oxford Handbook of Political Institutions*. Oxford: Oxford University Press.

Marker, P., McNamara, K., & Wallace, L. (2002). *The significance of information and communication technologies for reducing poverty*. London: DFID. Retrieved December 18, 2008 from http://www.dfid.gov.uk/pubs/files/ictpoverty.pdf

Markus, M. L. (1990a). Critical mass contingencies for telecommunications consumers . In Carnevale, M., Lucertini, M., & Nicosia, S. (Eds.), *Modeling the innovation: Communications, automation and information* (pp. 103–112). Amsterdam: Elsevier.

Markus, M. L. (1990b). Toward a critical mass theory of interactive media . In Fulk, J., & Steinfield, C. W. (Eds.), *Perspectives on new information technology in organizations*. Newbury Park, CA: Sage.

Marriott, L. (2006). What services are driving mobile internet usage? Retrieved July 18, 2009 from www.clickz.com/experts/ad/mobile/article.php/3606606

Mauro, P. (1995). Corruption and Growth. *The Quarterly Journal of Economics, 110*(3), 681–712. doi:10.2307/2946696

Mayer, R. (2003). The promise of multimedia learning: Using the same instructional design methods across different media. *Learning and Instruction, 13*, 125–139. doi:10.1016/S0959-4752(02)00016-6

McConatha, D., Praul, M., & Lynch, M. (2008). Mobile learning in higher education: an empirical assessment of a new educational tool. *The Turkish Online Journal of Educational Technology, 7*(3), 15–23.

McCullough, B. (1997). Automating field data collection in construction organizations. In [Minneapolis, MN.]. *Proceedings of ASCE Construction Congress, V*, 957–963.

McEvoy, S. P., Stevenson, M. R., McCartt, A. T., Woodward, M., Haworth, C., Palamara, P., et al. (2005). Role of mobile phones in motor vehicle crashes resulting in hospital attendance: a case-crossover study [Electronic Version]. *British Medical Journal*, bmj.38537.397512.397555. Retrieved July 12, 2005 from http://www.bmj.com/cgi/content/abstract/bmj.38537.397512.55v1

McIvor, R., McHugh, M., & Cadden, C. (2002). Internet technologies: Supporting transparency in the public sector. *International Journal of Public Sector Management*, *15*(3), 170–187. doi:10.1108/09513550210423352

McNeely, B. (2005). Using technology as a learning tool, not just the cool new thing. Retrieved July 18, 2009 from www.educause.edu/UsingTechnologyasaLearningTool%2CNotJusttheCoolNewThing/6060

Meade, N., & Islam, T. (2006). Modelling and forecasting the diffusion of innovation – A 25-year review. *International Journal of Forecasting*, *22*, 519–545. doi:10.1016/j.ijforecast.2006.01.005

Medhi, I., Sagar, A., & Toyama, K. (2007). Text-Free User Interfaces for Illiterate and Semi-Literate Users. *Information Technologies and International Development*, *4*(1), 37–50. doi:10.1162/itid.2007.4.1.37

Meissner, A., Mathes, I., Baxevanaki, L., Dore, G., & Branki, C. (2003). The cosmos integrated IT solution at railway and motorway construction sites - Two Case Studies. *ITcon*, *8*, 283–291.

Melitski, J., Holzer, M., Kim, S., Kim, C., & Rho, S. (2005). Digital government worldwide: An e-government assessment of municipal websites throughout the world. *International Journal of E-government Research*, *1*(1), 1–19.

Menzel, K., Keller, M., & Eisenblatter, K. (2004). Context sensitive mobile devices in architecture, engineering and construction. *ITcon*, *9*, 389–407.

Merriam, S. B. (1998). *Qualitative Research and Case Studies Applications in Education*. San Francisco: Jossey-Bass Publications.

Mertha, A. C. (2005). China's "Soft" Centralization: Shifting Tiao/Kuai Authority Relations. *The China Quarterly*, *184*(1), 791–810. doi:10.1017/S0305741005000500

MIIT. (2008). *The increasing numbers of mobile phone users*. Retrieved October 5, 2009 from http://it.people.com.cn/GB/1068/42899/8137255.html

Miles, M. B., & Huberman, A. M. (1994). *Qualitative Data Analysis: An Expanded Source Book*. Thousand Oaks, CA: Sage.

Miller, D. (2007). Mobile Phones and Development: The future in new hands? iD21 insights. Retrieved May 20, 2009, from www.id21.org/insights/insights69/insights69.pdf

Mittelstaedt, R. A., Grossbart, S. L., Curtis, W. W., & Devere, S. P. (1976). Optimal stimulation level and the adoption process. *The Journal of Consumer Research*, *3*, 84–94. doi:10.1086/208655

m-Learning.org. (2009). *Ultralab*. Retrieved July 14, 2009 from http://www.naec.org.uk/ultralab/ww3/projects/mlearning/

Mobile Phones UK. (2008). *Mobile Phones UK web site*. Landmark Internet Ltd. Retrieved on 2008-01-19., of phones from 2005 and earlier at "The Complete SAR List For All Phones (Europe)". On-Line-Net - Web Design & Internet Services (as SARValues.com). Retrieved on 2008-01-19. (a listing of US phones from 2005 and earlier, using the US 1 g standard, is also available at the SARValues site)

Mobilearn (2003). The Mobilearn Project Vision. Retrieved May 3, 2009 from http://www.mobilearn.org/vision/visiton.htm

Modis, T. (1997). Generic re-engineering of corporations. *Technological Forecasting and Social Change*, *56*, 107–118. doi:10.1016/S0040-1625(97)00076-0

Mody, B. (Ed.). (1997). Communication and Development: Beyond Panaceas. *The Journal of International Communication*, *4*(2), 138.

Mody, B. (Ed.). (2003). *International and Development Communication. A 21st Century Perspective*. Thousand Oaks, CA: Sage.

MOE. (2008). *Report on the investment in education*. Retrieved October 4, 2009 from http://www.moe.edu.cn/edoas/website18/84/info1229326947658184.htm

Moisio, R. J. (2003). Negative consequences of mobile phone consumption: Everyday irritations, anxieties and ambiguities in the experiences of Finnish mobile phone consumers. *Advances in Consumer Research. Association for Consumer Research (U. S.), 30*, 340–345.

Moore, D. S. (1997). Remapping resistance: 'Ground for struggle' and the politics of place . In Pile, S., & Keith, M. (Eds.), *Geographies of Resistance* (pp. 87–106). London: Routledge.

Moore, M. H. (1995). *Creating Public Value: Strategic Management in Government*. Cambridge: Harvard University Press.

Moreno, R., & Mayer, R. (2007). Interactive multimodal learning environments. *Educational Psychology Review, 19*, 309–326. doi:10.1007/s10648-007-9047-2

Morris, N. (2001). *Bridging the Gap: An examination of Diffusion and Participatory Approaches in Development Communication*. Washington, DC: The Change Project/USAID.

Motlik, S. (2008). Mobile Learning in Developing Nations. *International Review of Research in Open and Distance Learning, 9*(2), 1–7.

Moulder, J. E., Foster, K. R., Erdreich, L. S., & McNamee, J. P. (2005). Mobile phones, mobile phone base stations and cancer. *International Journal of Radiation Biology, 81*(3), 189–203. doi:10.1080/09553000500091097

Moulder, J.E., Erdreich, L.S., Malyapa, R.S., Merritt, J., & Pickard, W.F., Vijayalaxmi. (1999). Cell phones and cancer: what is the evidence for a connection? *Radiation Research, 151*(5), 513–531. doi:10.2307/3580028

Mulholland, P., Collins, T., & Zdrahal, Z. (2005). Bletchley Park Text: Using mobile and semantic web technologies to support the post-visit use of online museum resources. *Journal of Interactive Media in Education*. Retrieved June 28, 2009 from http://jime.open.ac.uk/2005/24.

Musso, J., Weare, C., & Hale, M. (2000). Designing web technologies for local governance reform: Good management or good democracy. *Political Communication, 17*(1), 1–19. doi:10.1080/105846000198486

Naismith, L., Lonsdale, P., Vavoula, G., & Sharples, M. (2004). *Literature Review in Mobile Technologies and Learning. Future lab Series, Report 11*. University of Birmingham.

Naismith, L., Lonsdale, P., Vavoula, G., & Sharples, M. (2004). *Literature review in mobile technologies and learning*. Bristol, UK: NESTA Futurelab. Retrieved March 3, 2009, from http://elearning.typepad.com/thelearnedman/mobile_learning/reports/futurelab_review_11.pdf

Natavidad, Sc. D. (n.d.) Summative evaluation of The ELSA text2teach project: Final report. Retrieved July 2, 2009, from pdf.usaid.gov/pdf_docs/PDACK830.pdf.

National Bureau of Statistics of China. (2000). *The Fifth National Census*. Retrieved October 4, 2009 from http://www.stats.gov.cn/tjgb/rkpcgb/qgrkpcgb/t20020331_15434.htm

National Health Service. (2008). Mobile effect on sleep. *Behind the Headlines*. National Health Service (2008-01-21). Retrieved on 2008-01-22.

Navarro, E. A., Segura, J., Portolés, M., & Gómez-Perretta de Mateo, C. (2003). The Microwave Syndrome: A Preliminary Study in Spain. *Electromagnetic Biology and Medicine, 22*(2), 161–169. doi:10.1081/JBC-120024625

NetworkComputing (2009). *A wireless umbrella for Delhi airport*. Retrieved July 2009 from http://www.networkcomputing.in/LT-ECC-Division-EDGE-001Jun009.aspx

Neubauer, (2007). Feasibility of future epidemiological studies on possible health effects of mobile phone base stations. *Bioelectromagnetics, 28*(3), 224–230. doi:10.1002/bem.20298

News, B. B. C. (2004a, June). Mobile usage shows gender split. Retrieved September 2009 from http://news.bbc.co.uk/2/hi/technology/3766643.stm

News, B. B. C. (2004b, May). Monthly spend on mobiles rises. Retrieved September 2009 from http://news.bbc.co.uk/2/hi/technology/3749739.stm

Nielsen, J. R. (2003). An analytical framework for studying: compliance and legitimacy in fisheries management. *Marine Policy*, *27*, 425–432. doi:10.1016/S0308-597X(03)00022-8

Norton, S. W. (1992). Transaction costs, telecommunications, and the microeconomics of macroeconomic growth. *Economic Development and Cultural Change*, *41*(1), 175–196. doi:10.1086/452002

NRPB (National Radiological Protection Board). (2003). *Health effects from radiofrequency electromagnetic fields: report of an independent advisory group on non-ionising radiation*. Chilton, UK: National Radiological Protection Board (Health Protection Agency), 2003. Retrieved from http://www. hpa.org.uk /radiation /publications /documents_ of_ nrpb /pdfs /doc_14_2.pdf

Nyiri, K. (2002). *Towards a philosophy of m-learning*. IEEE International Workshop on Wireless and Mobile Technologies in Education, August 29-30, 2002, Teleborg Campus.

Oblinger, D., & Oblinger, J. (2005). *Is it age or IT: first steps toward understanding the net generation*. Retrieved July 18, 2009 from www.educause.edu/IsItAgeorIT%3AFirstStepsTowardUnderstandingtheNetGeneration/6058

Omoto, A. (Ed.). (2005). *Processes of Community Change and Social Action*. Mahwah, NJ: LEA.

Orlena, C. Z., & Finkelstein, A. C. W. (1994). An Analysis of the Requirements Traceability Problem. In *Proceedings of 1st International Conference on Requirements Engineering* (pp. 94-101). IEEE Computer Society Press.

Ostrom, E. (2000). Collective action and the evolution of social norms . *The Journal of Economic Perspectives*, *14*(3), 137–158. doi:10.1257/jep.14.3.137

Oxford Internet Institute (2006). Breaking barriers to e-government online.

Ozcan, Y. Z., & Koçak, A. (2003). Research Note: A Need or a Status Symbol? Use of Cellular Telephones in Turkey. *European Journal of Communication*, *18*(2), 241–254. doi:10.1177/0267323103018002004

Pagani, M. (2004). Determinant of Adoption of Third Generation Mobile Multimedia Services. *Journal of Interactive Marketing*, *18*(3), 46–59. doi:10.1002/dir.20011

Palen, L., Salzman, M., & Youngs, E. (2001). Discovery and integration of mobile communications in everyday life. *Personal and Ubiquitous Computing*, *5*(2), 109–122. doi:10.1007/s007790170014

Palmer, M., & O'Kane, P. (2007). Strategy as practice: Interactive governance spaces and the corporate strategies of retail TNCs. *Journal of Economic Geography*, *7*, 515–535. doi:10.1093/jeg/lbm015

Palmer, M., Owens, M., & De Kervenoael, R. (in press). Paths of the least resistance: Understanding how motives form in international retail joint venturing. *The Service Industries Journal, 30*(8).

Panagopoulos, D. J., Chavdoula, E. D., Karabarbounis, A., & Margaritis, L. H. (2007). Comparison of bioactivity between GSM 900 MHz and DCS 1800 MHz Mobile Telephony Radiation. *Electromagnetic Biology and Medicine*, *26*(1), 33–44. doi:10.1080/15368370701205644

Panagopoulos, D. J., Karabarbounis, A., & Margaritis, L. H. (2004). Effect of GSM 900 MHz mobile phone radiation on the reproductive capacity of Drosophila melanogaster. *Electromagnetic Biology and Medicine*, *23*(1), 29–43. doi:10.1081/JBC-120039350

Papa, M., Singhal, A., & Papa, W. (2006). *Organizing for Social Change. A dialectic journey of theory and praxis*. New Delhi, India: Sage.

Paranavitana, S. (1956). *Sigiri graffiti: Being Sinhalese verses of the Eighth, Ninth and Tenth centuries (archaeological survey of Ceylon)*. Oxford University Press.

Peansupap, V., & Walker, D. (2005). Factors affecting ICT diffusion: A case study of three large Australian construction contractors. *Engineering, Construction, and Architectural Management*, *12*(1), 21–37. doi:10.1108/09699980510576871

Peters, B. G. (2005). *Institutional Theory in Political Science: The 'New Institutionalism* (2nd ed.). London: Continuum.

Piacentini, M., & Banister, E. M. (2009). Managing anti-consumption in excessive drinking culture. *Journal of Business Research, 62*, 279–288. doi:10.1016/j.jbusres.2008.01.035

Pieters, R. (1993). Consumers and their garbage a framework and some experiences from the Netherlands with garbage separation programs. *European Advance in Consumer Research, 1*, 541–546.

Pile, S., & Keith, M. (1997). *Geography of resitance*. London: Routledge.

Pinkwart, N., Hoppe, H. U., Milrad, M., & Perez, J. (2003). Educational scenarios for the cooperative use of Personal Digital Assistants. *Journal of Computer Assisted Learning, 19*(3), 383–391. doi:10.1046/j.0266-4909.2003.00039.x

Pistorius, C. W. I., & Utterback, J. M. (1997). Multi-mode interaction among technologies. *Research Policy, 26*, 67–84. doi:10.1016/S0048-7333(96)00916-X

Pitula, K., & Radhakrishnan, T. (2008). A Multimedia Tool to elicit Information Needs in Rural Communities. In *Proceedings of HCI for Community and International Development Workshop at CHI 2008*, Florence, Italy.

Plant, S. (2002). On the mobile, the effects of mobile telephones on social and individual life. Report for Motorola. Retrieved April 2009 from http://www.motorola.com/mot/doc/0/267_MotDoc.pdf

Pollock, S. (2003). *Literary cultures in history: reconstructions from South Asia*. University of California Press.

Pondy, L. R. (1966). A systems Theory of Organizational Conflicts. *Academy of Management Journal, 9*(3), 296–320. doi:10.2307/255122

Porter, A. L., Roper, A. T., Mason, T. W., Rossini, F. A., & Banks, J. (1991). *Forecasting and Management of Technology*. New York: Wiley.

Prasad, P., & Prasad, A. (2000). Stretching the iron cage: The constitution and implications of routine workplace resistance. *Organization Science, 11*(4), 387–403. doi:10.1287/orsc.11.4.387.14597

Prensky, M. (2001). Digital Natives, Digital Immigrants. *On the Horizon, 9*(5). Retrieved May, 25 2009, from http://www.marcprensky.com/writing/default.asp

Pressman, R. S. (2005). *Software engineering: A practitioner's approach*. Boston: McGraw-Hill.

Preston-Martin, S., Thomas, D. C., & White, S. C. (1988). Prior exposure to medical and dental x-rays related to tumors of the parotid gland. *Journal of the National Cancer Institute, 80*, 943–949. doi:10.1093/jnci/80.12.943

Prochaska, J. O., DiClemente, C. C., & Norcross, J. C. (1992). In search of how people change: Applications to addictive behaviors. *The American Psychologist, 47*(9), 1102–1114. doi:10.1037/0003-066X.47.9.1102

Pullchino, J. (2006). Mobile Learning Research Report. *The E-Learning Guild*. Retrieved July 10, 2009 from http://www.elearningguild.com/

QIA. (2008). *Mobile learning*. Retrieved June 28, 2009, from http://www.excellencegateway.org.uk/page.aspx?o=ferl.aclearn.page.id958

Qiang, C. Z.-W. (2007). *China's Information Revolution: Managing the Economic and Social Transformation*. Washington, D.C.: The World Bank. doi:10.1596/978-0-8213-6720-9

Quinn, C. (2000). M-Learning. Mobile, Wireless, In-Your-Pocket Learning. *Linezine*. Retrieved May 14, 2009 from http://www.linezine.com/2.1/features/cqmmwiyp.htm

Radiofrequency electromagnetic field exposure and non-specific symptoms of ill health: A systematic review. (n.d.). Röösli, Environmental Research. Available online 21 March 2008 at http://dx.doi.org/10.1016/j.envres.2008.02.003

Rahim, S., & Pennings, A. (1987). *Computerization and Development in Southeast Asia*. Singapore: Asian Mass Communication and Information Center (AMIC).

Ramaley, J., & Zia, L. (2005). The real versus the possible: closing the gaps in engagement and learning. Retrieved July 18, 2009 from www.educause.edu/TheRealVersusthePossible%3AClosingtheGapsinEngagementandLearning/6064

Randall, E. (2008). Mobile Technology Changes the Social Hierarchy. Is the text message more important than the messenger? Retrieved April 2009 from http://educationalissues.suite101.com/article.cfm/florida_can_be_flexible_about_education_outcomes

Rebolj, D., & Menzel, K. (2004). Mobile computing in construction. *ITcon*, *9*, 281–283.

Rebolj, D., Magdic, A., & Cus-Babic, N. (2000). *Mobile computing in construction*. Slovenia: Maribor.

Redelmeier, D. A., & Tibshirani, R. J. (1997). Association between cellular telephone calls and motor vehicle collisions. *The New England Journal of Medicine*, *336*, 453–458. doi:10.1056/NEJM199702133360701

Regel, (2006). UMTS Base Station-Like Exposure, Well Being and Cognitive Performance. *Environmental Health Perspectives*, *114*(8). doi:10.1289/ehp.8934

Reinhardt, J., Garrett, J. H. Jr, & Akinci, B. (2005). Framework for providing customized data representations for effective and efficient interaction with mobile computing solutions on construction sites. *Journal of Computing in Civil Engineering*, *19*(2), 109–118. doi:10.1061/(ASCE)0887-3801(2005)19:2(109)

Repacholi, M. H. (1997). Radiofrequency field exposure and cancer: What do the laboratory studies suggest? *Environmental Health Perspectives*, *105*(suppl 6), 7565–1568. doi:10.2307/3433671

Rettie, R. (2007). Mobile Phones as Network Capital: Facilitating Connections. *Mobilities*, *3*(2), 291–311. doi:10.1080/17450100802095346

Rezwan (2009, August 25). India: A wave of suicides among farmers. *Global Voices Online*. Retrieved September 12, 2009 from http://globalvoicesonline.org/2009/08/25/india-a-wave-of-suicides-among-farmers/

Rheingold, H. (2002). *Smart Mobs: the Next Social Revolution*. New York: Perseus.

Ribeiro, F. L., & Lopes, J. (2002, September). An approach to e-business in construction. In D. Greenwood (Ed.), *Proceedings of ARCOM 18th Annual Conference* (Vol. 2, pp. 475-484) University of Northumbria, UK.

Robertson, R. (1992). *Globalization: Social theory and global culture*. London: Sage.

Robinson/Sunna. S. (2007, May 17). Seeds of Despair. *Time.com*. Retrieved September 12, 2009 from http://www.time.com/time/magazine/article/0,9171,1622180,00.html

Rogers, E. M. (1983). *Diffusion of innovations* (3rd ed.). New York: Free Press.

Rogers, E. M. (1986). *Communication Technology: The new media in society*. New York: The Free Press.

Rogers, E. M. (1995). *Diffusion of Innovations* (4th ed.). New York: The Free Press.

Rolland, C., & Proix, C. (1992). A Natural Language Approach for Requirements Engineering. In *Proceedings of CAiSE 1992 on Advanced Information Systems Engineering*, Manchester, UK.

Röller, L.-H., & Waverman, L. (2001). Telecommunications Infrastructure and Economic Development: A Simultaneous Approach. *The American Economic Review*, *91*(4), 909–923. doi:10.1257/aer.91.4.909

Roodman, D. (2005). *XTABOND2: Stata module to extend xtabond dynamic panel data estimator*. Center for Global Development, Washington. Retrieved from http://econpapers.repec.org/software/bocbocode/s435901.htm

Roodman, D. (2009). How to do xtabond2: An introduction to difference and system GMM in Stata. *Stata Journal*, *9*(1), 86–136.

Roster, C. A. (2001). Letting Go: The Process and Meaning of Dispossession in the Lives of Consumers . In Meyers-Levy, J., & Gilly, M. (Eds.), *Advances in Consumer Research* ((pp. 425–430). Provo, UT: Association for Consumer Research.

Rothengatter, T., & Manstead, A. S. R. (1997). The role of subjective norm in predicting the intention to commit traffic violations . In Rothengatter, T., & Carbonell Vaya, E. (Eds.), *Traffic & Transport Psychology, Theory and Application* (pp. 389–394). London: Pergamon.

Rothstein, B. (2005). *Social Traps and the problem of trust*. Cambridge: Cambridge University Press. doi:10.1017/CBO9780511490323

Rouvinen, P. (2006). Diffusion of digital mobile telephony: Are developing countries different? *Telecommunications Policy*, *30*, 46–63. doi:10.1016/j.telpol.2005.06.014

Rushby, N. (2005). Editorial. *British Journal of Educational Technology*, *36*(5), 709–710. doi:10.1111/j.1467-8535.2005.00556.x

Sachs, J. D., & Warner, A. M. (1995). Economic Reform and the Process of Global Integration. *Brookings Papers on Economic Activity*, *1*, 1–118. doi:10.2307/2534573

Sack, R. D. (1986). *Human territoriality: Its theory and history*. New York: Cambridge University Press.

Sadetzki, S., Chetrit, A., Jarus-Hakak, A., Cardis, E., Deutch, Y., & Duvdevani, S. (2008). Cellular Phone Use and Risk of Benign and Malignant Parotid Gland Tumors—A Nationwide Case-Control Study. *American Journal of Epidemiology*, *167*(4), 457–467. doi:10.1093/aje/kwm325

Saku, T., Hayashi, Y., & Takahara, O. (1997). Salivary gland tumors among atomic bomb survivors, 1950–1987. *Cancer*, *79*, 1465–1475. doi:10.1002/(SICI)1097-0142(19970415)79:8<1465::AID-CNCR4>3.0.CO;2-A

Salford, L. G., Brun, A. E., Eberhardt, J. L., Malmgren, L., & Persson, B. R. R. (2003). Nerve Cell Damage in Mammalian Brain after Exposure to Microwaves from GSM Mobile Phones. *Environmental Health Perspectives*, *111*(7), 881–883. doi:10.1289/ehp.6039

Salz, P. A. (2006a). Learning to go mobile. Retrieved July 18, 2009 from www.econtentmag.com/Articles/ArticleReader.aspx?ArticleID=15299&AuthorID=157

Salz, P. A. (2006b). Mobile delivery discoveries. Retrieved July 18, 2009 from www.econtentmag.com/Articles/ArticleReader.aspx?ArticleID=15719&AuthorID=157

Samarajiva, R. (2000). The role of competition in institutional reform of telecommunications: Lessons from Sri Lanka. *Telecommunications Policy*, *24*(8-9), 699–717. doi:10.1016/S0308-5961(00)00057-4

Samuel, J., Shah, N., & Hadingham, W. (2005). *Mobile communications in South Africa, Tanzania, and Egypt: Results from community and business surveys*.

Santini, R., Santini, P., Danze, J. M., LeRuz, P., & Seigne, M. (2003). Survey Study of People Living in the Vicinity of Cellular Phone Base Stations. *Electromagnetic Biology and Medicine*, *22*(1), 41–49. doi:10.1081/JBC-120020353

Sayer, L.A. & Downing, A.J. (1981). *Driver knowledge of road safety factors in three developing countries*. TRRL Supplementary Report 713. Crowthorne: Transport and Road Research Laboratory.

Schechner, R. (1993). *The future of ritual: Writings on culture and performance*. London: Routledge. doi:10.4324/9780203359150

Schein, E. H. (2004). *Organizational culture and leadership*. San Francisco: Jossey-Bass.

Schoder, D. (2000). Forecasting the success of telecommunication services in the presence of network effects. *Information Economics and Policy*, *12*, 181–200. doi:10.1016/S0167-6245(00)00006-8

Schoemaker, M. J., Swerdlow, A. J., Ahlbom, A., Auvinen, A., Blaasaas, K. G., & Cardis, E. (2005). Mobile phone use and risk of acoustic neuroma: results of the Interphone case-control study in five North European countries. *British Journal of Cancer*, *93*(7), 842–848. doi:10.1038/sj.bjc.6602764

Scholte, J. A. (2005). *Globalization. A critical introduction*. New York: Palgrave.

Schramm, W. (1954). *The process and effects of mass communication*. Urbana, IL: University of Illinois Press.

Schramm, W. (1964). *Mass media and national development. The role of information in the developing countries*. Stanford, CA: Stanford University Press.

Schramm, W., & Lerner, D. (Eds.). (1976). *Communication and change. The last ten years - and the next.* Honolulu, HI: University Press of Hawaii.

Schreier, N., Huss, A., & Röösli, M. (2006). The prevalence of symptoms attributed to electromagnetic field exposure: a cross-sectional representative survey in Switzerland. *Sozial- und Praventivmedizin, 51*(4), 202–209. doi:10.1007/s00038-006-5061-2

Schüz, J., Böhler, E., Berg, G., Schlehofer, B., Hettinger, I., & Schlaefer, K. (2006). Cellular phones, cordless phones, and the risks of glioma and meningioma (Interphone Study Group, Germany). *American Journal of Epidemiology, 163*(6), 512–520. doi:10.1093/aje/kwj068

Schüz, J., Jacobsen, R., Olsen, J. H., Boice, J. D., McLaughlin, J. K., & Johansen, C. (2006). Cellular Telephone Use and Cancer Risk: Update of a Nationwide Danish Cohort. *Journal of the National Cancer Institute, 98*(23), 1707–1713. doi:10.1093/jnci/djj464

Sen, A. (2000). *Development as freedom*. New York: Anchor Books.

Sen, A. (2004). Cultural liberty and human development . In Fukuda-Parr, S. (Ed.), *Human development report: Cultural liberty in today's diverse world.* New York: United Nations Development Programme.

Seppala, P., & Alamaki, H. (2003). Mobile Learning in Teacher Training. *Journal of Computer Assisted Learning, 19*, 330–335. doi:10.1046/j.0266-4909.2003.00034.x

Servaes, J. (1997). Development Communication: What the 'Masters' Say [special issue]. *The Journal of Development Communication, 8*(2), 179.

Servaes, J. (1999). *Communication for Development. One World, Multiple Cultures*. Cresskill, NJ: Hampton Press.

Servaes, J. (2007a). Harnessing the UN system into a common approach on Communication for Development. *The International Communication Gazette, 69*(6), 483-507. Retrieved from http://gaz.sagepub.com

Servaes, J. (Ed.). (2003). *Approaches to Development. Studies on Communication for Development.* Paris: UNESCO Publishing House.

Servaes, J. (Ed.). (2007b). *Communication for Development. Making a Difference.* Background paper for the World Congress on Communication for Development, Rome, 25-27 October 2006.

Servaes, J. (Ed.). (2008). *Communication for Development and Social Change.* Los Angeles: Sage.

Servaes, J., & Liu, S. (Eds.). (2007). *Moving targets. Mapping the paths between communication, technology and social change in communities.* Penang, Malayasia: Southbound.

Servaes, J., Jacobson, T., & White, S. (Eds.). (1996). *Participatory Communication for Social Change.* New Delhi: Sage.

Shaanxi (Shaanxi Provincial Government). (2009). Shaanxi Province people's economy and societal development informatization special item program (Shaanxi sheng "shiyi wu" guomin jingji he shehui fazhan xinxihua zhuanxiang guihua). Retrieved June 12, 2009, from http://www.cia.org.cn/information/syw_dqzx_15.htm

Shade, L. R. (2008). Feminizing the mobile: Gender scripting of mobiles in North America . In Goggin, G. (Ed.), *Mobile Phone Cultures* (pp. 42–52). London: Routledge.

Shah, H. (2007). *Meta-Research of Development Communication Studies, 1997-2005: Patterns and Trends Since 1958.* Paper presented at ICA Conference, San Francisco, May.

Shankar Narayanan, R. (2008). Delhi International Airport – Moving on fast track. *ECC Concord, 31*(1), 31–34.

Sharma, G., & Sturges, P. (2007). Using ict to help the poor access public services: An action research programme. *Information Development, 23*(1), 15–24. doi:10.1177/0266666907075623

Sharma, S.K. & Kitchens, F.L. (2004). Web services architecture for m-learning. *Electronic Journal on e-Learning, 2*(1), 203-216.

Sharples, M. (2000). The Design of Personal Mobile Technologies for Lifelong Learning. *Computers & Education, 34*, 177–193. doi:10.1016/S0360-1315(99)00044-5

Sharples, M. (2005). Learning as conversation: Transforming education in the mobile age. In *Proceedings "Seeing Understanding, Learning in the Mobile Age"*, Budapest, April 28–30, 2005, 147-152.

Shaw, D., & Newholm, T. (2002). Voluntary simplicity and the ethics of consumption. *Psychology and Marketing, 19*(2), 167–185. doi:10.1002/mar.10008

Sheng, H., & Trimi, S. (2008). M-government: Technologies, applications and challenges. *Electronic Government: An International Journal, 5*(1), 1–18. doi:10.1504/EG.2008.016124

Sheth, J. N. (1981). Psychology of innovation resistance: The less developed concept (LDC) in diffusion research. *Research in Marketing, 4*, 273–282.

Shi, Y. (2007). The accessibility of Chinese local government Web sites: An exploratory study. *Government Information Quarterly, 24*, 377–403. doi:10.1016/j.giq.2006.05.004

Shin, J., & Park, Y. (2007). Building the national ICT frontier: The case of Korea. *Information Economics and Policy, 19*, 249–277. doi:10.1016/j.infoecopol.2007.01.003

Shiu, A., & Lam, P.-L. (2008). Causal relationship between telecommunications and economic growth in china and its regions. *Regional Studies, 42*(5), 705–718. doi:10.1080/00343400701543314

Shrivastava, M., Yadav, L. L., & Shrivastava, A. (2005). *From distance education to e-learning: Changing the roles of tutor and learner*. Retrieved June 28, 2009, from http://www.ignou.ac.in/ICDE2005/PDFs/theme2pdf/theme2_153.pdf

Siau, K., & Shen, Z. (2003). Mobile communications and mobile services. *International Journal of Mobile Communications, 1*(1), 3–14. doi:10.1504/IJMC.2003.002457

Sife, A. S., Lwoga, E. T., & Sanga, C. (2007). New technologies for teaching and learning: Challenges for higher learning institutions in developing countries. *International Journal of Education and Development using Information and Communication Technology, 3*(2), 57-67.

Simsekoglu, O., & Lajunen, T. (2008). Social psychology of seat belt use: A comparison of theory of planned behavior and health belief model. *Transportation Research Part F: Traffic Psychology and Behaviour, 11*, 181–191. doi:10.1016/j.trf.2007.10.001

Singh, A. (2006, November). Training and workforce planning in construction industry through e-learning and IT management systems. In *Proceedings of INCITE / ITCSED 2006 Conference* (Vol. 3, pp. 341-360). New Delhi, India.

Singh, J., & Pandya, S. (1991). Exploring the effects of Consumers' dissatisfaction level to complaint behaviours . *European Journal of Marketing, 25*(7), 7–21. doi:10.1108/EUM0000000000621

Singh, S. K. (2008). The diffusion of mobile phones in India. *Telecommunications Policy, 32*, 642–651. doi:10.1016/j.telpol.2008.07.005

Siriginidi, S. R. (2009). Achieving millennium development goals: Role of ICTS innovations in India. *Telematics and Informatics, 26*(2), 127–143. doi:10.1016/j.tele.2008.02.001

Slack, J. (1984). *Communication Technologies and Society*. Norwood, MA: Ablex.

Slater, D., & Kwami, J. (2005). *Embeddedness and escape: Internet and mobile use as poverty reduction strategies in Ghana* (Working Paper 4).

Smyth, G. (2006) Wireless Technologies Bridging the digital divide in Education. In *Proceedings, 4th World Conference on mLearning - Cape Town, South Africa*, 25-28 October 2005.

Smyth, R. (2005). Broadband videoconferencing as a tool for learner-centred distance learning in higher education. *British Journal of Educational Technology*, *36*(5), 805–820. doi:10.1111/j.1467-8535.2005.00499.x

Smyth, R. (in press). Enhancing learner-learner interaction using video communications in higher education: Implications from theorising about a new model. *British Journal of Educational Technology*.

Snowden, D. (1999). Story telling: an old skill in a new context. *Business Information Review*, *16*(1), 30–37. doi:10.1177/0266382994237045

Solomon, M. R., & Buchanan, B. (1991). A role-theoretic approach to product symbolism: Mapping a consumption constellation. *Journal of Business Research*, *22*(March), 95–110. doi:10.1016/0148-2963(91)90044-X

Solow, R. M. (1956). A Contribution to the Theory of Economic Growth. *The Quarterly Journal of Economics*, *70*(1), 65–94. doi:10.2307/1884513

Song, G., & Cornford, T. (2006). Mobile Government: Towards a Service Paradigm. In Proceedings of the 2nd International Conference on e-Government (pp. 208-218). Pittsburgh: University of Pittsburgh.

Spitz, M. R., Tilley, B. C., & Batsakis, J. G. (1984). Risk factors for major salivary gland carcinoma. A case-comparison study. *Cancer*, *54*, 1854–1859. doi:10.1002/1097-0142(19841101)54:9<1854::AID-CNCR2820540915>3.0.CO;2-1

Srivastava, L. (2005). Mobile phones and the evolution of social behaviour. *Behaviour & Information Technology*, *24*, 111–129. doi:10.1080/01449290512331321910

Standish-Group (2004). *CHAOS Third Quarter Research Report* (Technical Report).

Star InTech. (2006). Go mobile, get embedded. *Star InTech*, *13*, 19.

Statistical overview of the telecommunication sector as at end of 4th quarter 2008 (2009). Retrieved from http://www.trc.gov.lk/images/pdf/stodec1.pdf

Stiglitz, J. (1998). *Towards a New Paradigm for Development: Strategies, Policies, and Processes*. Prebisch Lecture at UNCTAD, Geneva, 19 October.

Strauss, A. (1987). *Qualitative research for social scientists*. Cambridge: Cambridge University Press. doi:10.1017/CBO9780511557842

Strauss, A., & Corbin, J. (1998). *Basics of Qualitative Research Techniques and Procedures for Developing Grounded Theory*. Thousand Oaks, CA: Sage.

Stroebe, V. (2000). *Social psychology and health*. Buckingham: Open University.

Suchman, M. C. (1995). Managing Legitimacy: Strategic and Institutional Approaches. *Academy of Management Journal*, *20*(3), 571–610. doi:10.2307/258788

Sui, Q. J., Zhang, Y. P., & Zhang, J. B. (2004). Review of mobile learning. *Journal of Education Exploration*, *8*, 66–67.

Suki, N., & Suki, N. (2007). Mobile phone usage for m-learning: comparing heavy and light mobile phone users. *Campus-Wide Information Systems*, *24*(5), 355–365. doi:10.1108/10650740710835779

Sun, Y. T. (2008). The practice and research of m-learning and mobile service. *Journal of Distance Education in China*, *8*, 68–70.

Sun, Y. T., & Chen, X. (2007). Research on the building of mobile campus in the open universities. *Journal of Educational Informatization in China*, *10*, 7–9.

Sun, Y. X. (2008). *Research on m-learning system architecture and application*. Unpublished master's thesis, Shanghai Jiaotong University, Shanghai, China.

Sungu, E., Sungu, H., & Bayrakci, M. (2008). E-inclusion: Providing services towards an information society for all. *8th International Educational Technology Conference.*

Sykes, G. M., & Matza, D. (1957). Techniques of Neutralization: A Theory of Delinquency. *American Sociological Review*, *22*, 664–670. doi:10.2307/2089195

Szmigin, I., & Foxall, G. (1998). Three forms of innovation resistance: The case of retail payment methods. *Technovation*, *18*(6/7), 459–468. doi:10.1016/S0166-4972(98)00030-3

Tajfel, H. (1974). Social identity and intergroup behaviour. *Social Sciences Information. Information Sur les Sciences Sociales*, *13*(2), 65–93. doi:10.1177/053901847401300204

Takebayashi, T., Varsier, N., Kikuchi, Y., Wake, K., Taki, M., & Watanabe, S. (2008). Mobile phone use, exposure to radiofrequency electromagnetic field, and brain tumour: a case-control study. *British Journal of Cancer*, *98*(3), 652–659. doi:10.1038/sj.bjc.6604214

Taylor, A. (2000). IT Projects: Sink or Swim? *The Computer Bulletin*, *41*(1), 24–26. doi:10.1093/combul/42.1.24

Taylor-Gooby, P. (2008). Choice and Values: Individualised Rational Action and Social Goals. *Journal of Social Policy*, *37*(2), 167–185. doi:10.1017/S0047279407001699

Telegraph (2008, April 19). New doubts raised over mobile phone safety, by Nic Fleming, Science Correspondent. Retrieved from http://www.telegraph.co.uk/news/uknews/1561587/New-doubts-raised-over-mobile-phone-safety.html

Tellis, G. J., Stremersch, S., & Yin, E. (2003). The international takeoff of new products: The role of economics, culture, and country innovativeness. *Marketing Science*, *22*(2), 188–208. doi:10.1287/mksc.22.2.188.16041

Teng, J. T. C., Grover, V., & Guttler, W. (2002). Information technology innovations: general diffusion patterns and its relationships to innovation characteristics. *IEEE Transactions on Engineering Management*, *49*(1), 13–27. doi:10.1109/17.985744

The Economist. (2007, 7 July). The eight commandments. *The Economist,* (pp. 26-29).

The Guardian. (2006, December 5). Jam mobile phones to combat exam cheating, report urges.

The National (2009, April 24). Apple underplays its success.

The National (2009a, February 5). UAE leads world in mobiles per person.

The National (2009b, March 15). Drive to cut road deaths by 40%.

The National (2009c, March 16). Far more drivers fined for phone calls.

The National (2010, August 11). Security scenario justifies concerns about BlackBerry.

The Times of India (2008, July 10). Delhi beats Mumbai to become busiest airport. Retrieved July 2009 from http://timesofindia.indiatimes.com/India/Delhi_is_countrys_busiest_airport/articleshow/3216435.cms July 10, 2008

The Times of India (2009, July 27). India will be third largest net user by 2013: Study. Retrieved July 2009 from http://lite.epaper.timesofindia.com

Thomke, S., & Fujimoto, T. (2000). The effect of front-loading problem-solving on product development performance. *Journal of Product Innovation Management*, *17*, 128–142. doi:10.1016/S0737-6782(99)00031-4

Thornton, P., & Houser, C. (2005). Using mobile phones in English education in Japan. *Journal of Computer Assisted Learning*, *21*, 217–228. doi:10.1111/j.1365-2729.2005.00129.x

Tomlinsion, J. (1999). *Globalization and Culture*. Cambridge, MA: Polity Press.

Tongia, R., & Subrahmanian, E. (2006). Information and Communications Technology for Development (ICT4D) - A Design Challenge? In *Proceedings of International Conference on Information and Communication Technologies and Development* (pp. 243-255).

Torenli, N. (2006). The 'other' faces of digital exclusion: ICT gender divides in the broader community. *European Journal of Communication*, *21*(4), 435–455. doi:10.1177/0267323106070010

Traxler, J. (2007). *Defining, discussing and evaluating mobile learning: the moving finger writes and having writ....* Retrieved May 10, 2009, from http://www.irrodl.org/index.php/irrodl/article/view/346/875

Traxler, J. (2008). *Mobility, Modernity and Development*. Paper presented at first m4d Karlstadt, Sweden. Retrieved July 2, 2009, from http://wlv/academia.edu/JohnTraxler/Papaers/83101/Mobility--Modernity--Development

Traxler, J. (2009). Learning in a Mobile Age. *International Journal of Mobile and Blended Learning*, *1*(1).

Trifonova, A., Georgieva, E., & Ronchetti, M. (2006). 'Has the time for university's mobile learning come? Determining students' readiness'. *WSEAS Transactions on Advances in Engineering Education*, *3*, 1790–1979.

Trimi, S., & Sheng, H. (2008). Emerging trends in m-government. *Communications of the ACM*, *51*(5), 53–58. doi:10.1145/1342327.1342338

Tuckman, B. W. (1965). Developmental sequences in small groups. *Psychological Bulletin*, *63*, 384–399. doi:10.1037/h0022100

TURKSAT. (2008). http://www.turksat.com.tr/english/index.php/e-Government-/e-Government-Gateway-Project/What-is-e-Government-Gateway-Project.html

Tutkun, C. (2007). *Devletin Kisa Yolu. In 12*. Ankara: Turkiye'de Internet Konferansi.

Tyler, T. R. (1990). *Why people obey the law*. New Haven: Yale University Press.

UNESCO. (2003). *International Experts Brainstorming Meeting on Development Communication* (New Delhi, 1-3 September 2003), Paris: UNESCO.

United Nations Development Programme. (2006). *Communication for Empowerment: developing media strategies in support of vulnerable groups*, Oslo, Norway. Retrieved from www.worldbank.org/wbi/news/docs/Severinoonfreepress.htm

United Nations Human Development Report 2007/08 (2008). Retrieved from http://hdr.undp.org/en/media/HDR_20072008_EN_Complete.pdf

United Nations. (2007). *The Millennium Development Goals Report 2007*. New York: UN.

Valberg, P. A., van Deventer, T. E., & Repacholi, M. H. (2007). Workgroup Report: Base Stations and Wireless Networks—Radiofrequency (RF) Exposures and Health Consequences. *Environmental Health Perspectives*, *115*(3), 416–424. doi:10.1289/ehp.9633

Valente, T. W. (1995). *Network models of the diffusion of innovations*. Creskill, NJ: Hampton Press.

Van Lamsweerde, A. (2001). Goal-Oriented Requirements Engineering: A Guided Tour. In *Proceedings of Fifth IEEE International Symposium on Requirements Engineering* (pp. 149-163).

Vaughan, P. (2007). Early lessons from the deployment of M-PESA, Vodafone's own mobile transactions service. In D. Coyle (Ed.), *The Transformational Potential of m-Transactions,* Policy Paper Series, No.6. Vodaphone, London. Retrieved April 4, 2009 from http://www.vodaphone.com/m-tranactions

Vavoula, G. N., & Sharples, M. (2002). KleOS: A Personal, Mobile, Knowledge and Learning Organisation System. In M. Milrad, H.U. Hoppe & Kinshuk (Eds.), *IEEE International Workshop on Wireless and Mobile Technologies in Education* (pp. 152-156). Los Alimatos, USA: IEEE Computer Society.

Venkatesh, V., Morris, M. G., Davis, G. B., & Davis, F. D. (2003). User acceptance of information technology: Toward a unified view. *Management Information Systems Quarterly*, *27*(3), 425–478.

Venkatraman, S., & Yoong, P. (2009). Role of mobile technology in the construction industry – A case study. *International Journal of Business Information Systems*, *4*(2), 195–209. doi:10.1504/IJBIS.2009.022823

Verzola, R. (2004). *Towards a political economy of information*. Quezon City, Philippines: Foundation for Nationalist Studies.

Vijayalaxmi, (2008). Genetic Damage in Mammalian Somatic Cells Exposed to Radiofrequency Radiation: A Meta-analysis of Data from 63 Publications (1990–2005). *Radiation Research*, *169*(5), 561–574. doi:10.1667/RR0987.1

Vincent, J., & Harris, L. (2008). Effective use of mobile communication in e-government: how do we reach the tipping point. *Information Communication and Society*, *11*(3), 395–413. doi:10.1080/13691180802025632

Virtanen, H., Keshvari, J., & Lappalainen, R. (2006). Interaction of radio frequency electromagnetic fields and passive metallic implants - A brief review. *Bioelectromagnetics*, *27*(6), 431–439. doi:10.1002/bem.20224

Vrechopoulos, A., Constantiou, I., Sideris, S., & Doukidis, G. (2003). The critical role of consumer behaviour in mobile commerce. *International Journal of Mobile Communications*, *1*(3), 329–340. doi:10.1504/IJMC.2003.003498

Wagner, E. (2005). Enabling Mobile Learning. *EDUCAUSE Review*, *40*(3), 40–53.

Walker, L., Williams, J., & Jamrozik, K. (2006). Unsafe driving behaviour and four wheel drive vehicles: observational study. *British Medical Journal*, *333*(7558), 71–74. doi:10.1136/bmj.38848.627731.2F

Walsh, S. P., White, K. M., Hyde, M. K., & Watson, B. (2008). Dialling and driving: Factors influencing intentions to use a mobile phone while driving. *Accident; Analysis and Prevention*, *40*(6), 1893–1900. doi:10.1016/j.aap.2008.07.005

Wang, G., Servaes, J., & Goonasekera, A. (Eds.). (2000). *The New Communications Landscape. Demystifying media globalization*. London: Routledge.

Ward, M., Thorpe, T., Price, A., & Wren, C. (2004). Implementation and control of wireless data collection on construction sites. *ITcon*, *9*, 297–311.

Watts, M. (1997). Black gold, white heat: State violence, local resistance and the national question in Nigeria . In Pile, S., & Keith, M. (Eds.), *Geographies of Resistance* (pp. 33–67). London: Routledge.

Weiber, R. (1995). Systemguter und klassische diffusiontheorie: Elemente einer diffusionstheorie kristische masse-systeme . In Stoetzer, M.-W., & Mahler, A. (Eds.), *Die diffusion von innovationen in der telekommunikation*. Berlin: Springer.

Weilenmann, A. (2001). Negotiating use: making sense of mobile technology. *Personal and Ubiquitous Computing*, *5*(2), 109–122. doi:10.1007/PL00000015

Wexler, S., Schlenker, B., Metcalf, D., Brown, J., Metcalf, D., Quinn, C., et al. (2007). *Mobile learning: What it is, why it matters, and how to incorporate it into your learning strategy*. Retrieved October 4, 2009 from http://www.elearningguild.com/assets/files/29/mobile_abstract.pdf

WFed [Wireless Federation] (2008): Middle East mobile subscription rates set to hit 15% growth in 2009. Retrieved April 2009 from http://wirelessfederation.com/news/category/mobile-subscription

Whetherall, M., Potter, J., & Antaki, C. (1988). *Analyzing Everyday Explanation*. Thousand Oaks, CA: Sage.

Whittenbraker, J., Gibbs, B. L., & Kahle, L. R. (1983). Seat belt attitudes, habits and behaviors: An adaptive amendment to the Fishbein model. *Journal of Applied Social Psychology*, *13*, 406–421. doi:10.1111/j.1559-1816.1983.tb01748.x

Whittington, R. (2007). Strategy practice and strategy process: family differences and the sociological eye. *Organization Studies*, *28*(10), 1575–1586. doi:10.1177/0170840607081557

WHO. (2000). *Fact Sheet on Electromagnetic Fields and Public Health: Mobile Phones and Their Base Stations*. Geneva: World Health Organization.

Wieraszko, A., Armani, J., Hanna, A., Maqsood, N., Raja, H., & Hogan, M. (2004). Changes in neurotransmitter turnover and second messenger levels in brain tissue exposed to magnetic fields. In P. Kostarakis, (Ed.), *Biological Effects of EMFs: Third International Workshop*, Kos Greece, October 4–8, (pp. 614–632). Ioannina, Greece: University of Ioannina.

Wijesekara, N. (1990). *Archaeological department centenary (1890-1990): Inscriptions*. Department of archaeology, Sri Lanka, Colombo.

wikipedia.org. (n.d.). Indian Construction Industry. Retrieved January 2009 from http://en.wikipedia.org/wiki/Indian_Construction_Industry

wikipedia.org. (n.d.). Indira Gandhi International Airport. Retrieved August 2009 from http://en.wikipedia.org/wiki/Indira_Gandhi_International_Airport#cite_note-8

Wilhelm, A. (2004). *Digital Nation. Toward an Inclusive Information Society*. Boston: The MIT Press.

Williams, H., & Torma, M. (2007). Trust and fidelity: from under the mattress to the mobile phone. In D. Coyle (Ed.), *The Transformational Potential of m-Transactions*, Policy Paper Series, No.6, Vodaphone, London. Retrieved April 4, 2008 from Available at: http://www.vodaphone.com/m-transactions

Wilska, T. A. (2003). Mobile phone use as part of young people's consumption styles. *Journal of Consumer Policy*, *26*(4), 441–463. doi:10.1023/A:1026331016172

Windmeijer, F. (2005). A Finite Sample Correction for the Variance of Linear Efficient Two-Step GMM Estimators. *Journal of Econometrics*, *126*(1), 25–51. doi:10.1016/j.jeconom.2004.02.005

Wiorkowski, J. J., & Gylys, V.A. (2006). An empirical real-time test for takeoff with applications to cellular telephony. *Review of Marketing Science,* 4, Article 1.

Witt, U. (1997). "Lock-in" vs. "Critical masses" - Industrial change under network effects. *International Journal of Industrial Organization*, *15*, 753–773.

Wittenbraker, Gibbs, & Kahle. (1983). Seat belt attitudes, habits and behaviors: an adaptive amendment to the Fishbein model . *Journal of Applied Social Psychology*, *13*, 406–421. doi:10.1111/j.1559-1816.1983.tb01748.x

Woo, K. S., & Fock, H. K. Y. (1999). Customer satisfaction in the Hong Kong mobile phone industry. *The Service Industries Journal*, *19*(3), 162–174. doi:10.1080/02642069900000035

Wood, K. (2003). Introduction to Mobile Learning (M Learning). Retrieved April 4, 2009 from http://ferl.becta.org.uk/display.cfm?page=65&catid=192&resid=5194&printable=1

Woodruffe, H. R. (1997). Compensatory Consumption: Why women go shopping when they're fed up and other stories . *Marketing Intelligence & Planning*, *15*(7), 325–334. doi:10.1108/02634509710193172

World Bank. (2008). *World Development Indicators Online*. Retrieved in 2008 from http://devdata.worldbank.org/dataonline/

World Bank. (n.d.). Retrieved from http://go.worldbank.org/K2CKM78CC0

Xiong, Z. G. (2005). *Research on mobile learning & the design of mobile learning resources*. Unpublished master's thesis, East China Normal University, Shanghai, China.

Yang, J. (2007). *Illiteracy continues to decline in China*. Retrieved October 4, 2009 from http://news.xinhuanet.com/english/2007-08/01/content_6460066.htm

Yin, R. (1989). *Case study research: Design and methods*. Thousand Oaks, CA: Sage Publishing.

Yin, R. (1994). *Case Study Research: Design and Methods*. Thousand Oaks, CA: Sage.

Yu, B., & Kushchu, I. (2004). The value of mobility for e-government. In *Proceedings of European Conference on E-Government (ECEG 2004)* (pp. 887-899). Trinity College, Dublin, June 17-18. Department of the Taoiseach, Dublin, Ireland. Reading, UK: Academic Conferences International.

Yu, S. Q. (2007). From knowledge transfer to cognitive constructivism, then to situated cognition - the development and prospects of mobile learning. *Journal of China Educational Technology*, *6*, 7–18.

Yu, S. Q. (2007). *The implementation plan for 'the application of handheld networked learning systems in subject teaching (AHNLSST)'*. Retrieved April 5, 2009, from http://www.360doc.com/showWeb/0/0/211018.aspx

Yunnan (Yunnan Provincial Government). (2009). Yunnan Province people's economy and societal development informatization special item program (Yunnan sheng guomin jingji he shehui fazhan xinxihua zhuanxiang guihua). Retrieved June 12, 2009, from Http://Www.Cia.Org.Cn/Information/Syw_Dqzx_14.Htm

Zawya (2008). RIM focused on Blackberry growth in the Gulf region. Retrieved January 2008 from www.zawya.com

Zelditch, M. (2001). Theories of Legitimacy. In J. T. Jost & B. Major (Eds.), *The psychology of legitimacy: emerging perspectives on ideology, justice, and intergroup relations* (Vol. 2, illustrated, pp. 33-53). Cambridge University Press.

Zhang, L. (2005). Integrate the social resources to accelerate the pace of e-government construction (Zhenghe shehui ziyuan jia kuai dianzi zhengwu jianshe bufa). Retrieved June 12, 2009, from http://www.nx.gov.cn/structure/zfxx/zfld/fsjhxx_5124_1.htm

Zhu, W. P. (2008). *The research and development of title English learning software based on J2ME*. Unpublished master's thesis, Sichuan Normal University, Chendu, China.

Zinelis, S. A. (2008). Short-term exposure to mobile phone base station signals. *Environmental Health Perspectives*, *116*(2), A62. doi:10.1289/ehp.10733

About the Contributors

Ahmed Gad Abdel-Wahab (Ph.D. University of Stirling, Scotland) is a professor of management at Mansoura University. He has a master's degree in business from the University of Mansoura. He is an author of two books in management and organizational behaviour that are used at the University of Mansoura. He has published a number of articles in Egypt in the field of organizational behaviour such as leadership, team building and managerial values. He has also published a number of articles in the field of information and communication technology in international journals such as Behaviour & Information Technology (BIT): Taylor & Francis; Electronic Journal of Communication: State University of New York; The Electronic Journal of Information Systems in Developing Countries (EJISDC): City University of Hong Kong; International Journal of Business Data Communications and Networking (IJBDCN): University of Auckland, New Zealand. Professor Abdel-Wahab's articles about the possibility of applying telecommuting, e-medicine, and telemedicine in Egypt have opened the door for researchers in Egypt to do further investigations on these topics and take further steps toward the application of such new concepts in Egypt.

Ahmed A. El-Masry graduated from the Faculty of Commerce of Mansoura University, Egypt where he got his BCom and MBA. He got his PhD in finance from Manchester Business School, UK. He is currently Associate Professor of Finance at Plymouth Business School, UK. In addition he is Affiliate Professor at Grenoble Graduate School of Business, France and Visiting Professor at Mansoura University, Egypt. His research and teaching interests have been fostered by several years of teaching and mentoring at Mansoura University from 1987 to 1999 and then Plymouth University from 2003 to present. He teaching interests are corporate finance, international finance, international financial management, financial research methods, investment & valuation, assets management, international securities markets, portfolio management, and valuation of securities. His PhD thesis on foreign exchange exposure and firm values built upon the most of his past research publications and his research has now focused upon capital structure; dividend policy; credit rating and credit scoring models; market efficiency & anomalies; corporate governance and online reporting; and Islamic finance and banking. Dr El-Masry is now doing research projects on e-government, m-government, m-banking and corporate internet reporting in developing countries. His research has been adapted to provide undergraduate and postgraduate students with research opportunities for their dissertation projects. In addition, his research has been published in Expert Systems with Applications, Journal of Operational Research Society, Managerial Finance and *Cross-Cultural Management: International Journal*, amongst others. Furthermore, he has been selected as an Editor-in-Chief of a new journal: *Journal of International Business and Finance (JIBF),* published by Serials Publications. He is also a reviewer in several international journals such as *British Accounting Review, European Journal of Operational Research, Frontiers in Finance and Economics, International Journal of Business Research, International Journal of Business Strategy, Journal of International Business and Economics, Journal of International Finance and Economics, Review of Business Research, and Review of Financial Economics.* In addition, he edited three issues for *Managerial Finance Journal* in vol. 32, issue 2, 2006 (Derivative use and risk management), vol. 33, issue 9, 2007 (foreign exchange rate exposure and its determinants) and vol. 34, issue 12, 2008 (corporate governance and ownership structure). In addition, he is an editorial board member for the International Academy for Business and Economics, and an associate editor of the International Journal of Customer Relationship Marketing and Management (a new journal published by IGI Global). Furthermore, Dr El-Masry currently acts as external examiner for several universities such as Glasgow Caledonian University Business School and Middlesex University Business School. He was also a chair of special sessions on Islamic banking and finance, IABE conference, Las Vegas, Nevada, USA, in October 2006 and 2007 and is a Jury Committee Member, for Best Website and Annual Reports Competition, Egyptian Institute of Directors, Ministry of Investment in collaboration with the United Nations and the European Union, EGYPT, 2008 to present. Finally Dr El-Masry participated in several international conferences in Switzerland, Spain, Ireland, Egypt, Greece, UK and USA.

***

Kaisu Puumalainen, D.Sc. (Tech), is Professor in technology research in the School of Business at Lappeenranta University of Technology, Finland. Her primary areas of research interest are diffusion of innovations, organizational innovation, strategic orientations, international marketing, and small business. She has supervised several doctoral dissertations and published about 40 refereed articles, e.g. in Journal of the Academy of Marketing Science, International Journal of Research in Marketing, Journal of Business Research, European Journal of Marketing, R&D Management, and Technological Forecasting and Social Change.

Lauri Frank, Dr.Sc (Econ) is Senior Assistant in electronic business with the department of Computer Science and Information Systems at the University of Jyväskylä. He also holds an Adjunct Professorship at the School of Business at the Lappeenranta University of Technology in the field of Technology Research. His research interests are in the diffusion and management of technological information and communications (ICT) innovations, the evolution of telecom operator software markets, and the uptake of the information society. Within these research areas, he has been involved in several research projects in various roles. Within the context of telecommunications related research, he is also actively participating in the Cost605 Telecommunications Economics network. Previously he has worked as a researcher for the universities of Lappeenranta and Joensuu, and for TeliaSonera.

Sanna Sundqvist, D.Sc. (Tech), is Professor of marketing, School of Business, Lappeenranta University of Technology, Finland. Her research interests deal with the international diffusion of innovations, market orientation (especially in an international context), and consumers' adoption behavior. She has published in the Journal of World Business, Journal of Business Research, International Journal of Research in Marketing, Journal of the Academy of Marketing Science, European Journal of Marketing, and the Canadian Journal of Administrative Sciences, among others.

Anni Tuppura, D.Sc. (Econ.), is a Senior Lecturer at the Lappeenranta University of Technology, School of Business (Finland). She has received her PhD from the Lappeenranta University of Technology in 2007. As a part of her doctoral studies she attended economic growth and development related courses in United Nations University in Maastricht (UNU-MERIT). Primarily her research focuses on strategic management, especially market entry timing, and corporate social responsibility. She has published three refereed journal articles.

Moon-Soo Kim is an associate professor of School of Industrial & Management Engineering, Hankuk University of Foreign Studies (HUFS) in Korea. He has experienced as a project manager at ETRI (Electronics and Telecommunications Research Institute) for 5 years before joining the university. His research focuses on technology strategy and its various application fields. He has been involved in studies on management of technology, telecommunication service management, and e-Business planning & strategy, etc. Dr. Kim has published papers in several international journals such as International Journal of Innovation Management, Scientometrics, Technological Forecasting & Social Change, Telecommunications Policy, Technology Analysis & Strategic Management, Omega and Journal of Scientific & Industrial Research etc., and also in several domestic journals. He holds a Ph.D. from Seoul National University in Korea.

Sungjoo Lee is a full-time lecturer in the Department of Industrial & Information Systems Engineering at Ajou University. She holds her B.S. and Ph.D. in industrial engineering, both from Seoul National University (SNU) in Korea. After spending six months working as a senior researcher in the Ubiquitous Computing Innovation Center, SNU in Korea, she moved to the UK to continue her research on technology intelligence and spent one year at the Centre for Technology Management, University of Cambridge as a visiting scholar. Telecommunications industry has been a key area of ongoing interest, in terms of both research and practice, and her current research focuses on technology forecasting and strategic management.

Vanita Ahuja is a project management consultant, teacher and trainer based in New Delhi, the capital city of India. She has done her graduation in Architecture and post graduation in Construction Management from IIT, the premier institutes of India and she has done her PhD from Australia. Her research area was studying factors that would enhance use of ICT for building project management, with special focus on SMEs. She has +17 years of industry experience as Construction planning and scheduling manager. She is also involved in research and academics. Her papers are published in peer reviewed International journals, she is a leading project management trainer in India and is also teaching the subject at post graduate courses.

Kristina Pitula is completing her doctoral degree in Computer Science at Concordia University with a research focus on methodologies for eliciting software requirements to design culturally appropriate, sustainable systems in an ICT4D context. To this research she brings 15 years of industrial experience in software development and 5 years of social development work in underdeveloped countries.

Daniel Sinnig received a PhD degree from Concordia University, in Canada in 2008. He also holds a Master's degree in Software Engineering from the University of Rostock, Germany and Master's degree in Human Computer Interaction from Concordia University. Daniel is the author of numerous publications and has taught several lectures on Software Engineering and User Interface Design. Daniel is a member of the ACM and an officer of the IFIP 13.2 working group on Methodologies for User-Centered Systems Design.

Thiruvengadam Radhakrishnan graduated from Guindy Engineering College, Madras with B.E.(Hons). He obtained his M.Tech and and PhD degrees from IIT, Kanpur, India. He has taught at IIT Kanpur and IIT Madras before coming to Concordia University where he is currently a full professor and Chair. He has supervised over 80 graduate students, co-authored four Computer Science text books, obtained three patents, and published in scholarly journals and conferences. His current interests are in Human Centered Computing, Medical Informatics, and Software Engineering for ICT for Development (ICT4D).

Ayse is a PhD student in Technology & Learning Sciences in the University of Oxford. Previously, she received her Msc degree from the same university. Ayse worked for international organisations such as Ernst & Young and UNSSC and get involved as a consultant for European Commission and UNDP projects. Ayse attended numerous international conferences as a speaker, published several articles about the use of Information Communication Technologies for learning and teaching. A detailed list of her publications can be found on https://ayshe.kok.googlepages.com.

Yuqin Yang (yuqin.yang@nie.edu.sg) is a research associate at the Learning Sciences Lab of the National Institute of Education, Nanyang Technological University. She got her Master degree from South China Normal University, Guangzhou, China. Since 2006, she has been conducting research on teachers' information seeking behaviour, and the design and development of web-based courses. She has published more than twenty journal and conference papers. Her research interests include information seeking behaviour of users, ubiquitous learning environment and mobile learning, and technology-enhanced learning and teaching, in particular, computer-assisted collaborative learning. She now focuses on the research on affordances of Facebook for collaborative learning.

Qiyun Wang (Qiyun.wang@nie.edu.sg) is an associate professor at the Learning Sciences and Technologies Academic Group in the National Institute of Education, Nanyang Technological University, Singapore. He has published a large number of journal papers and book chapters in recent years. His research interests include online learning, interactive learning environment design, constructivist learning, and the use of web 2.0 tools for teaching and learning. More information about him can be found on the web site of http://qywang.myplace.nie.edu.sg.

Susan Smith Nash has been involved in the design, development and administration of online courses and programs since the early 1990s. Her current research interests include the use of learning objects, mobile learning, leadership in e-learning organizations, and energy and sustainability technology transfer. Her articles and columns have appeared in magazines and refereed journals. She received her Ph.D. from the University of Oklahoma in 1996, and in addition to e-learning, Nash has also been involved in international economic development training, interdisciplinary studies, international energy education (renewables and non-renewables), and sustainable business and career training at the University of Oklahoma and Excelsior College (Albany, NY). Her book, *Leadership in the e-Learning Organization,* was co-authored with George Henderson, and published by Charles Thomas and Sons.

Her most recent books include Klub Dobrih Dejanj (Good Deeds Society) and E-Learner Survival Guide (Texture Press). Nash is also author of a book on Moodle teaching techniques. Her edublog, E-Learning Queen (www.elearningqueen.com) has received numerous awards and recognitions. Currently, Nash is an adjunct professor for the University of Oklahoma. Nash is Director of Education and Professional Development at AAPG, an international professional association of geoscientists, with offices in London, Dubai, Singapore, and Tulsa, OK, with more than 36,000 members worldwide, where she has launched a full suite of online and mobile learning synchronous and asynchronous courses and programs.

Andrew Vincent Berriman, ***MBA (Macquarie), MIMC,*** has more than 30 years experience in the IT and Telecommunications industry preceded by 6 years military service in secure telecommunications. The commercial experience in the IT and Telecommunications industry has covered the full spectrum of corporate and management roles including technical, marketing, sales, consulting and finance with Control Data, Sigma Data, Wang, Solution 6, CR Group, NVS and ServicePoint. Andrew's expertise includes hands-on experience with business start-ups, capital raising, mergers and acquisitions and a public company listing. Andrew has served on the board of two Australian publicly listed IT companies for a total of ten years. Andrew has a postgraduate degree in Business Administration from the Macquarie Graduate School of Management in Sydney and is a Member of the Institute of Management Consultants in Australia. Andrew's most recent executive IT appointment was Chief Executive Officer of ServicePoint Australia Pty Limited.

Robyn Smyth is a Senior Lecturer working as an Academic Developer in the School of Rural Medicine at University of New England specialising in curriculum design, using rich media technology in higher and medical education and for student centred pedagogies. She has been an active researcher investigating video communications, their pedagogies and potential for use in student centred learning. Principally her research is focussed on the potential of rich media technologies to enhance distance learning in higher education, including for higher degree research. Currently Robyn is leading a sector-wide investigation into the sustainability, viability and scalability of rich media technologies in Australian universities.

Deborah Vale I completed my doctorate at the University of New England, specialising in marine faunal remains from coastal archaeological sites. After working as an archaeologist and palaeoichthyologist in Australia and Central America, I became interested in academic development and completed a number of small research projects for the Teaching and Learning Centre at UNE. I am currently employed as a project officer on two ALTC funded projects at UNE, including the Leading Rich Media, which is investigating how tertiary institutions use rich media technologies to improve student learning.

Professor Belinda Tynan is the Director of e-University for the University of New England. Professor Tynan has held numerous teaching positions in the UK, Australia and Japan. She has demonstrated experience in the management of large projects in the higher education sector, both in Australia and overseas. Her latest major initiative, DEHub, received $3.544 million in Commonwealth funding in 2008. In addition, Professor Tynan has authored and co-authored almost two dozen journal articles and book chapters, in addition to many peer-reviewed conference papers. Her research has a focus in elearning, covering areas such as music teaching, distance education, academic staff development and collaboration. Belinda is also the Treasurer and hosts the Secretariat of the Open and Distance Learning Association of Australia (ODLAA).

Richard Caladine PhD, is the Head of Learning, Innovation, Facilities and Technologies (LIFT) at University of Wollongong. He is responsible for the operation of audio visual and rich media educational systems, including a videoconference service consisting of over 60 endpoints and a 40 port videoconference Multi Conference Unit that connects seven NSW campuses of the University. As well he is responsible for the podcasting, webcasting and streaming services. He is actively involved with the training of staff in the pedagogically appropriate use of these systems and other educational technologies. He researches the use of rich media in higher education, has published many papers, book chapters and books all on the appropriate pedagogical use of educational technologies. Richard's current area of investigation is unified communications in education. He has developed and is currently testing models for several communications technologies including immersive environments and the combination of online collaboration with videoconference.

Jesper Schlæger is a PhD Fellow at the University of Copenhagen, Department of Political Science. Primary field of interest is the role of new information and communication technology in institutional change as a harbinger of change in the Chinese political system. His PhD project concerns e-government in Chengdu, China. During this project a long series of qualitative interviews of public servants at all administrative levels and e-government users has been performed. Jesper Schlæger has lived and done research for extensive periods in Western China. Previous research includes detailed examinations of changes in internet mediated coal allocation and development of property rights in the coal sector.

Ronan De Kervenoael is a Lecturer in Marketing at Sabanci University and network Lecturer at Aston University. He received his Ph.D. from Sheffield University, UK. He has held position at Lancaster University, ITSM de Monterrey, and Manchester Universities. Ronan has a particular interest in choice and anti-choice investigating both consumers and within the supply chain actors through their everyday practice. His wider research interests lie under the umbrella of consumer behavior and retailing; the principal theme being the socio-spatial context of consumption. He is currently particularly interested in the Information Communication Technology change situation of emerging markets and associated strategies.

Mark Palmer is a Senior Lecturer in Marketing at Aston University. He received his Ph.D. from the University of Ulster, Northern Ireland. His research interests focus on retail TNC strategy as practice and corporate resistance within international supply chains. The central idea advanced by his research is the contestability of corporate strategizing and international market development. In particular his work draws upon the insights of Karl Polanyi's analysis of the human economy. This work has been published among others in the Journal of Economic Geography, European Journal of Marketing, Journal of Marketing Management, Environment & Planning A, The Service Industries Journal, The European Management Journal, Leadership and Organization Development Journal, Studies in Higher Education, Management Learning and the Journal of Strategic Marketing. This research has received financial support from a range of sources including, a Sainsbury's Ph.D Scholarship Award, The George Spencer Research Trust in conjunction with the British Shops and Stores Association, The Higher Education Learning and Management (HELM) Centre at Aston Business School, The British Academy and The Centre for Learning, Innovation and Professional Practice (CLIPP) at Aston University.

N. Meltem Cakici is a Lecturer in Marketing at Gediz University, Izmir. She received her B.S. degree in Management from Bogazici University and Ph.D. in Marketing from Yeditepe University, Istanbul. She worked as a post doctoral fellow on a Tübitak project at Sabanci University, Istanbul. Her areas of interest are consumer behavior, marketing research, international marketing and retailing. She is paying particular attention on consumers' decision making process, cross-cultural studies on purchase and consumption patterns, ethical consumption and corporate social responsibility issues. Emerging themes and concepts in marketing, marketing's role in strategy development and marketing's relationship with other disciplines are also part of her research agenda.

Heli Virta is a senior lecturer at the School of Business at Lappeenranta University of Technology in Finland. She previously worked as an economist at the Bank of Finland. She has a Ph.D. in economics from Helsinki School of Economics. Her main research interests are currently in development economics (particularly the role of institutions in economic development) and trade theory. Heli Virta teaches courses on economics, econometrics and mathematics.

Kaisu Puumalainen, D.Sc. (Tech), is a professor in technology research in the School of Business at Lappeenranta University of Technology, Finland. Her primary areas of research interest are diffusion of innovations, organizational innovation, strategic orientations, international marketing, and small business. She has supervised several doctoral dissertations and published about 40 refereed articles, e.g. in Journal of the Academy of Marketing Science, International Journal of Research in Marketing, Journal of Business Research, European Journal of Marketing, R&D Management, and Technological Forecasting and Social Change.

Anni Tuppura, D.Sc. (Econ.), is a senior lecturer at the Lappeenranta University of Technology, School of Business (Finland). She has received her PhD from the Lappeenranta University of Technology in 2007. As a part of her doctoral studies she attended economic growth and development related courses in United Nations University in Maastricht (UNU-MERIT). Primarily her research focuses on strategic management, especially market entry timing, and corporate social responsibility. She has published three refereed journal articles.

Ronan De Kervenoael is a Lecturer in Marketing at Sabanci University and network Lecturer at Aston University. He received his Ph.D. from Sheffield University, UK. He has held position at Lancaster University, ITSM de Monterrey, and Manchester Universities. Ronan has a particular interest in choice and anti-choice investigating both consumers and within the supply chain actors through their everyday practice. His wider research interests lie under the umbrella of consumer behavior and retailing; the principal theme being the socio-spatial context of consumption. He is currently particularly interested in the Information Communication Technology change situation of emerging markets and associated strategies.

Canan Devletkusu is a clinical psychologist and a research assistant in Psychology at Dogus University. She received her B.A degree on Psychology and B.S. degree on Political Science and International Relations from Bogazici University, Istanbul. She continued her education on Clinical Psychology and received her M.A. degree from Dogus University, Istanbul. She worked in several projects primarily on adolescents' health and substance abuse. Besides her continuing studies in clinical psychology, Canan has a particular interest in human motivations behind technology use and adoption. Political and public spheres emerging as the contexts for human behaviour are related to her investigation area.

Khaled Sabry has a PhD in Information Systems from Brunel University, UK. He is an Assistant Professor in the e-School of Business & Quality Management, Hamdan Bin Mohammed eUniversity, UAE. Research interests in Dynamic and Interactive Information Systems and Learning Systems. Postal address: PO Box 84087, Al-Ain, UAE. E-mail: khal_sabry@yahoo.co.uk.

Ahmed Al-Nakeeb has a PhD in Quality Management from University of Glamorgan, UK. He is an Assistant Professor in the Business Administration College–Al-Ain University, UAE. Research interests in various Management topics. Postal address: PO Box 64141, Al-Ain, UAE. E-mail: sinbad2001@btopenword.com.

Khalid Al-Rawi has a PhD in Business Administration from University of Strathclyde, UK. He is a Professor in the Business Administration College–Al-Ain University, UAE. Research interests in Organization Behavior and Marketing Services. Postal address: PO Box 64141, Al-Ain, UAE. E-mail: kalrawi47@hotmail.com.

Adrian David Cheok, Director of the Mixed Reality Lab, National University of Singapore, is an Associate Professor in the Department of Electrical and Computer Engineering at NUS and a Full Professor in Keio University, Graduate School of Media Design. Amongst numerous high quality academic journals, keynotes, exhibitions, he was featured in worldwide broadcasts on his research such as CNN, National Geographic, and Discovery Channel. In addition, he was the recipient of numerous prestigious awards such as the A-STAR Young Scientist of the Year, Hitachi Fellowship, SCS Singapore Young Professional of the Year, Microsoft Research Award in Gaming and Graphics, and was recently awarded Young Global Leader 2008 by the World Economic Forum.

Owen Noel Newton Fernando is a Research Fellow in the KEIO-NUS Connective Ubiquitous Technology for Embodiments (CUTE) Center of the National University of Singapore. Prior to joining Keio-NUS CUTE Center he was attached to the Mixed Reality Laboratory (MXR) of the same university. He received his BSc in Computer Science from the University of Colombo, Sri Lanka, and MSc and PhD in Computer Science and Engineering from the University of Aizu, Japan. In 2008, he completed a term of six months as a Visiting Research Scholar at the Academic Center for Computing and Media Studies of Kyoto University. He was awarded the Distinguish Laksakura fellowship (Herath Foundation) in 2001, a Japanese Government (Gakushu-shoreihi: MEXT) Scholarship in 2002, and a Japanese Government (Monbukagakusho: MEXT) Scholarship in 2004. He has produced technical publications and his work has been demonstrated at a wide variety of conferences. Furthermore, he is active in several research areas, including mobile computing, entertainment computing, human-computer interaction, and mixed & virtual reality.

Nimesha Ranasinghe is a PhD research scholar at the Mixed Reality Lab, National University of Singapore. He received his Bachelor of Science Honors degree in Information Technology from University of Moratuwa, Sri Lanka in 2007. He has superlative working experience on ATM, Kiosk systems and embedded research projects at Interblocks Ltd and Dialog Telekom Ltd in Sri Lanka as a Software

Engineer. His research interests are mainly focuses on Mixed Reality, Human Computer Interaction, Interactive Communication Media, and Novel Computer Interfaces.

Kening Zhu is a NGS PhD student of Mixed Reality Lab. He received his bachelor degree on Computer Science from Huazhong University of Science and Technology, China. His research interests include computer graphics, shape-changing interface and the combination of traditional art and digital technology. He is involved in many projects in Mixed Reality Lab, such as Petimo, Poetry Mix-up and so on. Currently Phd Research Scholar with the Department of Architecture of National University of Singapore. Graduated with an honors degree in Archaeology she holds several postgraduate qualifications related to Cultural Heritage Conservation, Museum Studies and Cultural Heritage Management. A professional Archaeologist and Cultural Heritage Conservator, she has several years of working experience related to UNESCO World Heritage Sites. Mixed and Virtual Reality and Cultural space are her current research orientations.

Jan Servaes (PhD, 1987, Catholic University of Louvain, Belgium) is Professor and Director of the SBS Center 'Communication for Sustainable Social Change (CSSC)' at the University of Massachusetts at Amherst (USA, www.csschange.org),). Honorary Guest Professor at Huazhong University of Science and Technology (HUST) in Wuhan, China, Researcher at the 'Brussels Center for Journalism Studies' (BCJS), Belgium, Editor-in-Chief of 'Telematics and Informatics: An Interdisciplinary Journal on the Social Impacts of New Technologies' (Elsevier), Editor-in-Chief of 'Communication for Development and Social Change: A Global Journal' (Hampton Press), Editor of the Southbound Book Series 'Communication for Development and Social Change', and Editor of the Hampton Book Series 'Communication, Globalization and Cultural Identity'. He chaired the Scientific Committee for the World Congress on Communication for Development (Rome, 25-27 October 2006), organized by the World Bank, FAO and the Communication Initiative. Servaes has taught International Communication and Development Communication in Australia (Brisbane), Belgium (Brussels and Antwerp), the USA (Cornell), The Netherlands (Nijmegen), and Thailand (Thammasat, Bangkok). He has been President of the European Consortium For Communications Research (ECCR, www.ecrea.eu) and Vice-President of the International Association of Media and Communication Research (IAMCR, www.iamcr.org), in charge of Academic Publications and Research, from 2000 to 2004.

Sohayla Mohamed El-Sherbiny Attalla is a senior lecturer at the Faculty of Medicine, Mansoura University, Egypt. She has a M.D. in Forensic Medicine and Clinical Toxicology, Faculty of Medicine , Mansoura University, Egypt,2006, diploma in Health Care Quality Management, Cambridge Training College Britain, 2008 and diploma in health organization administration, Faculty of Commerce, Mansoura University, Egypt and. Her main research interests are currently in environmental pollution and toxicity and their relation to chronic health problems. She also interested in medicolegal implication of health problems (the relation between medical malpractice and ethics). She is member of TIAFT (the international association of forensic toxicology), TAAFCT (the Arabic association of forensic and clinical toxicology), quality assurance and accreditation unit in Faculty of Medicine, Mansoura University, the Egyptian association of environmental toxicology, the Egyptian association of natural toxins, the Poineers society for development of human resources and health improvement society.

Index

Symbols

3G 122, 124, 130, 134, 137, 151, 208

A

acquired immunodeficiency syndrome (AIDS) 220
alliance legitimacy 178
analysis of variance (ANOVA) 6, 7
Arab Advisors Group (AAG) 196
Asian Development Bank 227
Association of Southeast Asian Nations (ASEAN) 227, 232
audio conferencing 104
audio files 94, 95
audio lectures 92, 94, 98
average revenues per user (ARPU) 9, 11

B

Bebo 93
Bioelectromagnetics 245, 247, 248, 249, 250, 251
Blackberry 196, 197, 203, 204
blended mobile learning 91, 99
blogs 92, 98
BlogWall 205, 206, 207, 209, 210, 212, 213, 214, 215, 216
British Broadcasting Corporation (BBC) 196
Buddhism 206
business management 135
buzzwords 79

C

cell phone markets 207
cell phones 71
cellular services 18, 19, 21, 22, 26, 27, 28, 29, 30, 35
cellular technologies 21, 26, 27, 28, 29, 30, 31, 35
China Internet Network Information Center (CNNIC) 80, 83, 88
China Mobile 121
China Telecom 121
China Unicom 121
civil servant resistances 135, 151
civil servants 134, 135, 136, 140, 142, 143, 144, 145, 146, 147, 148, 149, 150, 151, 158, 159
code division multiple access (CDMA) technologies 27
college graduates 81
common resource pools 36
Communication for Development and Social Change (CDSC) 218, 219, 224, 230, 233
communication sectors 205, 208
computer aided learning (CAL) 85
computer based learning applications 105
construction industry 36, 37, 38, 39, 40, 41, 47, 49, 50, 51
construction organizations 38, 39, 41, 47, 49, 51
construction project management 36, 37
construction projects 36, 37, 39, 40, 46, 47, 51
critical mass 1, 2, 3, 4, 5, 7, 8, 9, 10, 12, 13, 15, 16
critical mass points 2, 3, 4, 5, 7, 13

D

day-to-day activities 135
deoxyribonucleic acid (DNA) 240, 241, 245, 249

developed countries 1, 2, 12, 13, 14, 105, 107
developing countries 1, 2, 3, 13, 15, 52, 61, 64, 67, 68, 69, 70, 71, 73, 75, 103, 104, 105, 106, 108, 109, 110, 113, 114, 116
diffusion models 18, 19, 23, 32, 33
digital content 52
digital era, the 207
digital media players 91
digital strategies 134
digital subscriber line (DSL) 105
digital technologies 135, 143
DVD players 182

E

ebooks 92, 98
economical imbalances 206
economic development 161, 162, 164, 165, 166, 167, 168, 169, 170
economic growth 161, 162, 163, 165, 167, 168, 169, 170, 171
educational services 67, 69, 71
Educational Testing Service (ETS) 86
e-government 127, 134, 135, 136, 137, 145, 146, 149, 153, 154, 155, 157, 158
e-learning 68, 73, 81, 83, 86, 89, 91, 92, 93, 101
electromagnetic fields (EMF) 237, 238, 239, 240, 244, 245, 251
electromagnetic hypersensitivity (EHS) 239
Electronic and Telecommunications Research Institute (ETRI), the 20
end-user involvement 53
entrepreneurship 1
e-tools 57, 59, 61, 62, 64
Euromonitor 4

F

Facebook 93, 99
family life cycles 182
fixed lines 161, 162
Flickr 93
focus groups 53, 56, 62, 63
Food and Drug Administration (FDA) 238, 248

G

general method of moments (GMM) 161, 165, 166, 169, 171
geographical resistance 135, 145, 146
geographic information systems (GIS) 39, 92
geologists 91, 93, 98
Global Market Information Database (GMID) 163
global positioning system (GPS) 40, 93, 174, 182, 186, 195, 197, 199, 200, 201, 202
Global System for Mobile Communications (GSM) 8, 72, 122, 209, 238, 244, 245, 247, 248, 250, 251
Global System for Mobile Communications (GSM) cell towers 238
government organisations 117, 127, 131
gross domestic product (GDP) 4, 5, 6, 8, 10, 11, 12, 13, 40, 82, 161, 162, 163, 164, 165, 166, 167, 168, 169
gross national income (GNI) 5

H

Hancock, Alan 227, 232
handheld computers 91
handheld devices 68
higher education (HE) 103, 104, 109, 112, 115, 116
high-frequency electromagnetic fields (HF-EMF) 238
high-income countries 161
high-income economies 2
human immunodeficiency virus (HIV) 220, 229
human interactions 196

I

information and communications technologies (ICT) 18, 20, 32, 33, 34, 35, 37, 39, 41, 48, 49, 52, 65, 75, 82, 107, 112, 117, 118, 119, 120, 121, 122, 123, 124, 127, 128, 129, 131, 134, 135, 137, 142, 143, 144, 145, 146, 148, 149, 151, 156, 158, 159, 173, 176, 197, 203, 218, 223, 224, 227, 228, 229, 230, 231, 235, 236

information and communications technologies (ICT) skills 137
information and communications technologies (ICT) tools 197
Information and Communication Technologies for Development (ICT4D) 52, 53, 54, 56, 59, 62, 63, 65, 70
Information and Communication Technologies for Development (ICT4D) domain, the 52, 63
Information and Communication Technologies for Development (ICT4D) projects 52, 53, 56, 62, 63
information management 37, 38, 43
information systems (IS) 70, 73
innovation 206
instructional strategies 91, 92
interactive systems 206
interactive voice response (IVR) applications 55
interactive voice response (IVR) systems 55
inter-governmental organizations (IGO) 222
International Agency on Research on Cancer (IARC) 241
International Country Risk Guide (ICRG) 163, 164, 170
International Telecommunication Union (ITU) 162, 163
Internet access 91, 94, 98
Internet based services 104
investment legitimacy 178
iPods 80, 134
IT investments 38
IT solutions 37
IT systems 38, 40

J

Java 196
JD Power and Associates 196

K

Korea Mobile Communication Services (KMCS) 18, 19

L

landline telephones 68, 161, 162, 169
laptops 68, 134
learning environments 67
least developed countries (LDC) 69
LG Telecom 19
local area networks (LAN) 37, 38, 44
Lotka-Volterra Competition (LVC) model 19, 21, 22, 23, 24, 26, 29, 31, 32, 35
low-income countries 161, 162, 168, 169
low-income economies 2

M

management and information systems (M&IS) 70
market legitimacy 178
Millennium Development Goals (MDG) 228, 229
Ministry of Industry and Information Technology (MIIT), the 79, 89
Ministry of Information Industry (MII) 82
Ministry of Posts and Telecommunications (MPT), the 82
mobile access 136
mobile applications 136, 137, 139
mobile communications 18, 36, 37, 38, 39, 40, 41, 46, 47, 118, 132
mobile communication technologies 37, 38
mobile data communications 37
mobile devices 80, 81, 84, 85, 86, 87, 88, 92, 98, 103, 104, 107, 108, 109, 110, 111, 112, 113, 114
mobile government (m-government) 117, 118, 119, 120, 121, 122, 123, 124, 125, 126, 127, 128, 129, 130, 131, 132, 133, 134, 135, 136, 137, 139, 140, 142, 143, 145, 146, 148, 149, 150, 151, 152, 154, 156, 158, 159
mobile handsets 208
mobile industry 206
mobile infrastructures 161
mobile learning (m-learning) 67, 68, 69, 70, 71, 72, 73, 74, 75, 76, 77, 79, 80, 81, 82, 83, 84, 85, 86, 87, 88, 89, 90, 91, 92, 93, 94, 96, 98, 99, 100, 101, 102, 196
mobile learning systems (MLS) 85
mobile marketing 196, 203
mobile markets 18, 21

mobile networks 67, 71, 76, 122, 123
mobile numbers 207
mobile phone applications 67, 69
mobile phone density 121
mobile phone penetration 161, 162
mobile phones 2, 10, 16, 67, 68, 69, 70, 71, 74, 75, 76, 78, 80, 82, 84, 86, 91, 94, 102, 104, 106, 115, 121, 122, 124, 131, 134, 135, 137, 142, 143, 155, 157, 158, 161, 162, 163, 170, 173, 174, 175, 176, 179, 180, 181, 182, 183, 184, 185, 186, 188, 189, 191, 192, 194, 195, 196, 197, 198, 199, 200, 201, 202, 203, 205, 206, 207, 208, 209, 237, 238, 239, 240, 241, 242, 243, 245, 246, 247, 248, 249, 250, 251
mobile phone technologies 206
mobile phone usage 173, 174, 175, 182, 185, 186
mobile phone users 174, 237, 243
mobile policies (m-policies) 134
mobile service markets 18, 19, 21
mobile service operators 18, 19
mobile services (m-services) 18, 19, 21, 31, 134, 149, 150, 206, 216, 217
mobile subscription markets 196
mobile technologies 2, 67, 68, 69, 71, 77, 91, 93, 103, 104, 105, 106, 107, 108, 109, 110, 111, 112, 113, 118, 127, 135, 140
mobile technologies (m-technologies) 173, 175, 187, 195, 196, 197, 198, 200, 202, 205, 206, 207, 208, 209, 215, 216
mobile technology (m-technology) consumption 175
mobile technology solutions 91
mobile telecommunications 2, 14, 15, 19, 20, 22, 31, 32
mobile telephone base stations 237, 238
mobile telephones 209
mobile telephone technologies 1
mobile telephony 2, 15, 17, 161, 162, 163, 165, 167, 168, 169
mobile users (m-users) 173
mobility literature 134
model-based approaches 73
MP3 players 80
multi-generation products 22
multimedia education technologies (MET) 85
multimedia messaging services (MMS) 62

N

national development agencies 53
National Health Service (NHS) 245
netbooks 134
neurobehavioral test battery (NBTB) 239, 246
New Public Management 125
non-governmental organizations (NGO) 222, 228
non-profit sectors 54
North Atlantic Treaty Organization (NATO) 91

O

odds ratio (OR) 242
Organisation for Economic Co-operation and Development (OECD) 125, 162, 163
organizational developments 136, 148
Orkut 93

P

palmtops 68, 80
Participatory Rural Assessment (PRA) 62
personal communications services (PCS) 19, 20, 21, 22, 26, 27, 28, 29, 30, 31, 35
personal digital assistants (PDA) 37, 39, 40, 48, 68, 80, 84, 91, 134, 137, 174
personal legitimacy 182
Philippines, the 72
pocket PCs 37, 51
podcasts 92, 96, 97, 98
poetry 205, 206, 207, 209, 210, 213, 214, 215, 216
political environments 206
portable devices 91, 94
prepaid cards 105
private service providers 117, 119, 120, 123, 124, 125, 126, 131
public administration 134, 135, 140, 143, 144, 146, 147, 148, 149, 150, 158
public service vehicles 174

R

radiofrequency electromagnetic radiations (RFR) 239, 240

radio frequency identification devices (RFID) 40, 41, 46
radio frequency (RF) 237, 238, 241, 242, 243, 245, 246, 251
relational legitimacy 178
requirements engineering 52, 53, 58, 63
rich media 103, 104, 105, 108, 109, 112, 113, 114
rich media technologies 103, 104, 105, 108, 110, 112, 113, 114
ringtones 196
Road and Transport Authority (RTA) 197

S

safety education 174
service providers 18, 19, 27, 28, 29, 30, 31, 117, 119, 120, 121, 123, 124, 125, 126, 131
severe acute respiratory syndrome (SARS) 118
short message service (SMS) 80, 84, 85, 88, 89, 118, 122, 124, 128, 129, 130, 131, 175, 184, 192, 205, 206, 207, 208, 209, 210, 211, 214, 215, 216
simulation models 23
Skype 92, 98
small and medium enterprises (SME) 36, 38, 41, 47, 48
smart phones 37
social change 218, 219, 220, 222, 224, 227, 230, 233, 235
social development projects 52
social impacts 67, 70, 75
social interactions 104
social justice 79
social legitimacy 178
social networking 91, 98, 104, 105, 107, 109, 110, 112
socio-economic situations 52, 54, 61
software development 52, 53
software engineers 53
software requirements 53, 54, 58, 62, 63, 64
software requirements specifications (SRS) 58, 59
Soviet Union 219
stakeholders 53, 54, 55, 57, 58
structured digital storytelling (SDS) 53, 54, 55, 56, 57, 59, 61, 63, 64
Superman 225
support groups 91, 99

T

technological changes 134, 148
technological evolutions 23
technology acceptance models (TAM) 73
technology-process-culture (TPC) triangle, the 39
technoverts 196
Telecomm Regulatory Authority (TRA) 197
telecommunications 1, 2, 3, 14, 15, 16, 17, 105, 106, 114, 161, 162, 163, 164, 165, 166, 167, 168, 169, 170, 171
telecommunication services 2, 15
telecommunications infrastructures 161, 162, 167
telecommunications markets 105
telephone hotlines 118
third world countries 219, 222, 225, 226, 227, 232
transient evoked otoacoustic emissions (TEOAE) 243
TV broadcasts 118
Twitter 99

U

United Arab Emirates (UAE) 195, 196, 197, 198, 201, 203, 204
United Nations Development Programme (UNDP) 5, 228
United Nations Educational, Scientific and Cultural Organization (UNESCO) 224, 227, 231, 232, 233
United Nations (UN), the 228, 233
U.S. military professionals 91

V

video display units (VDU) 244
video-oculography (VOG) 243
virtual higher education (HE) 104
virtual private networks (VPN) 37, 38, 45
voice calls 206, 208
voice services 206

W

Web 2.0 98, 99
Web-based learning management systems 68
web-based solutions 91
Web browsers 196
WebCT 68
WFed 196, 204
Wi-Fi 105
WiMAX 105, 134
wireless application protocols (WAP) 62
wireless communications 1, 4, 5, 6, 7, 8, 11, 12, 14, 16
wireless devices 237
wireless local area networks (WLAN) 37, 38, 39, 47
wireless protocols 105
wireless technologies 68
WordNet 210, 216
word sense disambiguation (WSD) 210, 211
World Bank, the 104, 163, 171, 227
World Cellular Information Service 5
World Health Organisation (WHO), the 197, 241
World War II (WWII) 219

X

x-ray machines 237

Y

YouTube 93, 99, 101

Breinigsville, PA USA
05 November 2010

248705BV00003B/5/P